The Gastrointestinal System

Po Sing Leung
Editor

The Gastrointestinal System

Gastrointestinal, Nutritional
and Hepatobiliary Physiology

 Springer

Editor
Po Sing Leung
School of Biomedical Sciences
Faculty of Medicine
The Chinese University of Hong Kong
Hong Kong, People's Republic of China

ISBN 978-94-017-8770-3   ISBN 978-94-017-8771-0 (eBook)
DOI 10.1007/978-94-017-8771-0
Springer Dordrecht Heidelberg New York London

Library of Congress Control Number: 2014939026

© Springer Science+Business Media Dordrecht 2014
This work is subject to copyright. All rights are reserved by the Publisher, whether the whole or part of the material is concerned, specifically the rights of translation, reprinting, reuse of illustrations, recitation, broadcasting, reproduction on microfilms or in any other physical way, and transmission or information storage and retrieval, electronic adaptation, computer software, or by similar or dissimilar methodology now known or hereafter developed. Exempted from this legal reservation are brief excerpts in connection with reviews or scholarly analysis or material supplied specifically for the purpose of being entered and executed on a computer system, for exclusive use by the purchaser of the work. Duplication of this publication or parts thereof is permitted only under the provisions of the Copyright Law of the Publisher's location, in its current version, and permission for use must always be obtained from Springer. Permissions for use may be obtained through RightsLink at the Copyright Clearance Center. Violations are liable to prosecution under the respective Copyright Law.
The use of general descriptive names, registered names, trademarks, service marks, etc. in this publication does not imply, even in the absence of a specific statement, that such names are exempt from the relevant protective laws and regulations and therefore free for general use.
While the advice and information in this book are believed to be true and accurate at the date of publication, neither the authors nor the editors nor the publisher can accept any legal responsibility for any errors or omissions that may be made. The publisher makes no warranty, express or implied, with respect to the material contained herein.

Printed on acid-free paper

Springer is part of Springer Science+Business Media (www.springer.com)

*To my wife, Wan Chun Hu, and my daughter,
Choy May Leung and my son, Ho Yan Leung*

Preface

Human gastrointestinal (GI) physiology is a study of our GI system that addresses the regulation and integration of major physiological functions of motility, secretion, digestion, absorption and blood flow, as well as immunity. The coordination of these processes is vital for the maintenance of GI health; thus, any dysregulation will result in GI disease. In fact, GI physiology is a fundamental subject that is indispensable not only for undergraduate but also for graduate students of any biomedical courses, including, but not limited to, medical, pharmacy, nursing, human biology, Chinese medicine, and natural science, as well as other health-related subjects.

From my GI teaching experience over the past decade, it has always been my wish to produce a succinct monograph that can serve as a companion book for biomedical students not only for their initial studies, but also for their career paths. From the students' viewpoint, one of the most common problems they encounter is the lack of any available textbooks that cover both the basic science and provide relevant clinical correlations. Owing to the lack of exposure to patients with real clinical problems, students often cannot see the whole picture of the patient during the diagnostic process. In view of this, students have often shared with me their thoughts on what they want from a textbook; it should cover the basic science comprehensively, but with equal emphasis on relevant clinical problems. In addition, the textbook should be user-friendly and easy-to-understand. This is of particular importance as a well-formatted textbook can facilitate the understanding of the material covered and thereby save the students' time. Furthermore, students would appreciate the provision of relevant multiple-choice questions that would reinforce their understanding of, and ability to apply, the basic concepts, as well as honing their examination skills.

The overall objectives of this *Gastrointestinal System* book are to present basic concepts and principles of normal GI physiology and, most importantly, to convey an understanding of how to apply this knowledge to the understanding of abnormal GI physiology in the clinical context. The ultimate goal is to let the readers have an integrated systems-based approach in order to be able to grasp knowledge on GI disease and its management. The understanding of basic GI concepts and principles would be guided by scenario-based clinical case-studies, critical for bedside care and also for preparation for professional examinations, and for being able to deal with future developments in clinical care. In this book, the aim is to achieve these various objectives by covering the breadth of GI system. The contents are, therefore, designed to fall systematically into three core sections, namely Gastrointestinal Physiology (Part I), Nutritional Physiology (Part II) and Hepatobiliary Physiology (Part III) with closely relevant scenario-based clinical case presentations at the end of each chapter to help students learn to apply their growing knowledge of basic GI science, in the clinical setting. Last but by no means least, we provide a wide range of multiple-choice questions (Part IV) so that students can evaluate their understanding of the basic science in each area of the GI system and to develop the students' ability to apply their knowledge to solving clinical problems.

Finally, we would like to take this opportunity to express my sincere gratitude to Dr. Thijs van Vlijmen, the Publishing Editor, and to Miss Sara Germans, the Publishing Assistant of Springer, for their support and encouragement. We would also like to express my appreciation to Leo Ka Yu Chan, Medical Student of this University, and Sam Tsz Wai Cheng, PhD student in my Department, for technical assistance.

Hong Kong, People's Republic of China Po Sing Leung
January 2014

Prologue

A major challenge facing medical students and subspecialty trainees is mastery of basic organ system physiology and its application to understanding and treating human disease. This is especially true of the gastrointestinal and liver system where symptoms of disease are frequently vague and non-specific. As I tell my medical students at the beginning of the GI lectures, "Most of us have experienced diarrhea, upset stomach, abdominal pain, and nausea that your patients will present with, and your task as physicians will be to determine whether their severity and duration warrant intervention and, if so, to identify the cause." In addition, knowledge in this field continues to expand at an ever increasing rate. The ideal textbook for this audience would combine state-of-the-art science with relevant application to disease processes and treatment in a logical and understandable format. This *Gastrointestinal System* book, edited by PS Leung and written by DD Black, EB Chang and MD Sitrin clearly satisfies these requirements. The informative, up-to-date chapters cover all aspects of gastrointestinal and hepatobiliary physiology and span pathophysiology, diagnosis and treatment of disease. Clinical correlations using cases that clearly illustrate the concepts presented are included in each chapter along with a reading list for more in-depth study. Multiple-choice questions at the end of the book help the reader assess his or her understanding of the material and identify areas for additional review. The editor and authors are all outstanding academicians who have contributed significantly to our knowledge of both basic physiology and clinical medicine. This book belongs in every student's and trainee's armamentarium for mastering this fascinating area of medicine.

Memphis, TN, USA
Leonard R. Johnson
Thomas A. Gerwin

Contents

Part I Gastrointestinal Physiology

1   **Regulation of Gastrointestinal Functions** 3
    Eugene B. Chang and Po Sing Leung

2   **Gastrointestinal Motility** 35
    Eugene B. Chang and Po Sing Leung

3   **Gastric Physiology** 63
    Eugene B. Chang and Po Sing Leung

4   **Pancreatic Physiology** 87
    Eugene B. Chang and Po Sing Leung

5   **Intestinal Water and Electrolyte Transport** 107
    Eugene B. Chang and Po Sing Leung

Part II Nutritional Physiology

6   **Digestion and Absorption of Carbohydrates and Proteins** 137
    Michael D. Sitrin

7   **Digestion and Absorption of Dietary Triglycerides** 159
    Michael D. Sitrin

8   **Digestion and Absorption of Other Dietary Lipids** 179
    Michael D. Sitrin

9   **Absorption of Water-Soluble Vitamins and Minerals** 211
    Michael D. Sitrin

Part III Hepatobiliary Physiology

10 **Structure, Functional Assessment, and Blood Flow of the Liver** 237
    Dennis D. Black

11	**Protein Synthesis and Nutrient Metabolism**	271
	Dennis D. Black	
12	**Biotransformation, Elimination and Bile Acid Metabolism**	295
	Dennis D. Black	

Part IV Review Examination

13	**Multiple Choice Questions**	327
	Dennis D. Black, Eugene B. Chang, Po Sing Leung, and Michael D. Sitrin	

Index 357

Contributors

Dennis D. Black, M.D. Department of Pediatrics, University of Tennessee, Memphis, TN, USA

Eugene B. Chang, M.D. Department of Medicine, University of Chicago, Chicago, IL, USA

Po Sing Leung, Ph.D. School of Biomedical Sciences, Faculty of Medicine, The Chinese University of Hong Kong, Hong Kong, People's Republic of China

Michael D. Sitrin, M.D. Department of Medicine, University at Buffalo, The State University of New York, Buffalo, NY, USA

Part I
Gastrointestinal Physiology

Chapter 1
Regulation of Gastrointestinal Functions

Eugene B. Chang and Po Sing Leung

1 Introduction of the Gastrointestinal System

The **Gastro-Intestinal (GI) system** is divided into **two parts:** the **luminal GI** and **hepato-biliary-pancreatic GI**.

The **luminal (or tubular) GI** consists of the alimentary (digestive) canal or GI tract, which extends from the mouth to the anus (Fig. 1.1). The GI tract includes the **pharynx, esophagus, stomach, small intestine** (*duodenum, jejunum and ileum*) and **large intestine** (*colon, cecum and rectum*), as well as the **anus**. The GI tract is a muscular tube of about *5 m* long when one is alive; however, after a person dies and during autopsy or postmortem examination, the length of the tract can be doubled to *10 m*. This is due to the loss of muscle tone. The GI tract can contract and relax with different transit time in each segment of the tract, which, in turn, depends on its own specific function (i.e. **motility** or **secretion**) of each segment. The motor and secretory activities of the GI system are highly controlled and integrated by the **gut endocrine** and **enteric nervous systems** (see Section on "Neural and Hormonal Regulators of Gastrointestinal Function").

The **hepato-biliary-pancreatic GI** consists of the associated glands and organs of the GI system; they include the *salivary glands* (parotid, sublingual and submandibular glands), *pancreas, gallbladder* and *liver* which empty their secretions into the lumen of the GI (e.g. luminal digestive enzymes). The **salivary glands** secrete saliva for digestion and lubrication; the **pancreas** produces hydrolytic enzymes for the digestion of our daily foodstuff and bicarbonate for

---

E.B. Chang, M.D. (✉)
Department of Medicine, University of Chicago, Chicago, IL, USA
e-mail: echang@medicine.bsd.uchicago.edu

P.S. Leung, Ph.D. (✉)
School of Biomedical Sciences, Faculty of Medicine, The Chinese University of Hong Kong, Hong Kong, People's Republic of China
e-mail: psleung@cuhk.edu.hk

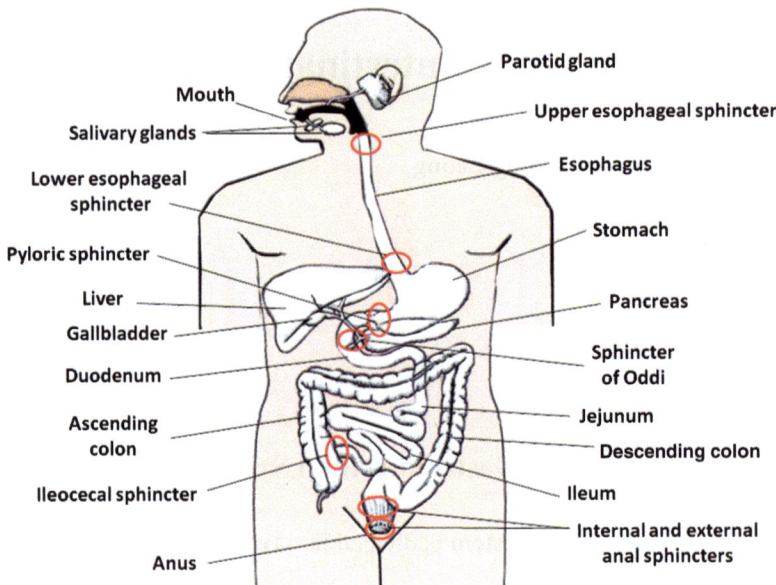

Fig. 1.1 A schematic diagram of the gastrointestinal system showing the digestive tract and the associated organs. The *circled* structures indicate the key locations that separate the digestive tract from each other by the structures called "sphincters"

the neutralization of our gastric contents; the **liver** secretes bile, which is stored temporarily in the gallbladder and subsequently delivered to the duodenum for fat digestion and absorption.

There are key locations that separate the different parts of the GI tract from each other by a structure called ***"sphincter"***, e.g. esophageal (upper and lower), pyloric, sphincter of Oddi, ileocecal, and anal sphincters (Fig. 1.1). Sphincters are made up of **smooth muscle** and they act as the "valve of a reservoir" for holding luminal content adequately before emptying the content into next segment by their highly coordinated activity. **Dysregulation of the activity of sphincters** results in **GI motility disorders** (e.g. *Gastroparesis/Dumping Syndrome* and *Achalasia/Gastro-Esophageal Reflux Disease (GERD)*). As a basic concept, dysfunction of either GI motility or secretion, or both, can lead to some common GI disorders including, but not limiting to, *GERD, Peptic ulcer Disease (PUD),* and *Diarrhea.*

1.1 General Structure of the Gastrointestinal Tract

The structure of the GI tract varies greatly from region to region, but common features exist in the overall organization of the wall of the tract (Fig. 1.2). **From inside out**, there are four characteristic layers.

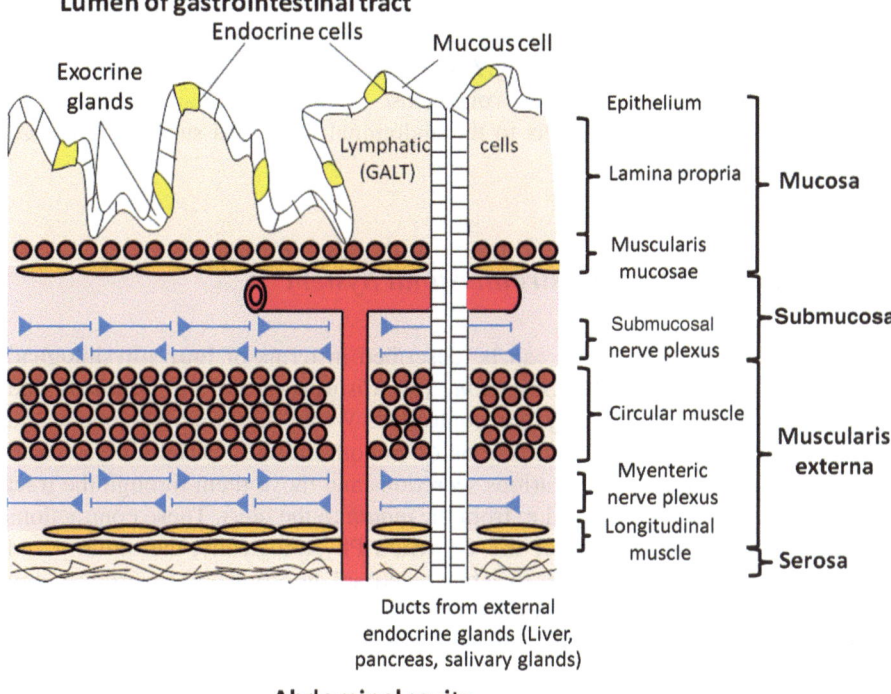

Fig. 1.2 A longitudinal section of the digestive tract wall illustrating the four major gut layers from inside outward, i.e. mucosa, submucosa, muscularis externa and serosa. GALT indicates gut-associated lymphoid tissue (Modified from Widmaier et al. [9])

The **first layer** *mucosa* is the innermost layer, and it consists of an epithelium, the lamina propria, and the muscularis mucosae. In the *epithelium*, it contains exocrine gland cells and endocrine cells, which secrete mucus and digestive enzymes into the lumen, and release GI hormones into the blood, respectively. The endocrine cells are interspersed among the epithelial lining and they constitute the *gut endocrine system*. In the *lamina propria*, it contains small blood vessels, nerve fibers, and lymphatic cells/tissues, the latter being called *gut-associated lymphoid tissue,* as introduced in Sect. 1.2. In addition, a thin muscle layer called **muscularis mucosae** is also found and the activity of its muscle is responsible for controlling mucosal blood flow and GI secretion.

The **second layer** *submucosa* is a connective tissue with major blood and lymphatic vessels, along with a network of nerve cells, called the **submucosal nerve plexus**, passing through.

The **third layer** *muscularis externa* is a thick muscle and its contraction contributes to major gut motility (segmentation and peristalsis). This muscle layer typically consists of two substantial layers of smooth muscle cells: an **inner circular layer** and an **outer longitudinal layer**. A prominent network of nerve cells, called the **myenteric nerve plexus**, is also located between the circular and longitudinal smooth muscle layers. The myenteric nerve plexus and the submucosal plexus constitute the *enteric nervous system (ENS)*.

The **fourth layer** *serosa* is the outermost layer, which mainly consists of connective tissues and it connects to the abdominal wall, thus supporting the GI tract in the abdominal cavity.

1.2 Functions of the Gastrointestinal System

The function of the GI system can be described in terms of **four physiological processes: (1)** *Digestion*, **(2)** *Secretion*, **(3)** *Absorption* **and (4)** *Motility*, and the mechanisms by which they are controlled. While digestion, secretion, and absorption are taking place, contractions of smooth muscles in the GI tract wall mix the luminal contents with various secretions and move them through the tract from proximal to distal regions, i.e. from the *mouth* to the *anus*. These contractions are referred to as the **motility of the GI tract.** Physiologically, the motility and secretion are finely tuned in order to achieve optimal digestion and absorption; it in turn facilitates assimilation of nutrients, which is the primary role of our GI system. Put it simply, the overall function of the GI system is to take in nutrients and to eliminate waste. In fact, one can survive without the GI system (yet the liver is still essential for survival) if one is fed parenterally and some vital secretions (such as *digestive enzymes, intrinsic factor* and *insulin*) are replaced.

In general, the cells lining the luminal intestinal organs are exposed to hostile environments, including antigens from food and bacteria, digestive enzymes and various solutions at variable pH levels. In view of this fact, certain *nonimmunologic defense mechanisms* are present to protect against these potential hazards; they include *gastric acid secretion, intestinal mucin, epithelial cell permeability barrier* and *gut peristalsis*, which are critical for maintaining the ecology of intestinal flora. For example, abnormally high levels of bacteria in an individual with impaired small intestinal peristalsis can lead to diarrhea and/or steatorrhea (fecal fat excretion), a clinical condition being referred to as *Intestinal Blind Loop Syndrome.* Of note, the GI tract is also an important part of the immune system of the body. The so-called *Gut-Associated Lymphoid Tissue or GALT* (Fig. 1.3) consists of both organized aggregates of lymphoid tissue (e.g. *Peyer's patches*) and diffuse (or migrating) populations of immune cells (e.g. *intraepithelial lymphocytes*). **GALT has two primary functions: (1) protection** against the potential microbial pathogens, and **(2)** permission of **immunologic tolerance** to both the potentially immunologic dietary substances and bacteria that normally reside primarily in the lumen of the GI tract (called the *intestinal microflora*). The GALT and intestinal microflora play critical roles in regulating GI functions, thus having clinical relevance to gut diseases.

Immunologically, GALT is a part of the *Mucosa-Associated Lymphoid Tissues or MALT* of the body, which contributes significantly to our gut defense mechanism or GI immune system. Apart from GI tract, MALT can also be found in respiratory

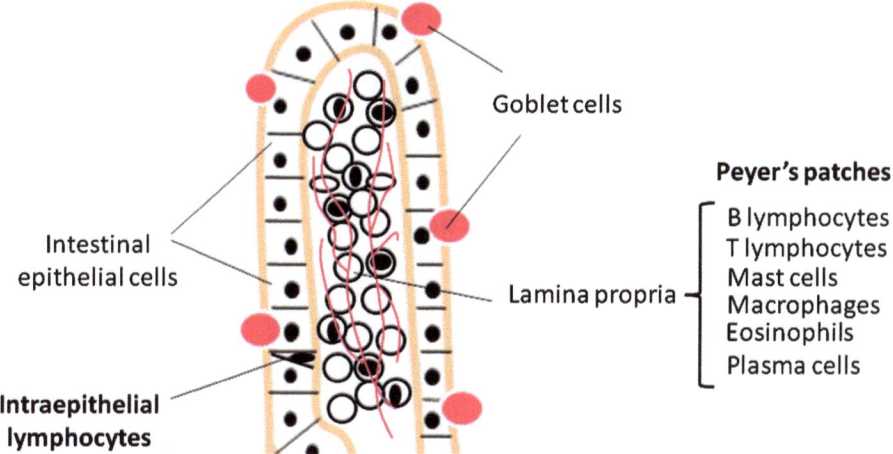

Fig. 1.3 A schematic diagram showing the Peyer's patches (lymphoid nodules) and intraepithelial lymphocytes (migrating immune cells) that constitute the major components of the gut-associated lymphoid tissue (Modified from Berne and Levy's Physiology, 2008)

and urinary systems. Even though GALT has some interactions with the systemic immune system, GALT is functionally and operationally distinct from the systemic system. There is evidence of communication between the GALT and other MALT such as the pulmonary epithelia, as manifested in asthma. The GALT system secretes antibodies in response to specific food or bacterial antigens, and triggers immunological reactions against them, thus finally leading to mucosal inflammation and damage. Activation of this local GI immune system is involved in some of the common GI disorders, including *celiac disease*, and *inflammatory bowel diseases (IBD)* such as *ulcerative colitis* and *Crohn's disease*.

2 Integration of Gastrointestinal System

Upon the ingestion of a food bolus into the GI tract, a sequence of regulatory mechanisms is elicited. This includes complex digestive, secretory, absorptive, and excretory processes. Most of these events occur automatically without conscious effort and, for the most part, they work in an integrated manner. Considering the substantial variations in the content, volume, and timing of our daily oral intake, it is remarkable that the gut is so accommodating and adaptable. Adequate nutrition is finely maintained to meet our metabolic needs so that abdominal symptoms are rarely experienced.

The luxury of taking our digestive system for granted is largely due to complex and seemingly redundant regulatory systems that make it possible to achieve great efficiency in the digestion and absorption of what we ingest. As an example of the highly integrated nature of gut functions, stimulation of the smooth muscles of the

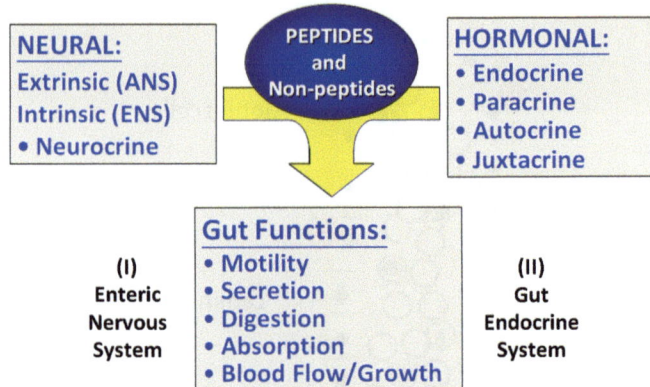

Fig. 1.4 A summary illustrating the basic neuro-hormonal mechanisms by which the gastrointestinal functions are regulated and integrated. There are two major body systems involved in these regulatory pathways, namely the enteric nervous system and the gut endocrine system

esophagus at different levels must be coordinated; this in turn leads to sequential stimulation to produce the unified propulsive contractile wave front (**peristalsis**) that is required for distally directing boluses of food. For food to enter the stomach, the lower esophageal sphincter (LES), which is usually closed to prevent reflux of gastric contents back into the esophagus, must be open at the precise arrival of the peristaltic wave. To achieve this, the gut uses intricate neural and hormonal signals to coordinate esophageal and gastric functions and to prepare the rest of the gut for processing of food. Occasionally, something goes awry and severe GI symptoms can develop. This will be discussed in greater detail in the following chapter.

2.1 *Neural and Hormonal Regulators of Gastrointestinal Function*

In view of the complex and automatic nature of gut functions, it is not surprising that the GI tract has its own endocrine system, i.e. *gut endocrine system* and its local nervous system, i.e. *enteric (intrinsic) nervous system (ENS)* of the body. Although its functions can be modulated by the **central nervous system (CNS)** and **autonomous (extrinsic) nervous system (ANS),** the gut can work on its own and independently from the CNS and ANS. This is because numerous regulatory systems are intrinsic and "hard-wired", making it possible for the gut and associated digestive organs to produce reflexive and measured responses to luminal contents and metabolic needs. Figure 1.4 summarizes the regulation and integration of GI physiological processes via neural and hormonal pathways of the GI system. In general, there are **three principal control mechanisms** involved in the regulation of GI function, namely *endocrine, paracrine,* and *neurocrine* pathways, depending on the methods by which the regulators are delivered to their target sites.

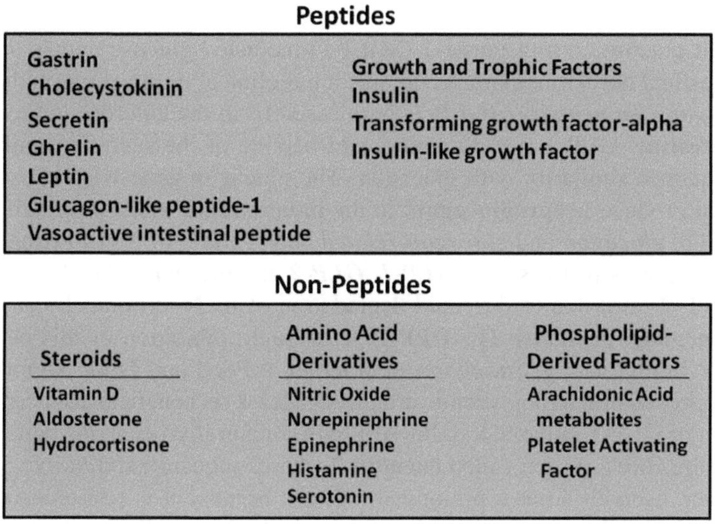

Peptides	
Gastrin	**Growth and Trophic Factors**
Cholecystokinin	Insulin
Secretin	Transforming growth factor-alpha
Ghrelin	Insulin-like growth factor
Leptin	
Glucagon-like peptide-1	
Vasoactive intestinal peptide	

Non-Peptides		
Steroids	**Amino Acid Derivatives**	**Phospholipid-Derived Factors**
Vitamin D	Nitric Oxide	Arachidonic Acid metabolites
Aldosterone	Norepinephrine	Platelet Activating Factor
Hydrocortisone	Epinephrine	
	Histamine	
	Serotonin	

Fig. 1.5 Some typical examples of gut regulators that are grouped into the categories of peptides and non-peptides

2.2 Peptide Hormones

The regulators of the gut is grouped into several different classes of compounds in the form of ***peptides*** and ***non-peptides*** (Fig. 1.5). As shown, peptide hormones such as gastrin, cholecystokinin (CCK), secretin, vasoactive intestinal peptide (VIP), gastric inhibitory peptide/glucose-dependent insulinotropic peptide (GIP), and motilin are recognized as extremely important regulators of the gut function. Many other peptide hormones have recently been identified, which are released by GI endocrine cells and influence gut functions.

Ghrelin is a 28-amino acid peptide, released from the **stomach**, which acts as a regulator of food intake or **appetite enhancer**. Plasma levels of ghrelin increase during fasting and fall during feeding. Interestingly, ghrelin levels are raised in dieters who try to lose weight; this observation may explain why it is difficult for most dieters to maintain their weight loss. In contrast, *leptin,* a peptide hormone released from **fat cells** or **GI tract,** acts as a circulating satiety factor or **appetite-suppressant**. Rather than regulating meal-to-meal food consumption as ghrelin, leptin helps maintain the usual level of adiposity (fat storage) of the body. Interestingly, leptin has been shown to affect nutrient absorption via direct inhibition of luminal amino acid uptake by the enterocytes. Similarly, several novel gut peptide candidates have also recently been discovered to control nutrient absorption by the intestine; one of the good examples is *angiotensin II*, which is a physiologically active peptide of the *renin-angiotensin system (RAS)*. In this context, locally generated angiotensin II from the enterocytes is found to inhibit sodium-dependent glucose co-transporter-1 (SGLT-1)-mediated glucose uptake across the small intestinal border membrane and thus it has clinical implications in diabetes.

Glucagon-like peptide-1 (GLP-1) is released from the enteroendocrine **L cells** in the **intestine.** GLP-1 is a 30-amino acid peptide of the secretin family with a 50 % sequence similarity with glucagon. The glucagon gene is composed of six exons that produce **preproglucagon.** In the **pancreatic α-cells,** preproglucagon is processed to *glucagon* and *glucagon-related polypeptide*. In the **intestinal L cells,** preproglucagon is processed to *GLP-1, GLP-2* and *glycentin*. GLP-1 has a short half-life (1–2 min) due to the rapid degradation of its N-terminus by an enzyme, called *dipeptidyl peptidase IV* (**DPP-IV**). As such, inhibition of this enzyme by a DPP-IV inhibitor (e.g. *Sitagliptin* or *Januvia*, a Food and Drug Administration (FDA) approved oral hypoglycemic drug from Merck) is beneficial as it can prolong the action of GLP-1 in the blood. Interestingly, a naturally-occurring peptide in the saliva of the Gila monster, called **exendin-4**, shares sequence similarity with GLP-1; however, exendin-4 has a prolonged half-life because it is resistant to DPP-IV degradation. The drug *exenatide (Byetta)* is a synthetic exendin-4, which is the first GLP-1 based drug for the treatment of diabetes approved by the FDA.

The GI tract is a major site for production and release of many of the peptide hormones, some acting exclusively on GI organs while some affecting extra-intestinal tissues, and a few having actions that are still unknown. These regulators can be found not only in endocrine cells throughout the GI tract but also in enteric neurons. Many of these agents are also found in the brain, leading some to believe that the gut has a "*visceral brain*" that regulates intestinal function in response to food or other stimuli. Abnormalities in the regulation by these agents or their overactive responses to external stimuli, such as stress, may contribute to certain diseases, including peptic ulcer disease, motility abnormalities, and irritable bowel syndrome.

Most peptide hormones of the gut are **single peptide chains**, insulin being a notable exception. About half of the bioactive gut peptide hormones are **amidated** (i.e. having an additional terminal amide or NH_2 group) at the carboxyl terminal end, a specific post-translational processing event that appears to be essential for biological activity and stability of the peptide. Based on structural similarities, most gut peptides can be grouped into a variety of **peptide families** (Fig. 1.6).

Gastrin-Cholecystokinin Family

In the gastrin-cholecystokinin family, there are various forms of gastrin and cholecystokinin (CCK) arising from post-translational modifications; they share an identical **C-terminal tetra-peptide sequence** (Fig. 1.7a), which is the biologically active domain of both peptides. Not surprisingly, each peptide can bind to the receptors of the other peptides, but with lesser affinity than to its own receptor,

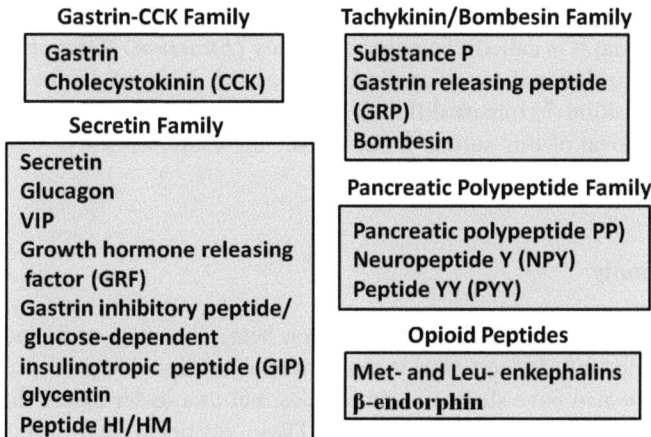

Fig. 1.6 Some representatives of the major gut peptide families including gastrin-cholecystokinin family, secretin family and tachykinin family

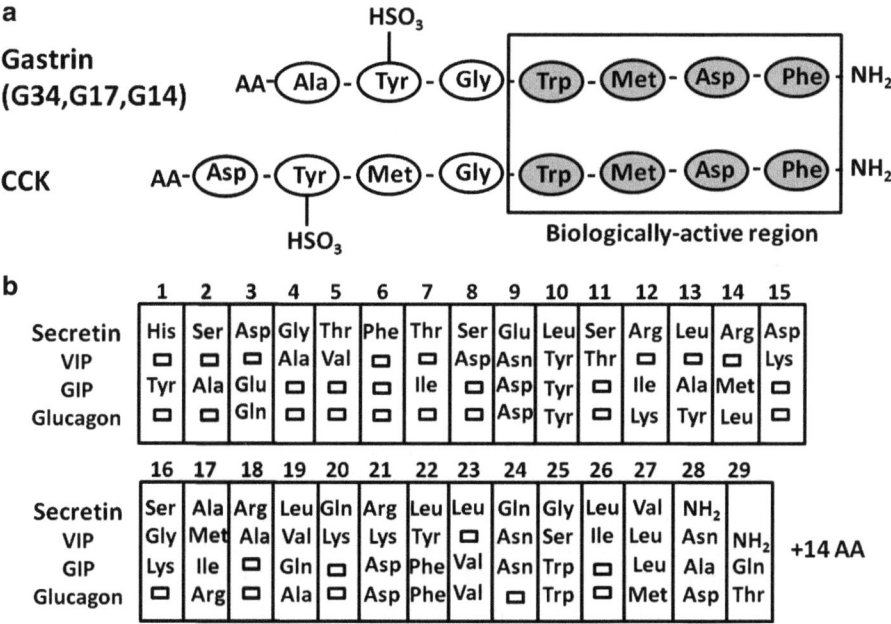

Fig. 1.7 Comparisons of peptide sequences of the (a) gastrin-CCK and (b) secretin families. In the gastrin family, gastrin and cholecystokinin (CCK) members share identical C-terminal tetrapeptide sequences, which is the biologically active domain of the family. Members of the secretin family have structural similarities, but to a lesser extent than the gastrin family. Boxes within lower panel indicate amino acids that are identical to those in secretin

i.e. *CCK_A* for CCK and *CCK_B* for gastrin, thus exhibiting similarities in physiological function; it is called **"*Structural-Activity (Function) Relationship*" (SAR)**. Other post-translational modifications, such as the sulfation of the CCK tyrosyl residue in position 7, impart different functional and binding characteristics. For instance, removal of this sulfate group causes the CCK peptide to have properties more similar to gastrin.

Secretin Family

The secretin family merits further discussion here, as several of its members have prominent and well-defined roles in the activation of digestive processes. Members of this family also have structural similarities, but to a lesser extent than members of the gastrin-CCK family (Fig. 1.7b). These members share some structural similarities throughout the amino acid sequence that is critical for the biological activity. As such, they have, not surprisingly, diverse and independent actions. **Gastric inhibitory polypeptide (GIP)** and **vasoactive intestinal polypeptide (VIP)**, for instance, each has nine amino acids that are identical to secretin, albeit in different positions. Consequently, the known physiological actions and target tissues of these peptides can vary significantly. For example, secretin and GIP both inhibit gastric acid secretion and gastric emptying, but, unlike secretin, GIP has little or no effect on pancreatic function. Unlike members of the gastrin-CCK family, most peptides in the secretin family require the entire molecule for biological activity.

Tachykinin Family

The tachykinins are a family of peptides which share a number of biological actions, especially on gut smooth muscle, and have a common C-terminal sequence Phe-X-Gly-Leu-Met-NH$_2$, where X is a variable amino acid residue. The members of this family include **substance P, neurokinin A, neurokinin B, gastrin releasing peptide (GRP)**, and **bombesin** (Fig. 1.6).

Finally, several other types of the regulatory peptides not typically associated with endocrine or neural tissues are worthwhile mentioning here, particularly those of growth and trophic factors, and cytokines. Many of these agents are synthesized locally by a variety of epithelial and mesenchymal cell types (Fig. 1.5). ***Transforming growth factors-α and β (TGF-α and TGF-β)***, for instance, are made and secreted by epithelial cells of the intestinal mucosa, as well as by cells in the lamina propria. ***Insulin-like growth factor (IGF)***, on the other hand, is largely made by lamina propria cells. Many of these agents have recently been shown to affect mucosal functions, including barrier function, water and electrolyte transport, and cellular growth and differentiation. Although they have prominent roles in mediating tissue responses during injury or inflammation, these peptides probably serve important physiological roles in maintaining mucosal functions as well.

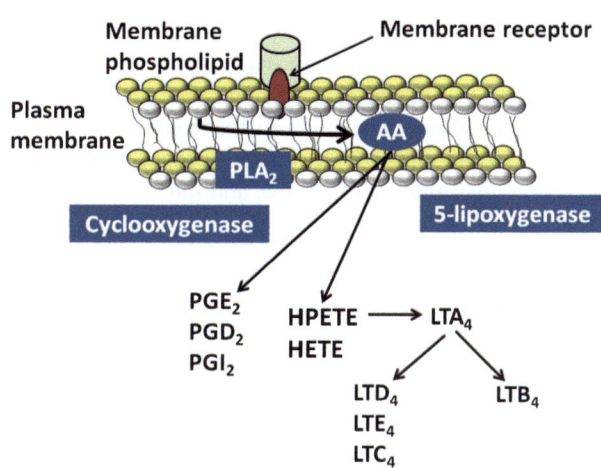

Fig. 1.8 Examples of some arachidonic acid-derived gut regulators. Prostaglandin and leukotriene metabolites of arachidonic acid (AA) metabolism are important regulators of gut function. PLA_2, Phospholipase A_2

2.3 Non-peptide Regulators

In addition to peptide hormones, many non-peptide regulatory agents are involved in the modulation and coordination of gut functions (Fig. 1.5). Factors such as **glucocorticoids** (e.g. cortisol) and **mineralocorticoids** (e.g. aldosterone), which affect intestinal fluid and electrolyte absorption, are steroids made by the adrenal gland. Vitamin D_3, or 1,25-dihydroxyvitamin D_3, is also a steroid that has significant effects on intestinal Ca^{2+} absorption and possibly growth and differentiation of gut mucosa.

Several non-peptide bioactive amines and neurotransmitters should also be worthwhile mentioning. These include substances derived from single amino acid sources, for example, *epinephrine* **(adrenaline)**, *dopamine*, and *norepinephrine* *(noradrenaline)*, which are derived from phenylalanine and tyrosine; *nitric oxide*, derived from L-arginine metabolism; and *histamine* from the decarboxylation of histidine. The catecholamines within the gut are exclusively made and secreted by enteric neurons, whereas nitric oxide is found in many different cell types, serving as an intracellular mediator as well as a cellular signaling agent. **Epinephrine and norepinephrine**, secreted by the adrenal glands and sympathetic nerves, have numerous effects on intestinal blood flow, water and electrolyte transport, and motility. **Nitric oxide** is believed to be a major *non-adrenergic and non-cholinergic (NANC)* neurotransmitter important for the regulation of intestinal motor functions and mesenteric blood flow. **Histamine**, which is secreted by the gut *enterochromaffin-like (ECL)* cells, is a major regulator of gastric acid secretion.

Several regulatory substances of gut functions originate from membrane-derived fatty acids, such as the synthesis of **platelet activating factor (PAF)**, as illustrated in Fig. 1.8, showing metabolites of arachidonic acid. These agents are made by virtually all cell types and have been recognized as being important intracellular mediators. However, they are made and secreted in large quantities by some cells and appear to have effects on numerous target tissues. Although they have been more thoroughly investigated in the context of intestinal inflammation, they also have important physiological roles in regulating gut functions such as gastric acid secretion, mesenteric blood flow, and intestinal motility.

3 Cellular Mechanisms of Gut Regulators

Gut hormones and neurotransmitters stimulate intracellular processes through a variety of mechanisms (Fig. 1.9). **Hydrophobic substances** such as corticosteroids and nitric oxide rapidly permeate the plasma membrane to stimulate intracellular receptors or targets. On the other hand, **hydrophilic mediators** such as peptide hormones, purinergic agonists, bioactive amines, and acetylcholine bind to and stimulate specific surface membrane receptors, which in turn initiate a number of intracellular signal transduction pathways so as to mediate alterations in cell functions or behaviors.

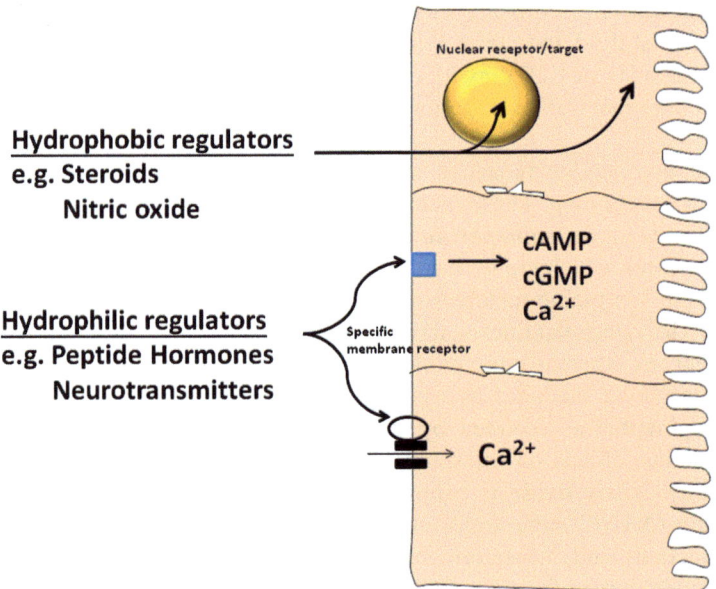

Fig. 1.9 Cellular sites of action for regulatory agents of the gut. Hydrophobic regulators permeate the plasma membrane to stimulate intracellular receptor or targets. Hydrophilic mediators generally bind to and stimulate specific surface-membrane receptors, initiating intracellular events that mediate alterations in cell functions or behaviors

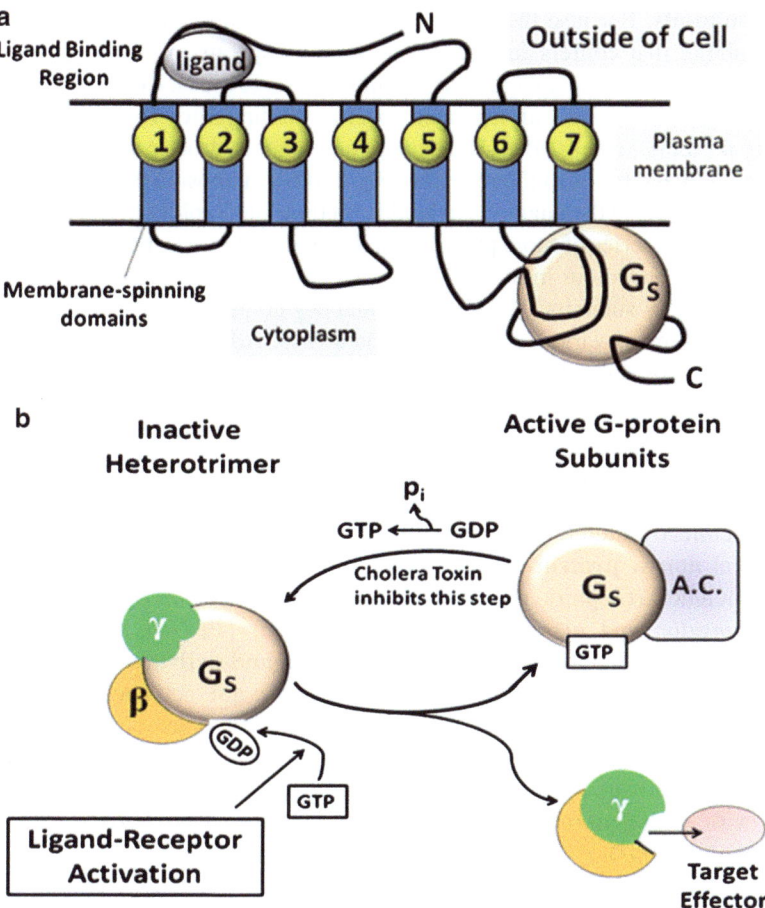

Fig. 1.10 G protein-mediated receptor action. (**a**) G proteins have structural and functional similarities, characterized by seven helical hydrophobic membrane-spanning domains and a cytoplasmic binding site for certain G proteins. (**b**) When a ligand binds to the G-protein receptor, GDP (normally occupying the guanine nucleotide binding site) rapidly exchanges for GTP, resulting in physiochemical alterations in G-protein structure and function

3.1 Types of Membrane Receptors

Several types of membrane receptors are now well characterized. Among these, the family of receptors associated with **guanosine triphosphate (GTP)-binding regulatory proteins or G proteins,** is perhaps the most numerous. These receptors share structural and functional similarities, typified by **seven hydrophobic, helical, membrane-spanning domains** (Fig. 1.10a), and a cytoplasmic binding site for certain G proteins. G proteins are heterotrimeric complexes formed from α, β

and γ subunits. Because the β and γ subunits are shared among G proteins, it is the α subunit that confers functional and binding specificity to the various types of G proteins. Each α subunit has an intrinsic GTPase activity at the guanine nucleotide binding site and separates specific binding sites for the receptor and effector proteins, such as adenylate cyclase and other hydrolases.

When a ligand binds to the receptor (Fig. 1.10b), guanosine diphosphate (GDP), normally occupying the guanine nucleotide binding site, rapidly exchanges for GTP. This results in a series of rapid physiochemical alterations in G-protein structure and function. The G protein immediately dissociates from the receptor and α subunit from the β and γ subunits. The α subunit and β-γ complex appear to activate other effector proteins of various biochemical pathways. Additionally, some G proteins appear to directly regulate ion channels such as *voltage-sensitive Ca^{2+} channels* and *inwardly rectified K^+ channels*. When GTP is hydrolyzed by the intrinsic GTPase activity of the α subunit, the G protein heterotrimeric complex reforms, and becomes ready to be activated again if further receptor ligand signals are still present. As will be discussed below and in subsequent chapters, the cycling of certain G proteins can be biochemically altered by bacterial toxins such as *cholera* and *pertussis toxins*.

Several **non-G protein-coupled receptors** have also been characterized. One group includes receptors for growth factors such as *epidermal growth factor (EGF), insulin,* and *platelet-derived growth factor (PDGF)*, which have only one membrane-spanning domain but possess **ligand-activated protein kinase activity**. For most growth and trophic factors of the gut, activated receptors have tyrosine kinase activity, which can stimulate several other signal transduction pathways, including phosphatidylinositol metabolism, the arachidonic acid cascade, and mitogen-activated protein (MAP) kinase. Another class of receptors includes those with **endogenous guanylate cyclase activity**, best characterized by the receptors for *atrial natriuretic factor (ANF)* and *guanylin*.

Guanylin is a recently identified gut peptide that stimulates secretion of water and electrolytes by gut epithelial cells, particularly **crypt cells** in which guanylin receptors are most highly expressed. This receptor is also the target of the heat-stable enterotoxin of **enterotoxigenic *Escherichia coli***, which stimulates increases in mucosal **cyclic guanosine monophosphate (cGMP)** levels and causes profuse watery diarrhea. Finally, some receptors appear to be ion channels, including the nicotinic acetylcholine receptor and the serotonin receptor subtype 5-HT_3. Most of these receptors, when activated, undergo conformational changes resulting in the formation of membrane "pores" that allow the entry of different ions.

3.2 Intracellular Effector Pathways

Several intracellular effector pathways mediate the actions of gut hormones and neurotransmitters. None is specific to a particular receptor. In some instances, such as activation of growth factor receptors, several pathways may be simultaneously

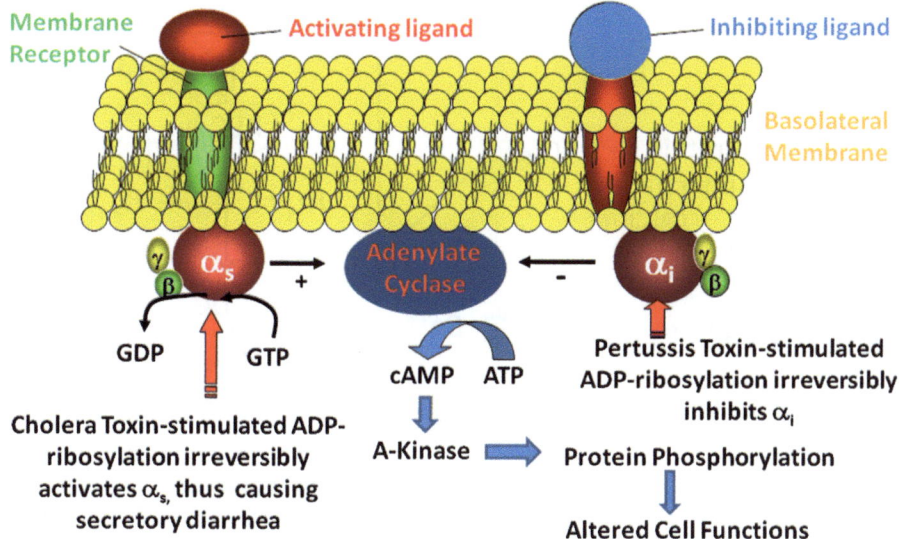

Fig. 1.11 Regulation of cellular cAMP levels. Adenylate cyclase activity is regulated by the α subunit of two G proteins, i.e. G_s and G_i. The ADP-ribosylation of the G_s-α, stimulated by cholera toxin, prevents the re-association of the inactive heterotrimeric G_s protein. This results in irreversible activation of G_s-α and adenylate cyclase activity. In contrast, ADP-ribosylation of G_i-α, stimulated by pertussis toxin, results in inactivation of the counter-regulatory pathway of adenylate cyclase and shifting the balance in favor of stimulation and increased cellular cAMP

stimulated. This may occur when different G proteins are coupled to a common receptor or when the activated pathway initiates other effector mechanisms essential for activating or amplifying requisite events of the cellular response.

Adenylate Cyclase

Adenylate cyclase is a membrane-bound enzyme with several isotypes. As illustrated in Fig. 1.11, the activity of adenylate cyclase I is regulated by the α subunits of two G proteins, **G_s and G_i**. On receptor-ligand binding, G_s-α dissociates from the β-γ subunits and activates adenylate cyclase, which catalyzes the formation of cyclic adenosine monophosphate (cAMP) from adenosine triphosphate (ATP). **Counter-regulatory hormones**, such as somatostatin and $α_2$-noradrenergic agonists, stimulate the dissociation of G_i-α from the G_i-protein complex, which inhibits adenylate cyclase activity. Thus the balance of these regulatory mechanisms of adenylate cyclase is an important determinant of steady-state levels of cellular cAMP. **Cholera toxin** stimulates the ADP-ribosylation of G_s-α, which prevents the re-association of the inactive heterotrimeric G_s protein. This causes irreversible activation of G_s-α and adenylate cyclase activity and is the mechanism by which cholera toxin causes profuse *watery diarrhea* in the absence of tissue injury.

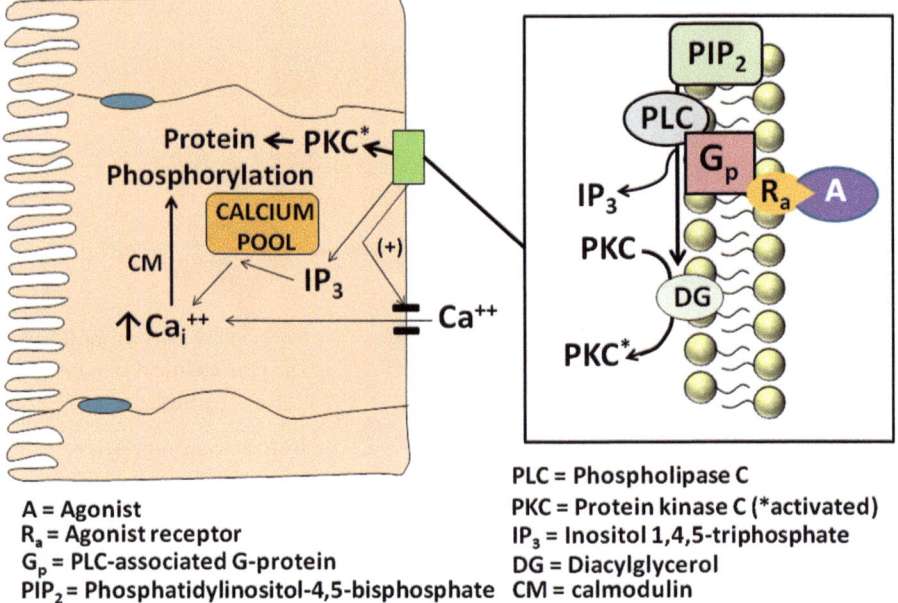

Fig. 1.12 The phosphatidylinositol pathway. This is a major pathway by which hormones and neurotransmitters lead to increases in cytosolic Ca^{2+} and activation of protein kinase C

Pertussis toxin, on the other hand, stimulates ADP-ribosylation of G_i-α, uncoupling the G_i protein from its receptor. This results in inactivation of the counter-regulatory pathway of adenylate cyclase, shifting the balance in favor of stimulation and increased cellular cAMP.

An **increase in cAMP** activates cAMP-dependent protein kinases (protein kinase A or A-kinase), which have numerous intracellular phospho-protein targets. The phosphorylation of these proteins initiates a series of events that ultimately leads to alterations in cellular function or behavior, such as stimulated ion secretion, altered motility, and changes in capillary and absorptive functions. These events will be discussed at greater length in subsequent chapters.

Phosphatidylinositol (PI) Pathway

Phosphatidylinositol 4, 5-bisphosphate (PIP_2) is a membrane phospholipid that serves as an important substrate for **phospholipase C**, a membrane-associated enzyme, when activated by receptor-associated G proteins. The hydrolysis of PIP_2 results in the formation of two products, **inositol 1,4,5-triphosphate (IP_3)** and **1,2-diacylglycerol (DAG)** (Fig. 1.12). IP_3 is a water-soluble product that freely diffuses into the cytosol and stimulates the increase of cytosolic Ca^{2+} by releasing it from intracellular calcium stores. This action is mediated by a specific receptor for IP_3, which regulates a membrane calcium channel for intracellular calcium stores.

Characteristically, a rapid, **initial-phase increase** in cytosolic Ca^{2+} is observed, which is believed to serve as a triggering event for the second and more sustained phase of increased Ca^{2+}. In most cells, the **second plateau phase** is dependent on the influx of extracellular Ca^{2+} through Ca^{2+}-activated calcium channels in the plasma membrane. Increases in cytosolic Ca^{2+} have numerous effects on cell function depending on the cell type. In many cells, the increase activates a multifunctional **calcium-calmodulin-dependent protein kinase, CaMK II kinase. Calmodulin** is a small, calcium-dependent regulatory protein that mediates many of the actions of increased cellular Ca^{2+}, including phosphorylation and activation of myosin light chain kinase (an important step in smooth muscle contraction), the stimulation of Ca^{2+}-dependent ATPase (important for restoration of resting Ca^{2+} levels), and the regulation of adenylate cyclase activity.

In contrast to IP_3, DAG is lipophilic and remains associated with the plasma membrane. It is a specific activator of **protein kinase C (PKC)**, which is cytosolically located but translocated to and activated at the plasma membrane by DAG. Although PKC activity is Ca^{2+}-dependent, DAG stimulation causes a shift in the calcium requirement such that PKC is fully active at resting cytosolic Ca^{2+} concentrations. PKC stimulates cell-specific serine and threonine phosphorylation of many protein substrates, which can cause alterations in cell functions, growth, and differentiation.

Guanylate Cyclases

There are at least **two classes of guanylate cyclase**: one is an integral part of the plasma membrane, and the other is soluble and located within the cytosol (Fig. 1.13). The latter is activated by a number of non-receptor-associated systems including nitric oxide, which stimulate rapid increases in cGMP from GTP in many cells. As will be discussed in subsequent chapters, increases in cGMP in smooth muscle cells cause muscle relaxation. By contrast, **membrane guanylate cyclases** are ligand-activated, and in most cases function as receptors themselves. G-protein coupling with guanylate cyclase receptors is not likely. Ligand-receptor interaction activates the guanylate cyclase activity of a family of guanylate cyclase receptors, which includes those for *atrial natriuretic factor (ANF)*, and the gut peptide *guanylin* or heat-stable enterotoxin of *E. coil* (ST_a). Among them, *guanylin* is a recently discovered peptide produced by the ileum and colon, and its name is derived from its ability to activate the enzyme guanylate cyclase. These agonists result in increases in cGMP, smooth muscle relaxation, and stimulation of water and electrolyte secretion in epithelial cells. These actions are mediated by cGMP-dependent protein kinases, which phosphorylate critical membrane proteins involved in these processes.

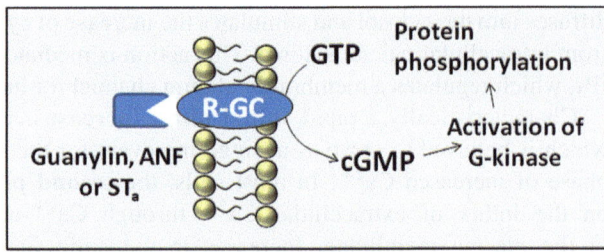

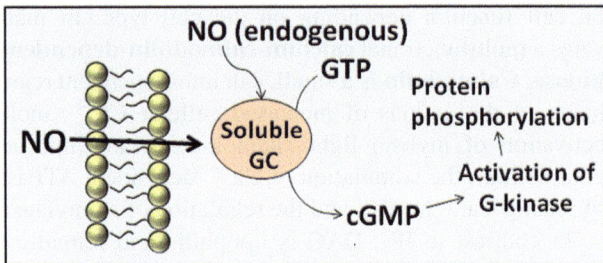

Fig. 1.13 Regulation of cellular cGMP. There are two types of guanylate cyclases that are found in intestinal cells, one integral with the plasma membrane and the other in the cytosol. A receptor-mediated mechanism activates the membrane-bound guanylate cyclase (*upper panel*), whereas the cytosolic guanylate cyclase, which is soluble, is stimulated by endogenous agents. e.g. nitric oxide which is able to permeate the plasma membrane (*lower panel*)

Tyrosine Kinases

The actions of many intestinal growth factors such as *EGF, PDGF* and *IGF* are mediated by the activation of cellular tyrosine kinase receptors. Tyrosine kinase receptors are trans-membrane proteins that can be single polypeptides or heterodimers. They have extracellular domains for specific ligand binding and cytoplasmic domains that possess intrinsic tyrosine kinase activity. The activation of the receptor often induces auto-phosphorylation of the receptor, which may negatively feedback on its activity. There may be considerable amplification of the initial signal by the integration of this system with other signal transduction pathways (Fig. 1.14). Several substrates appear to be the **targets of tyrosine kinase receptors**, including important regulatory proteins of other biochemical pathways such as phospholipase C-γ and phospholipase A_2. Additionally, **auto-phosphorylation of the receptor** appears to be an important event since it provides recognition and binding sites for SH2 (src homology) domain of the adaptor protein Grb2. This in turn causes the binding of another protein, SOS, at the SH3 domain of Grb2, required for activation of ras proto-oncogene. The latter initiates a cascade of events that activate Raf-1 and MAP kinases, involved in regulating numerous cellular and transcriptional processes. Details of this pathway are beyond the scope of this discussion.

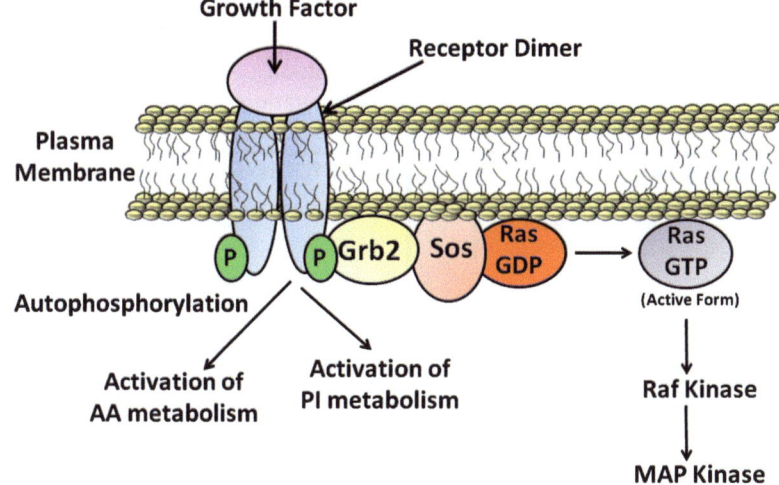

Fig. 1.14 Cellular activation by growth factors. The action of many intestinal growth factors such as EGF, PDGF, and IGF are mediated by the activation of cellular tyrosine kinase receptors

4 Gut Regulatory Pathways

4.1 Neural Hormonal Mechanisms

The regulation and integration of gut functions are mediated by a variety of pathways, each having unique characteristics suited for specialized tasks (Fig. 1.15). These pathways are present throughout most of the gut, although they assume more importance in certain regions relative to others. Likewise, the regulatory agents mediating the actions of each pathway can substantially differ. On the other hand, it is also clear that some regulatory agents are common to several pathways, albeit affecting different target tissues.

Endocrine Pathway

This represents the traditional concept of regulatory hormones **secreted into the blood stream** and delivered to distant target tissues. This system has the advantage of simultaneously affecting large numbers and regions of target tissues. *Insulin secretion from pancreatic beta cells* stimulated by increased blood glucose concentration is an example of this classic endocrine pathway. Insulin has a large systemic action, including effects on GI and hepatic metabolism, and functions. Another example is the *release of gastrin into the blood* stimulated by gastric

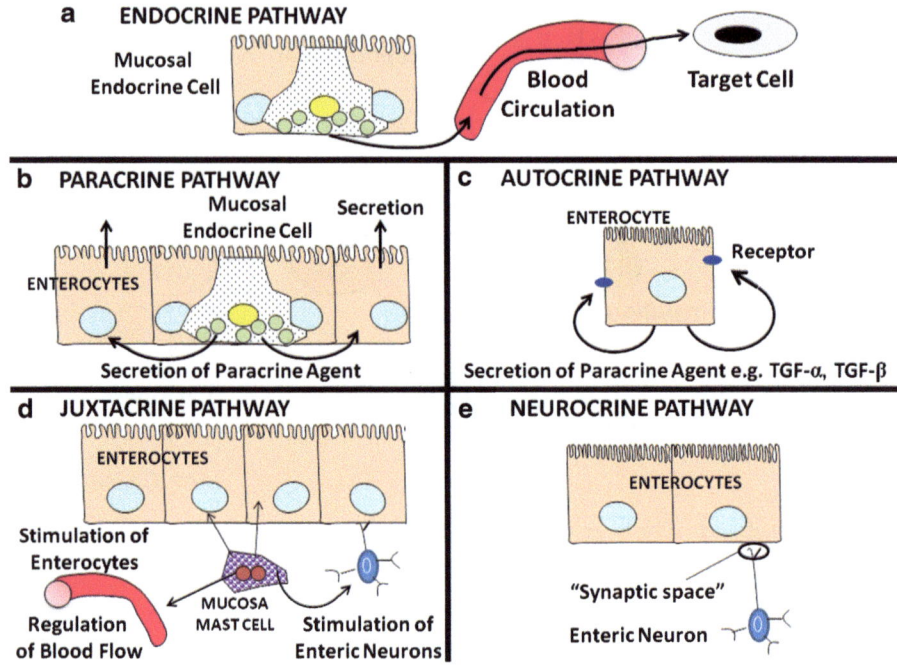

Fig. 1.15 Gut neural and hormonal mechanisms. The regulation and integration of gut functions are mediated by a variety of pathways, which are unique for specialized tasks. (**a**) Endocrine, (**b**) Paracrine, (**c**) Autocrine, (**d**) Juxtacrine, and (**e**) Neurocrine pathways

contents and increased intraluminal pressure. Gastrin stimulates gastric acid and pepsin secretion and activates gastric motility important for grinding up food and emptying gastric chyme into the duodenum. Gastrin also promotes increased tone of the lower esophageal sphincter to prevent reflux of gastric contents. Thus gastrin affects several target tissues simultaneously, which function coordinately to start the initial phase of digestion.

Paracrine Pathway

This represents the mode of action of numerous endocrine cells interspersed among epithelial cells of the intestinal and gastric mucosa. These cells are morphologically distinct, characterized by **epithelial polarity**, and have well-defined **microvillous membranes** at the luminal surface, **whereas nuclei and dense secretory granules** are located at the base of the cells. These cells are capable of sensing changes in the chemical and osmotic conditions of the intestinal lumen, which trigger the release of their regulatory substances stored in the basally located granules at the contra-luminal membrane. Additionally, neural structures and products of other endocrine cells may regulate the release of paracrine factors by stimulating specific receptors on the baso-lateral surfaces of these cells. The released paracrine agents traverse the interstitial space to adjacent target cells, altering their functions or cellular behaviors. This mechanism of action is particularly suitable for the intestine, as the effect is a localized, immediate, and measured response to changes in an ambient environment. Potential systemic effects of these agents are thus avoided and adjacent areas of the gut that may have other important and separately-regulated functions are not affected.

Autocrine Pathway

This pathway is closely related to paracrine action, except that the stimulated release of a hormone message serves to **auto-regulate** the functional or biological behaviors of the cell itself. In some instances, this appears to serve as a **negative feedback mechanism** to auto-regulate the hormonal secretory process. However, these autocrine signals may also serve to auto-regulate cellular functions under special circumstances or at certain stages of development, such as growth and differentiation. For example, TGF-α and TGF-β, which are made and secreted by intestinal epithelial cells at critical points in cellular development, may be involved in the differentiation process of immature crypt cells to more mature villus cells.

Juxtacrine Pathway

This pathway is a newly-described mechanism of action of several regulatory agents that are secreted by **cells in close proximity to their target tissues** within the intestinal mucosa. This includes cells of the **lamina propria**, which are now believed to be important in regulating mucosal functions such as epithelial fluid and electrolyte transport, capillary blood flow, enteric neural stimulation, and motility. For example, a layer of myofibroblasts underlies the intestinal epithelium. These cells elaborate arachidonic acid metabolites, platelet activating factor (PAF), purinergic agonists (adenosine, ATP), and other bioactive substances that modulate the electrolyte transport function of the overlying epithelial cells. It is also believed that these cells secrete **trophic/growth factors** that nurture the epithelial cells and play a role in their development to mature epithelial cells. Cells using this pathway are distinct from paracrine cells in that they are not typical endocrine cells, can exhibit **considerable motility** (e.g. immune cells of the lamina propria), and can **affect numerous target tissues**. Apart from playing a role in regulating physiological processes of the guts, many of these cells also have a prominent role in the intestinal responses to injury and inflammation (e.g. against pathogen invasion of the intestinal mucosa).

Peptides	Non-peptides
Substance P	Acetylcholine
Cholecystokinin (CCK)	Norepinephrine
Somatostatin	Serotonin
Vasoactive Intestinal Peptide (VIP)	Nitric Oxide
Gastrin Releasing Peptide (GRP)	Dopamine
Enkephalins	Purinergic agonist
Calcitonin gene-related peptide (CGRP)	(adenosine, ATP)
Neuropeptide Y (NPY)	

Fig. 1.16 Some typical examples of neuroregulators of the gut. These regulators are divided into peptides and bioactive non-peptide substances

Neurocrine Pathway

This pathway is involved in regulatory agents that are made and released from enteric neurons and function as **typical neurotransmitters**; that is, after release from terminal axons they traverse a narrow "synaptic space" and stimulate specific receptors of the target tissue. This form of regulation is quite *specific* in that the delivery of the neurotransmitters precisely and rapidly affects a designated target cell. On the other hand, the neural regulation of intestinal function can be quite *extensive*, as neurons are organized into intricate complexes capable of producing rapid and faithful patterns of stimulation that simultaneously affect numerous structures to coordinately regulate their functions. These patterns are *hard-wired*, reflexive responses involved in physiological functions such as swallowing, gastric emptying, and defecation, as well as coordination of secretory, motor, and capillary functions. Integrated neural responses to physiological stimuli will occur repetitively and proportionately to the intensity of the afferent stimulus. A variety of target cells of motor neurons is known, including *epithelial, muscle, endothelial,* and other *mesenchymal cells*.

4.2 Organization of Enteric Neurons

Neurotransmitters Involved

The neurotransmitters involved in neurocrine regulation of the gut can be divided into **peptides** and bioactive **non-peptide substances** (Fig. 1.16). Non-peptide agents include serotonin (5-hydroxytryptamine, 5-HT), nitric oxide, purinergic agonists, acetylcholine, and norepinephrine; whereas for gut neuropeptides, numerous have now been identified. In fact, many were originally identified as neurotransmitters of the brain, but some, such as cholecystokinin (CCK) and somatostatin, are also found in endocrine cells of the gut. Neuropeptides are largely concentrated in the intrinsic nerves of the gut.

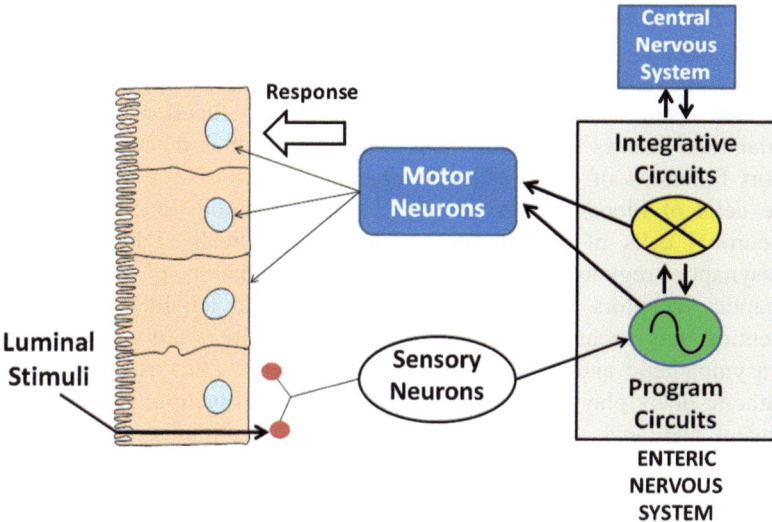

Fig. 1.17 Organization of the enteric nervous system. Enteric neurons have their own sensory neurons, interneurons and motor neurons which are capable of producing motor and secretory responses upon luminal stimuli. These responses can be modified by input from autonomic nerves from the central nervous system. (Modified from Chang EB, Binder HJ. Diarrheal diseases. American Gastroenterological Association Teaching Slide Collection 25 © slide 36, Bethesda, Maryland. Used with permission)

Enteric Nervous System

The intrinsic or enteric nervous system is composed of **afferent (sensory) neurons**, **interneurons**, and **efferent (motor) neurons**, whose cell bodies are located within the bowel wall. Enteric neurons are quite numerous and well organized. As shown in Fig. 1.17, enteric nerves are organized into hard-wired circuits that produce characteristic and rapid patterns of response. Interneurons, nerves that connect and integrate afferent signals to efferent motor neurons, play an important role in integrating the neural response, receiving inputs from sensory neurons and activating several neural targets of intestinal plexuses. Their functions can be modified by inputs from autonomic nerves from the central nervous system. But still, the interneurons are also quite capable of functioning independently despite loss of autonomic neural input.

Sympathetic Innervation

The gut receives extensive innervation by sympathetic and parasympathetic neurons. Most sympathetic fibers of the gut are **postganglionic**. Efferent sympathetic fibers to the *stomach* originate from the celiac plexus, whereas those to the *small intestine* originate from the celiac and the superior mesenteric plexuses.

Fibers to the ***cecum, appendix, ascending colon,*** and ***transverse colon*** arise from the superior mesenteric plexus. The ***remaining portion*** of the colon receives sympathetic fibers from the superior and inferior hypogastric plexuses. Sympathetic neurons to the gut have a variety of functions. Some regulate inter-glandular tissues or closely approximate intestinal mucosa by regulating transport functions of these cells. Sympathetic neurons also innervate smooth muscle cells of blood vessels, thereby regulating blood flow to the intestine, and neuronal cells of the intramural plexus, specifically inhibitory receptors of presynaptic regions of enteric neurons. Stimulation of α_2-noradrenergic presynaptic receptors inhibits the release of acetylcholine and serotonin, representing one mechanism by which sympathetic fibers are counter-regulatory to parasympathetic and serotoninergic stimulation. In other regions of the gut, sympathetic fibers play a role in regulating sphincter function, such as the lower esophageal and anal sphincters where sympathetic stimulation appears to be excitatory.

Parasympathetic Innervation

Parasympathetic innervation to the ***stomach, small intestine,*** and ***proximal colon*** arises from the **vagus nerve**, with its fibers running along blood vessels and ending in the myenteric plexus. ***The rest of the colon*** receives parasympathetic innervation from the pelvic nerves of the **hypogastric plexus**, which also end in the myenteric plexus. All parasympathetic fibers are **preganglionic** and most are **cholinergic and excitatory**. However, other neurotransmitters, such as *substance P, vasoactive intestinal peptide (VIP), and enkephalin-like compounds,* are also made and released by parasympathetic neurons. For example, VIP released from parasympathetic nerve endings causes relaxation of the lower esophageal, pyloric, and internal anal sphincters.

The **efferent response of vagal fibers** requires a continuous exchange of information between the gut and brain. This is mediated by afferent vagal fibers that connect with neurons of the **hypothalamus, globus pallidus**, and the **limbic systems**. In this way, regulation of intestinal functions can be modulated by inputs from the central nervous system. For example, signals from stretch receptors and chemoreceptors of the stomach and alterations in metabolic conditions, such as *glucosideapenia* (low glucose content inside cells), when perceived by the brain, prompt an immediate efferent response to stimulate gastric acid secretion and motility.

4.3 Determinants for the Integration of Gut Functions

You may wonder: why does the gut have such intricate and seemingly redundant regulation systems? Given the complexity and the number of regional and temporal

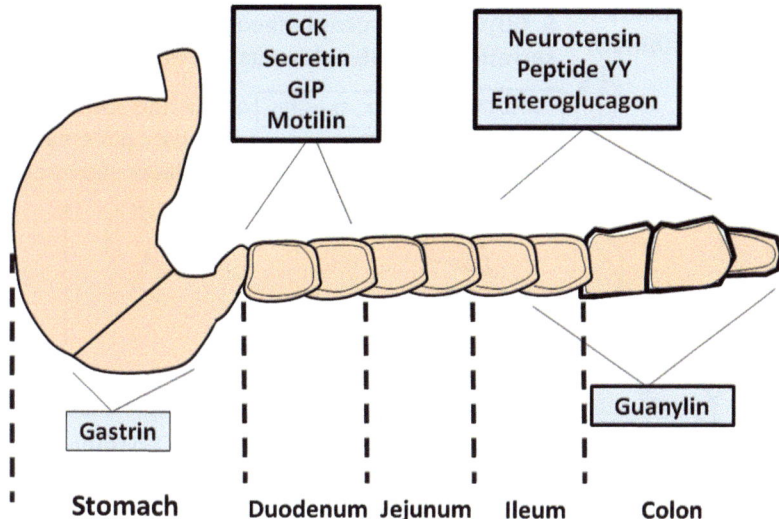

Fig. 1.18 Region-specific distribution of mucosal endocrine cells. The gut endocrine cells are not randomly but regionally specific distributed in the gut

events that must be coordinated for digestive and absorptive functions to proceed autonomously and efficiently, such regulation systems are necessary. Different levels of organization are utilized, as indicated in the examples that follow.

4.4 Regional Distribution of Mucosal Endocrine Cells

Most mucosal endocrine cells are distributed in a **region-specific** way (Fig. 1.18). This actually makes teleological sense, as the physiological roles of many gut peptides depend on their regional presence, either to produce a rapid, appropriate, and measured stimulation of their release or to regulate regional tissue responses. This basic principle can be illustrated by the predominant expression of the peptides *gastrin, CCK, GIP,* and *secretin* in the stomach and upper intestinal tract (Fig. 1.19). The entry of a food bolus into the *stomach* stimulates gastrin release from the antral G cells (mucosal endocrine cells located in the distal one-third of the stomach) into the blood. Through a classical endocrine effect, gastrin initiates gastric motility, increases the tone of the lower esophageal sphincter, and stimulates gastric acid secretion. As the gastric chyme enters the *duodenum*, mucosal endocrine cells are stimulated to release GIP, CCK, and secretin, which activate enteric neural pathways and/or stimulate via an endocrine mechanism. Pancreatic and biliary secretions begin the next phase of food digestion. In addition, these hormones provide feedback inhibition of gastric motility and acid secretion. As will be

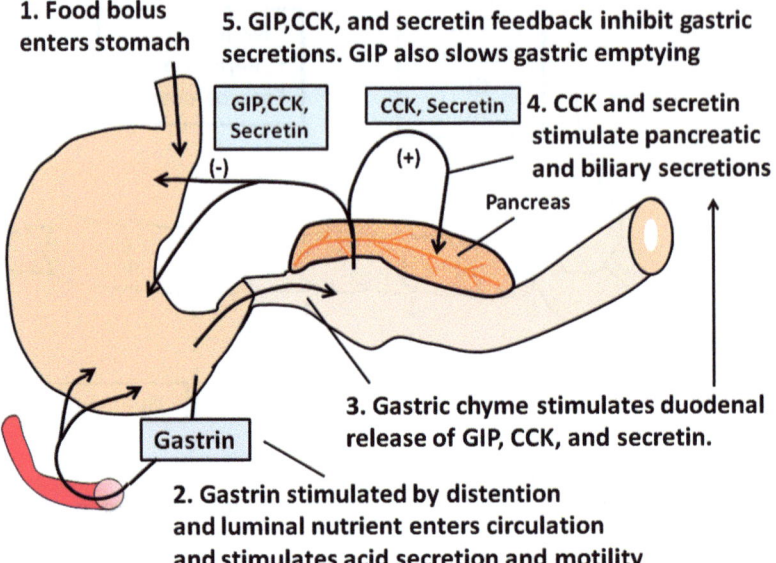

Fig. 1.19 Physiological roles of gut peptide hormones during digestion. The hormones are released in a sequential manner in the regulation of intestinal functions in response to a meal

discussed in the subsequent chapters, precise regulation of the timing of these events and the degree of tissue response are critical for preventing maldigestion of food and development of abnormal symptoms.

4.5 Systemic Integration of Associated Gut Functions

Upon the reception of physiological stimuli, the response of tissues and organs to the stimuli is regulated by the regulatory systems of the gut. This is achieved through the simultaneous and coordinated activation of several key co-dependent processes. As an example, neural regulation of any regions of the gut integrates the processes of intestinal water and electrolyte transport, motility, and blood flow. **Neurocrine regulation of the small intestine** is initiated by chemoreceptors and mechanical stretch receptors (Fig. 1.20). **Afferent neurons** stimulate *interneurons of intestinal neural plexuses* to activate several efferent motor pathways. Some neurons *directly* stimulate *intestinal epithelial cells* to secrete water and electrolytes into the lumen, necessary for providing the aqueous phase and proper chemical composition for luminal digestion and absorption. Concomitantly, *motor neurons* are activated to stimulate segmental contractions of the small intestine, for mixing intestinal contents with digestive juices. Finally, **motor neurons** stimulate *alterations in mesenteric and splanchnic blood flow* necessary for absorption of nutrients and their delivery to the liver.

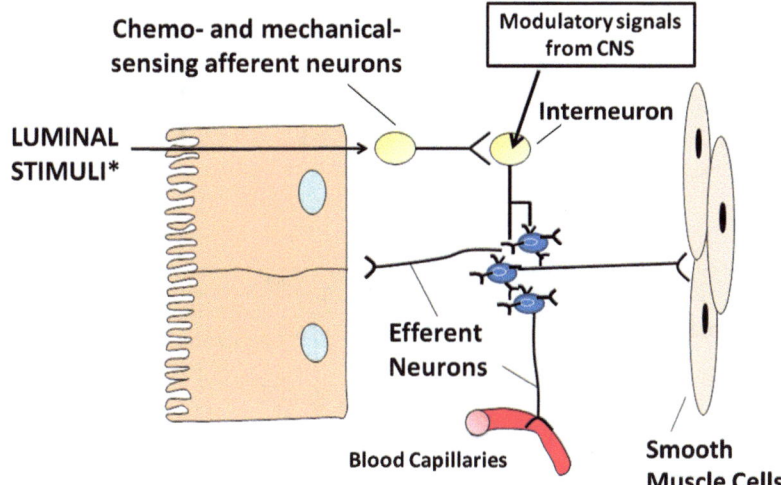

Fig. 1.20 Enteric regulation of gut functions. Neuroregulation of the gut integrates the processes of intestinal water and electrolyte transport, motility and blood flow in response to food stimuli. Luminal pressure and chemical signals activate chemo- and mechano-receptors of the sensing afferent enteric neurons and thus stimulate a response such as simultaneous alterations in motor, capillary and mucosal functions

4.6 Integration of Regulatory Systems of the Gut

Considerable "cross-talk" is essential between regulatory systems of the gut to prevent counterproductive and inefficient actions. An example of this is the relationship between *juxtacrine* and *neurocrine* cells of the intestinal mucosa (Fig. 1.21). Under physiological and pathophysiological situations, juxtacrine cells of the lamina propria (e.g. sub-epithelial myofibroblasts and other mesenchymal cells) are stimulated to secrete a variety of bioactive substances, including arachidonic acid metabolites. These agents can directly stimulate intestinal epithelial, endothelial, and smooth muscle cells. However, their effects are further amplified by activation of secreto-motor enteric neurons, which potentiate the actions of the mediators at the target tissues. Thus, the intensity and regional effects of juxtacrine agents can be significantly enhanced.

Summary

- The **gastrointestinal (GI) system** is massive in size compared to all other organ systems; it consists of *esophagus*, *stomach*, *small intestine*, *large intestine*, *gallbladder*, *liver*, and *pancreas*. In view of this, complex control mechanisms are needed to integrate these seven organs together into a functional correlate.
- The **major functions of the GI system** are *secretion*, *motility*, *digestion*, *absorption*, *metabolism*, and *defense mechanisms,* as well as serving *immunity*.

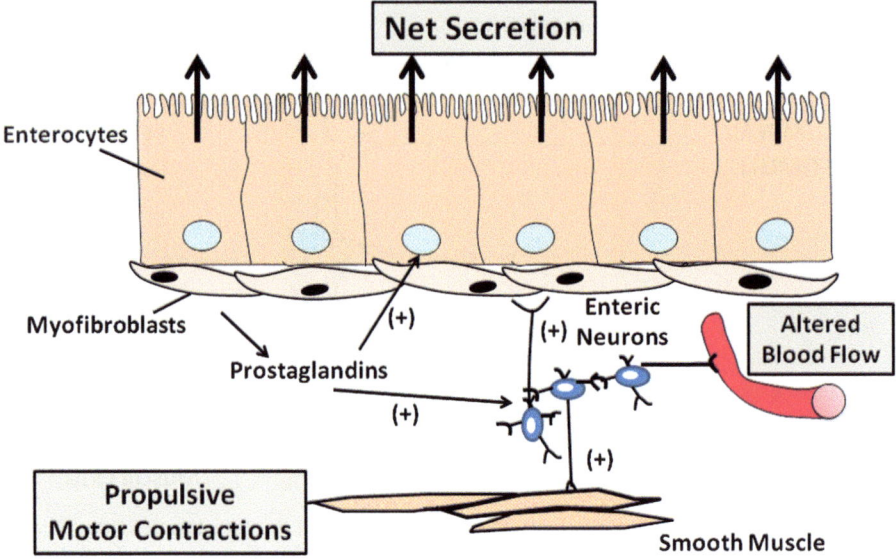

Fig. 1.21 Cross-talk between gut regulatory systems. Such an interaction is exemplified by the effects of prostaglandins on secretory function that can be amplified by secretomotor enteric neurons

- The **GI function is regulated** by the *extrinsic, intrinsic*, and *hormonal* mechanisms via *endocrine*, *paracrine* and *neurocrine* pathways. Each regulatory mechanism does not occur as an isolated event but are all **integrated into one control mechanism**. For example, extrinsic nerves alter hormone release, which are in turn affected by paracrine agents. One representative example is typified by the *control of gastric acid secretion*, where extrinsic nerves, local enteric reflexes, paracrines, and hormones act in a highly integrated manner.

Clinical Correlations

Case Study 1

A 50-year-old man presents with a 6-month history of severe watery diarrhea and dehydration. The diarrhea is characterized by intermittent loose stools or watery diarrhea and occasional flushing. As his symptoms become worse, his family doctor prescribes him with antibiotics and several antidiarrheal medications, but to no avail.

He is admitted to the hospital when found to be orthostatic by blood pressure and pulse (i.e. blood pressure falls and pulse increases, indicative of severe dehydration), hypokalemic (low serum K^+), and hypochloremic (low serum Cl^-). His serum electrolyte abnormalities are corrected by intravenous replacement, but his diarrhea persists in spite of fasting for 48 h. Stool output over a 24-h period averages 1–1.5 L. Stool examination reveals no blood, pathogens, leukocytes, or fat. Diagnostic studies, including endoscopy and barium studies of the bowel, are unrevealing.

A computerized axial tomography (CAT) scan of the abdomen, however, shows a pancreatic mass and numerous lesions in the liver, consistent with tumor metastases. Plasma VIP levels are markedly elevated.

Questions:

1. **What does this patient have?**
   Answer: This patient has a rare endocrine tumor originating from the pancreas called a **VIPoma**. It is a functional endocrine tumor that makes and secretes large amounts of gut peptides; they include **VIP (vasoactive intestinal polypeptide), motilin,** and **neurotensin**. Because these tumors are generally slow growing, patients can tolerate them for years before they become clinically manifested.

2. **What is the pathophysiological basis of the clinical symptoms?**
   Answer: VIP is an important gut peptide typically found in neuroendocrine cells (but also expressed in some enteric neurons) that regulates numerous intestinal functions including motility, intestinal water and electrolyte transport, and blood flow. Because this tumor produces abnormal blood hormone levels, agents like VIP, in particular, can have significant systemic effects. VIP causes vasodilation that is manifested by facial flushing and sometimes hypotension. It is also one of the most potent secretagogues (i.e. stimulants that cause net secretion of water and electrolytes) of the gut; this action is mediated by stimulation of adenylate cyclase and increased cAMP. *VIP-induced diarrhea* persists despite fasting, a clinical feature of secretory (opposed to "osmotic") diarrheal diseases. *Hypokalemia* occurs because of stool losses and the chronic effects of aldosterone, a mineralocorticoid that is secreted by the adrenal gland in response to hypovolemia (decreased blood volume). Aldosterone increases renal reabsorption of Na^+ at the expense of increased excretion of K^+. *Hypochloremia* may be a consequence of increased stool Cl^- losses.

3. **How would you treat this patient's diarrhea?**
   Answer: Because this tumor has already spread, surgical intervention is not feasible. However, treating the patient with the long-acting somatostatin analog, *Octreotide*, can inhibit the effects of the tumor. Octreotide has minor direct effects on intestinal mucosa cells that promote net absorption, probably mediated by activation of the inhibitory G-protein, G_i (thereby inhibiting VIP-stimulated adenylate cyclase activity). Octreotide's major effect appears to be the inhibition of hormone release (and possibly synthesis) by tumor cells. This is evidenced by plasma levels of the peptides, VIP and neurotensin, returning to normal following the administration of Octreotide, which is accompanied by a marked and rapid reduction of daily stool output to normal (<200 gm/day).

Case Study 2

A 30-year-old woman has increasing symptoms of abdominal bloating, weight loss, and diarrhea. The latter is characterized as foul smelling and bulky (typical of fat in the stool, or steatorrhea). Although the pancreas appears to be anatomically normal, exocrine pancreatic functions are abnormal, i.e. decreased secretion of pancreatic enzymes and bicarbonate. Her gallbladder also empties poorly after meals. Fecal fat levels are very high, consistent with fat maldigestion and malabsorption. A small bowel biopsy shows villus atrophy, inflammation, and crypt cell hypertrophy.

Questions:

1. **What does this patient have?**
   Answer: This patient has **Celiac disease (Celiac sprue)**, an inflammatory condition (enteropathy) of the small intestinal mucosa triggered by dietary exposure to gluten. **Gluten** is a water-insoluble protein moiety of cereal grains (primarily wheat) that stimulates an inflammatory or immune response in individuals who are genetically susceptible. The condition is characterized by the development of small intestinal mucosal inflammation, most severe in the duodenum and proximal jejunum. The histology shown above is typical of Celiac disease, showing *villus atrophy* (shortening or absence of villi), *crypt hypertrophy*, and a *lamina propria packed with immune and inflammatory cells*. As an aside, Celiac disease is a common disorder that likely has a genetic basis, but is often clinically silent. It is a relatively "new" disease, as humans only started eating wheat about 10,000 years ago.

2. **What is the pathophysiological basis of the patient's symptoms and abnormal gut functions?**
   Answer: As mentioned above, the duodenum is severely affected by Celiac disease. Many of the mucosal endocrine cells that regulate the early digestive phases of intestinal function are located in this region, including those that secrete CCK, GIP, and secretin. **Compromised secretion of these gut peptides**, after gastric chyme enters the duodenum, impairs pancreatic function because CCK and secretin are required for activating pancreatic enzyme and bicarbonate secretion, respectively. Similarly, gallbladder contraction is dependent on the release of CCK. **Compromised pancreatic and biliary secretions** prevent proper digestion of food. **Injured and inflamed small bowel mucosa** impairs proper absorption of nutrients, water, and electrolytes. Net secretion of water and electrolytes occurs because inflammation stimulates active mucosal secretion and crypts cells are hypertrophied (crypt cells are secretory cells). As a result, the patient develops *maldigestion*, *malabsorption*, *steatorrhea*, *weight loss*, and *abdominal bloating*.

3. **How would you treat this patient?**
   Answer: By going on a **gluten-free diet** (e.g. no more wheat-based bread or pasta), the patient should experience gradual improvement and weight gain. Pancreatic and digestive functions should also return to normal. **Glucocorticoid treatment** is recommended if deemed to do so; glucocorticoid is a potent drug with anti-inflammation and immunosuppression.

Case Study 3

A medical student traveling through South America develops acute, severe watery diarrhea without fever or chills. Stool examination shows no blood, leukocytes, or fat. However, stool cultures are positive for the pathogenic bacteria, enterotoxigenic *E. coli*.

Questions:

1. **How does this bacteria cause diarrhea in the student?**
   Answer: This strain of enterotoxigenic *E. coli* makes an enterotoxin called **heat-stable enterotoxin (ST_a)** that binds to the guanylin receptor on the luminal surface of intestinal epithelial cells. **Guanylin** is a gut peptide made by gut mucosal endocrine cells that stimulates net water and electrolyte secretion by a paracrine pathway. The guanylin receptor is a membrane guanylate cyclase that is also activated by binding with ST_a. Resulting increases in cellular cGMP stimulate cGMP-dependent protein kinases, setting off a cascade of biochemical events that ultimately leads to increased mucosal secretion while inhibiting absorption. Also compare with the mechanism of cholera-induced diarrhea (Chap. 5).

2. **What do you expect from an intestinal mucosal biopsy of this student?**
   Answer: It would be entirely normal (i.e. intact villus architecture). *E. coli* ST_a causes a functional impairment; however, the organism is not invasive and does not cause an inflammatory reaction.

3. **How would you treat this student?**
   Answer: You do not need any specific treatment and **oral rehydration solution** is sufficiently enough in most cases, as the disease is self-limiting in this sort of patients. However, **antibiotics** taken immediately after the development of symptoms can shorten the duration of illness. In the chapter on "Intestinal Physiology", you will learn why cholera is more severe and how oral rehydration solutions work.

Further Reading

1. Avau B, Carbone F, Tack J, Depoortere I (2013) Ghrelin signaling in the gut, its physiological properties, and therapeutic potential. Neurogastroenterol Motil 25:720–732
2. Baldwin GS (2012) Post-translational processing of gastrointestinal peptides. In: Johnson LR (ed) Physiology of the gastrointestinal tract, 5th edn. Academic, New York, pp 43–63
3. Berridge MJ (2009) Inositol trisphosphate and calcium signalling mechanisms. Biochim Biophys Acta 1793:933–940
4. Campbell JE, Drucker DJ (2013) Pharmacology, physiology, and mechanisms of incretin hormone action. Cell Metab 17:819–837
5. Ducroc R, Sakar Y, Fanjul C, Barber A, Bado A, Lostao MP (2012) Luminal leptin inhibits L-glutamine transport in rat small intestine: involvement of ASCT2 and B0AT1. Am J Physiol 299:G179–G185
6. Gomez GA, Englander EW, Greeley GH Jr (2012) Postpyloric gastrointestinal peptides. In: Johnson LR (ed) Physiology of the gastrointestinal tract, 5th edn. Academic, New York, pp 155–198
7. Kuemmerle JF (2012) Growth factors in the gastrointestinal tract. In: Johnson LR (ed) Physiology of the gastrointestinal tract, 5th edn. Academic, New York, pp 199–277
8. Oldham WM, Hamm HE (2008) Heterotrimeric G protein activation by G-protein-coupled receptors. Nat Rev Mol Cell Biol 9:60–71

9. Widmaier EP, Raff H, Stran KT (2011) The digestion and absorption of food. In: Vander's human physiology, 12th edn. McGraw-Hill, New York, pp 516–553
10. Wong TP, Debnam ES, Leung PS (2007) Involvement of an enterocyte renin-angiotensin system in the local control of SGLT1-dependent glucose uptake across the rat small intestinal border membrane. J Physiol 584:613–623
11. Wood JD (2012) Cellular neurophysiology of enteric neurons. In: Johnson LR (ed) Physiology of the gastrointestinal tract, 5th edn. Academic, New York, pp 629–669
12. Wood JD (2012) Integrative functions of the enteric nervous system. In: Johnson LR (ed) Physiology of the gastrointestinal tract, 5th edn. Academic, New York, pp 671–688
13. Yu JH, Kim MS (2012) Molecular mechanisms of appetite regulation. Diabetes Metab J 36:391–398

Chapter 2
Gastrointestinal Motility

Eugene B. Chang and Po Sing Leung

1 Introduction to Gut Musculature and Neural Innervations

Gastrointestinal (GI) motility is an essential function of digestive and absorptive processes of the gut, required for propelling intestinal contents, mixing them with digestive juices, and preparing unabsorbed particles for excretion. Pivotal to the discussion of mechanisms for producing defined patterns of motor responses are important and intimately related aspects of intestinal neurophysiology and motor regulation. Many principles presented in this section are intentionally oversimplified to provide a better conceptual framework for understanding both normal physiology and the pathophysiological mechanisms of various intestinal diseases arising from aberrations in intestinal motility and neurophysiology.

1.1 Gut Smooth Muscle and Its Organization

With the *exception* of the *upper one third of the esophagus* and the *external anal sphincter*, the muscular layers of the bowel wall are made up of **smooth muscle cells**. *Like striated muscles*, smooth muscle contractions of mammalian small intestine are preceded by changes in membrane potential differences. **Depolarization** of the membrane tends to cause the muscle cell to contract, whereas

---

E.B. Chang, M.D. (✉)
Department of Medicine, University of Chicago, Chicago, IL, USA
e-mail: echang@medicine.bsd.uchicago.edu

P.S. Leung, Ph.D. (✉)
School of Biomedical Sciences, Faculty of Medicine, The Chinese University of Hong Kong, Hong Kong, People's Republic of China
e-mail: psleung@cuhk.edu.hk

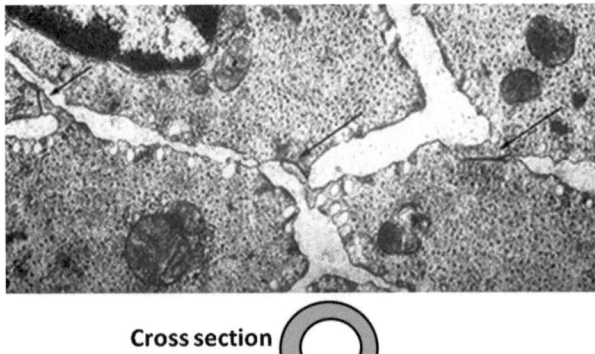

Fig. 2.1 Transverse section of intestinal smooth muscles. Gap junctions (*arrows*) are points of low electrical resistance between cells. They allow depolarizing signals to be instantaneously propagated to a large number of cells (Photomicrograph kindly provided by Dr. Giorgio Gabella)

hyperpolarization has the opposite effect. However, *in contrast to skeletal muscles*, smooth muscle cells have **no striations** because they lack well organized arrays of sarcomeres, are smaller, and are organized into discrete muscle bundles surrounded by connective tissue. Each bundle functions as a single unit; all its smooth muscle cells contract simultaneously to produce an efficient motor response. This is possible because of **gap junctions**, or points of low electrical resistance between cells (Fig. 2.1), through which a depolarizing signal can be instantaneously transmitted to all cells. This also makes unnecessary the neural activation of each individual cell, as each muscle bundle unit could potentially be regulated by a single neuron. In most instances, as will be later evident, each muscle bundle receives several types of neural input that are necessary to coordinate their activities with other motor and intestinal processes.

Most smooth muscles of the GI tract are organized into **two layers** that make up the muscular sheath of the intestinal wall. Cells of the **inner circular muscle layer** are *circumferentially organized* so that each level of the gut muscle fibers simultaneously contracts to produce an annular contraction. The circular muscle layer can be *further subdivided into two lamellae*, one **thick, outer layer** and another **thin, inner layer**. These layers differ electrophysiologically and in the neural innervations they receive: the **inner lamella** receives a greater proportion of fibers from the *submucosal plexus*, while the **outer** is predominantly innervated by the *myenteric plexus* (see below). Muscles of the outer lamella are also larger, less electron-dense, and have more gap junctions. Thus the two lamellae of the circular muscle layer are believed to have different functional responses; however, understanding of this area remains incomplete.

Longitudinal muscles of the gut are oriented such that their *axis runs along the length of the bowel*. With the *exception* of the *colon* where they are organized into separate **thick cords (teniae coli)**, the longitudinal muscles form a continuous outer muscle sheet encircling the bowel wall. Contraction and relaxation of the longitudinal muscle layer cause alterations in the length of the bowel, important for complex motor functions such as *peristalsis* (see Sect. "3.2" below).

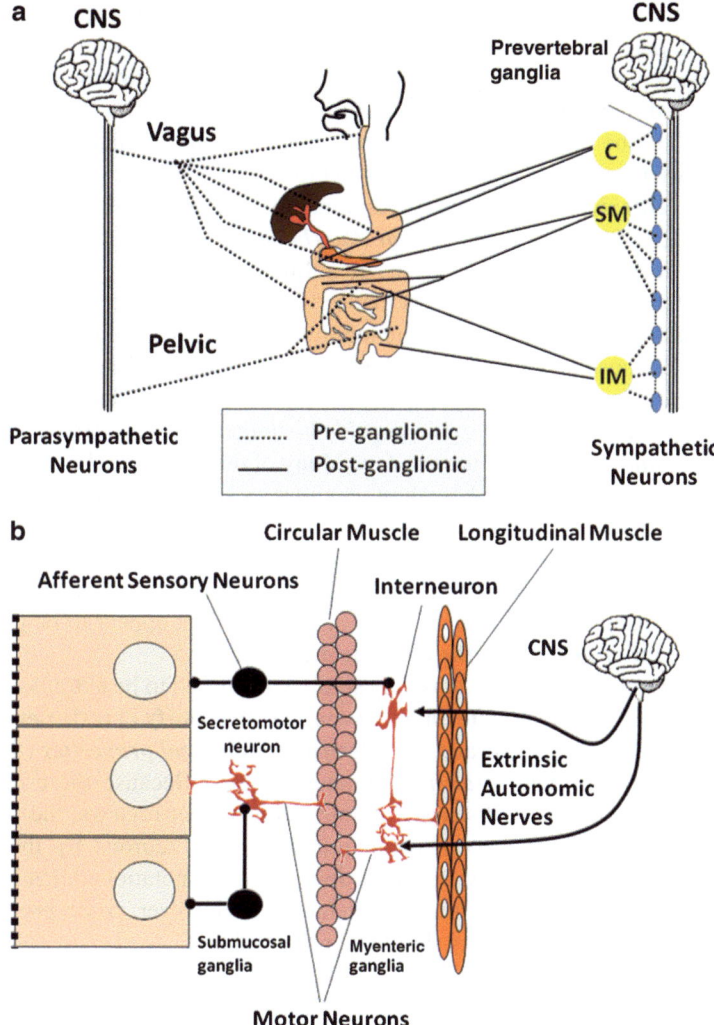

Fig. 2.2 Neural innervation of the autonomous (extrinsic) nervous system and enteric (intrinsic) nervous system of the gut. Gut functions are regulated by (**a**) autonomic nerves and (**b**) enteric neural network. *C* celiac, *SM* superior mesenteric and *IM* inferior mesenteric plexuses, *CNS* central nervous system

1.2 Innervation of Gut Muscle

The gut receives **extrinsic sympathetic and parasympathetic innervation**, but it also has its **own enteric nervous system** capable of *independent* function in many instances. As shown in Fig. 2.2a, **preganglionic parasympathetic motor neurons** reach the gut either via the *vagus nerve* or the *pelvic nerve*, the latter supplying efferent fibers to the distal colon and rectum. Parasympathetic fibers mainly synapse with postganglionic parasympathetic or other enteric neurons located in the intestinal wall plexuses. In contrast, **preganglionic efferent fibers** emanate from the spinal cord and synapse at *paravertebral ganglia*, from which postganglionic fibers project to the intestine, following the celiac, superior mesenteric and interior mesenteric arteries in their respective distributions.

Intrinsic enteric neurons are predominantly found within **two major gut plexuses** (Fig. 2.2b). The *myenteric or Auerbach's plexus*, located between the circular and longitudinal muscle layers, is a thin layer array of ganglia, ganglion cells, and inter-ganglionic nerve tracts that serve to interconnect the plexus. Enteric neurons of this plexus largely innervate longitudinal muscles and the outer lamella of the circular muscle layer. A large number of peptides and nonpeptide neurotransmitters are expressed by myenteric plexus neurons. These include *vasoactive intestinal peptide (VIP)*, *nitric oxide*, *peptide histidine isoleucine (PHI)*, *substance P*, and *neurokinin A (NKA)*, to name a few. Many of these neurons have projections into **adjacent muscle layers**, where they are either excitatory or inhibitory, but some are interneurons involved in integrative functions. The *submucosal or Meissner's plexus* is the other neuronal array found between the submucosal layers and circular muscle. This plexus has neurons that are functionally distinct from those of the myenteric plexus and, relative to intestinal motor function, appear to be projecting mainly to the **inner lamella of the circular muscle layer**.

Enteric neurons have been extensively studied and found to be extremely diverse and complex. At least **three criteria** have been used to **classify enteric neurons**, *i.e. morphology* (Dogiel type I, II and III), *electrophysiological properties* (S and AH neurons), and *type of neurotransmitters* they express. Because these classifications have been extensively discussed by numerous recent reviews, the following discussion will mostly focus on distinction of enteric neurons by the content of neurotransmitters. This provides some functional connotations meaningful for understanding their role in regulating gut motility. However, basic principles of smooth muscle electrophysiology should be reviewed first.

2 Smooth Muscle Electrophysiology

2.1 Myogenic Control System

Slow Waves and Spike Potentials

Two types of electrical activity are found in gut smooth muscle cells; they are *slow waves* which contribute to the *basic electrical rhythm* and *spike potentials*. Except for the esophagus and proximal stomach, most parts of the bowel have a **spontaneous rhythmical fluctuation** in **resting membrane potential** or **slow waves**, generally *between -60 and -65 mV*. This fluctuation represents an interval of cyclic depolarization followed by repolarization and is called the **basic electrical**

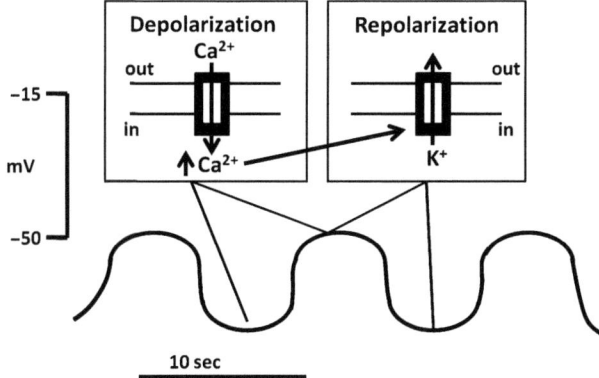

Fig. 2.3 Basic electrical rhythm or BER. The BER is a rhythmical fluctuation in resting membrane potential, characterized by intervals of cyclic depolarization followed by repolarization. BERs arise from specialized pacemaker cells. Depolarization phases are not always accompanied by muscular contraction. Inserts illustrate ion fluxes that are involved in depolarizing phases

rhythm (BER). The frequency of BER oscillations varies from one part of the bowel to another (Fig. 2.3), and depolarization is not necessarily accompanied by muscular contractions.

The **origin of the BER** appears to be *specialized "pacemaker" cells (interstitial cells of Cajal)*. These stellate, muscle-like cells interconnect, have projections to muscle cells, and receive neural fibers from nearby myenteric plexuses. In the **stomach and small intestine**, they are located at the *inner circular muscle layer*, near the myenteric plexus. In the **colon**, dominant pacemaker cells appear to be located in the *submucosal border of the circular muscle layer*. As in the conduction system of the heart, dominant pacemakers, i.e. ones having greater cyclic frequency, capture and entrain subsidiary pacemakers. Thus, in any region of the gut, the electrical characteristics of the dominant regional pacemaker will be not only rapidly propagated circumferentially but also transmitted orad and aborad. However, a *descending gradient* of pacemaker frequency is known to exist in most regions of the gut, with the possible exception of the colon. Thus, when propagation of a pacemaker BER along the long axis of the gut weakens, more distal subsidiary pacemakers will assume the pacemaker role. Studies have suggested that an aboral ascending gradient may be present in the colon. As will be discussed below, this hierarchical arrangement in BER frequency in various regions of the gut may have important functional significance.

Generation of Slow Waves

Because of the tight circumferential electrical coupling of smooth muscle cells, the BER propagation is extremely rapid and can also be measured in adjacent muscle cells in any given gut segment. The **mechanism for generation of the BER** is believed to be the consequence of a **balance** between the *depolarizing*

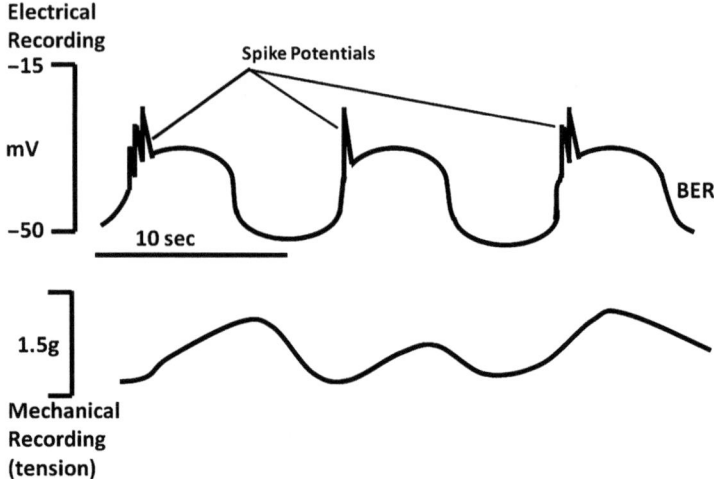

Fig. 2.4 Relationship between BER and smooth muscle contraction. The likelihood and magnitude of smooth muscle contractions depend on the number of rapid changes in membrane potential (spike potentials) which are superimposed on the depolarizing phase of the BER

inward flux of calcium through dihydropyridine-insensitive Ca^{2+} channels and the *repolarizing efflux of K^+* from the cell through Ca^{2+}-activated K^+ channels (Fig. 2.3, insert). The cycle begins with the slow buildup of Ca^{2+} ions in the cell *(a depolarizing process)* that progressively inactivates further Ca^{2+} influx and simultaneously activates Ca^{2+}-activated K^+ channels, leading to K^+ efflux *(a repolarizing process)*.

Although depolarizations of the BER in stomach and colon can occasionally lead to small motor contractions, most contractions occur after rapid changes in membrane potential differences called **spike potentials**, which are superimposed on the depolarizing phase of the BER (Fig. 2.4). These "action potentials" are quite different from those seen in mammalian nerve or striated muscle. Their depolarizing phase is caused by the activation of Ca^{2+}-dependent K^+ channels or increased activity of nonselective cation or Cl^- channels. This is often initiated by stimulation of cells with excitatory ligands. The resulting influx of K^+ depolarizes the cells, causing an activation of voltage dependent Ca^{2+} channels. However, this is only one component contributing to the observed increases in cytosolic Ca^{2+} necessary for producing muscular contraction. In **circular muscle cells**, the other component involves *inositol 1,4,5-triphosphate (IP_3)-stimulated Ca^{2+} release* from intracellular stores, with IP_3 generated from ligand-stimulated hydrolysis of membrane phosphatidylinositol. In **longitudinal muscle cells**, IP_3 appears to have little role in activating increases in $[Ca^{2+}]_i$. Little IP_3 is generated after ligand stimulation, and addition of IP_3 to permeabilized cells has little effect in releasing Ca^{2+}, consistent with the lack of IP_3 receptors regulating internal calcium pools. In these cells, *excitatory ligands* appear to activate Cl^- channels, depolarizing the cells and stimulating the opening of voltage-sensitive Ca^{2+} channels. However, the end

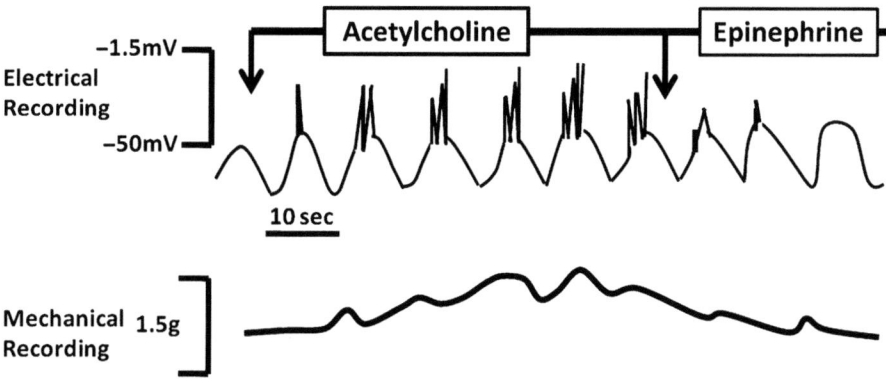

Fig. 2.5 Regulation of smooth muscle contraction. The appearance and frequency of spike potentials are greatly affected by hormonal agents and neurotransmitters. The greater the number of spike potentials there are per BER cycle, the greater is the degree of muscular contraction

result in both circular and longitudinal muscles may be the same. Initial increases in cytosolic calcium augment additional and more sustained increases through the activation of plasma membrane or sarcoplasmic Ca^{2+} channels.

Stimulated increases in Ca^{2+} are required for electromechanical coupling. This explains *why not all phases of smooth muscle BER are associated with contraction.* Even at the peak of membrane depolarization, if the threshold potential for initiating these events is not achieved, contraction will not occur. Spike potentials superimposed on phasic BER depolarization increase the probability that the threshold potential will be reached (Fig. 2.5). The appearance and frequency of spike potentials are greatly affected by hormonal agents and neurotransmitters. The greater the number of spike potentials per BER cycle, the greater is the degree of muscular contraction. *Epinephrine*, for example, markedly decreases spike potentials and inhibits contraction, whereas *acetylcholine* stimulates them and promotes contraction. The major function of the BER, therefore, is to dictate when contractions can occur in a certain area in the bowel. Its basic rhythm does not appear to be affected by hormones and neurotransmitters. If there are no spike potentials, contractions will generally not occur. On the other hand, if an agent promotes numerous spike potentials, the frequency of muscular contractions cannot exceed that of the BER.

2.2 Electromechanical Coupling

Contraction of Smooth Muscles

How do stimulated increases in cytosolic Ca^{2+} cause contraction of smooth muscles? Increased cytosolic Ca^{2+} binds to the calcium-regulatory protein

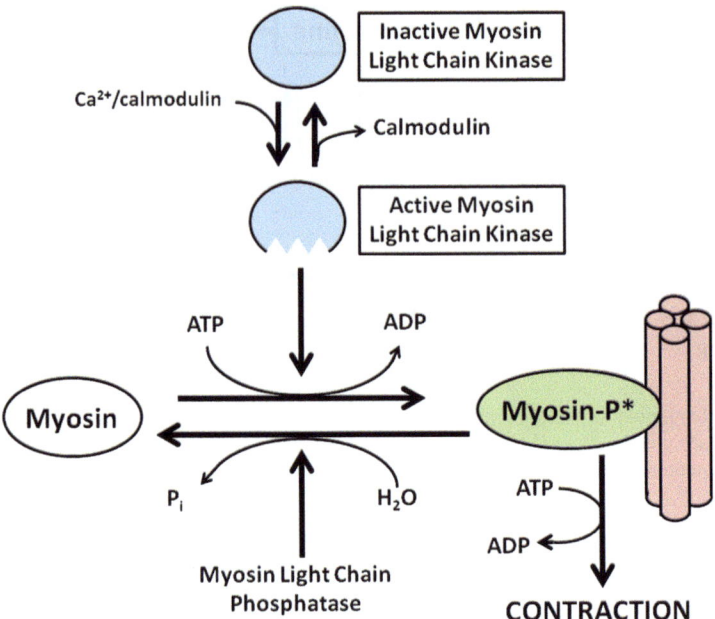

Fig. 2.6 Ca^{2+}-dependent smooth muscle contraction. Increased cytosolic Ca^{2+} binds to calmodulin, a calcium-regulatory protein important for activating myosin light chain (MLC) kinase. Activating MLC kinase stimulates the phosphorylation of myosin, a critical step in initiating the actin-myosin interaction essential for muscular contraction

calmodulin, forming a complex that activates **myosin light chain kinase (MLC kinase)** (Fig. 2.6). MLC kinase stimulates the phosphorylation of myosin, a critical step in initiating the actin-myosin interaction essential for muscular contraction. **Protein kinase C**, activated through the generation of diacylglycerol, also appears to have a role, especially in maintaining the duration of contraction.

Relaxation of Smooth Muscles

Relaxation of smooth muscles involves **restoration of basal Ca^{2+} levels, decreased MLC kinase activity,** and **commensurate increase in MLC phosphatase action**. Several mechanisms appear to be involved in decreasing cytosolic Ca^{2+}, including the activation of Ca^{2+}-ATPase to pump Ca^{2+} out of the cells or back into internal Ca^{2+} stores. In addition, repolarization of the plasma membrane inhibits further entry of Ca^{2+} via voltage-activated calcium channels. MLC phosphatase causes the dephosphorylatioin of myosin, with cessation of the actin-myosin interaction.

3 Neural and Hormonal Regulation of Gut Motility

3.1 Neurohumoral Regulators

Gut motility is regulated by the balance of effects received from excitatory and inhibitory regulatory agents. These agents can regulate motor functions in several different ways (Fig. 2.7). Some neurotransmitters and hormones have direct effects on smooth muscle, including agents such as *acetylcholine, epinephrine, norepinephrine, tachykinins* (such as *substance P*), and *opioid peptides* such as *enkephalins* and *dynorphins*.

Inhibitory Agents

Inhibitory agents generally work by stimulating **increases in cyclic adenosine monophosphate (cAMP)** or **cyclic guanosine monophosphate (cGMP)**, secondary messengers that appear to counteract the effects of increased cytolic Ca^{2+} in several ways. For example, the stimulation of cAMP- and cGMP-dependent protein kinases have been shown to inhibit IP_3 formation and its effects on Ca^{2+} release from internal calcium stores. These kinases may also stimulate mechanisms involved in the **re-uptake or lowering of cytosolic calcium**, including stimulation of Ca^{2+}-ATPase of plasma membrane or cellular organelles. As a third mechanism,

Excitatory Agents	Mode of Action
Acetylcholine Serotonin Opioid Peptides Substance P	Increased Ca_i^{2+}
CCK Bombesin	Increased release of acetylcholine and substance P
Opioid Peptides	Inhibit adenylate cyclase

Inhibitory Agents	Mode of Action
VIP β-Adrenergic agonists Glucagon PHI Nitric oxide	Increased cAMP or cGMP
Somatostatin NPY α-Adrenergic agonists	Inhibit release of ACH and substance P

Fig. 2.7 Regulation of intestinal motor function. Agents involved in the regulation of intestinal motor function may be excitatory (*upper panel*) or inhibitory (*lower panel*). The mode of action of each representative agent is shown on the *right* side of each box

increases in cyclic nucleotides may play a role in stimulatory mechanism, such as Ca^{2+}-dependent K^+ channels, which repolarize the cell and inhibit the actions of excitatory stimuli. Inhibitory agents that stimulate increases in smooth muscle cell cAMP include *vasoactive intestinal peptide (VIP), glucagon,* and *peptide histidine isoleucine (PHI) or peptide leucine methionine (PHM),* in humans. *Nitric oxide* is inhibitory because it stimulates soluble guanylate cyclase and increased cellular cGMP levels.

Excitatory Agents

Most excitatory agents of intestinal motility work by stimulating **increases in cytosolic Ca^{2+}** and/or **inhibiting the formation of cyclic nucleotides**. For instance, *acetylcholine*'s excitatory actions are mediated by muscarinic receptor-stimulated increases in cytosolic Ca^{2+}. Other agents such as *μ- and δ-opioid receptor agonists* also inhibit adenylate cyclase activity, in addition to stimulating increases in cytosolic Ca^{2+}. *Serotonin (5-hydroxytryptamine, 5-HT)*, on the other hand, simultaneously stimulates increases in cytosolic Ca^{2+} through 5-HT_2 receptors and increases in adenylate cyclase activity and cAMP via stimulation of 5-HT_4 receptors. These seemingly counterproductive actions probably serve as a form of cellular feedback regulation.

Other Regulatory Agents

Other regulatory agents of intestinal motility **act indirectly**, either by **stimulating or inhibiting the release of other neuropeptides or hormones**. As an example of this modulatory role, *cholecystokinin (CCK)* and *bombesin* stimulate intestinal motility by enhancing the release of excitatory peptides such as acetylcholine and substance P. By contrast, neuropeptide Y and somatostatin inhibit the release of these agents. As another layer of complexity, agents such as *somatostatin* can further inhibit motility by inhibiting the release of opioid peptides that normally would inhibit VIP release of nitric oxide production. The net effect, therefore, is decreased inhibitory tone on release of VIP secretion, causing inhibition of motility.

3.2 Types of Gut Motility

Digestive and absorptive functions of the gut are very much dependent on certain characteristic patterns of motility which is present in various parts of the bowel. Each serves important functional roles, including *mixing, propulsion*, and *separation* of luminal contents. These actions are possible because of the coordinated interaction of excitatory and inhibitory neurons. As schematically shown in Fig. 2.8, stimulation of intramural baroreceptors by the presence luminal contents

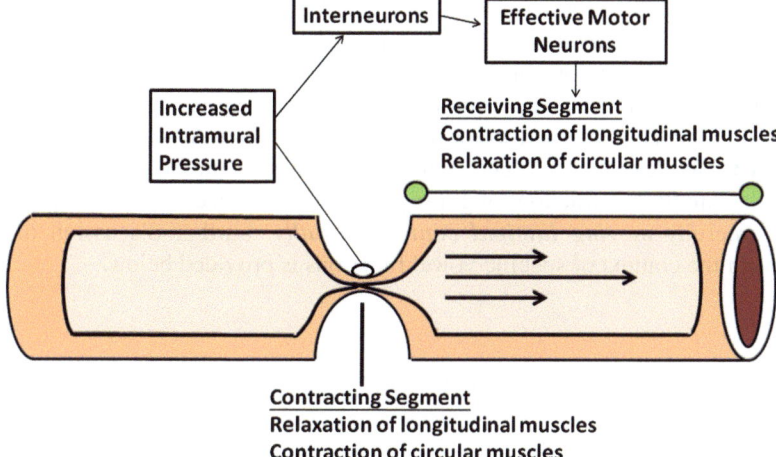

Fig. 2.8 The mechanism of peristaltic reflex. Contraction and relaxation of each adjacent segment of the gut involved are highly integrated

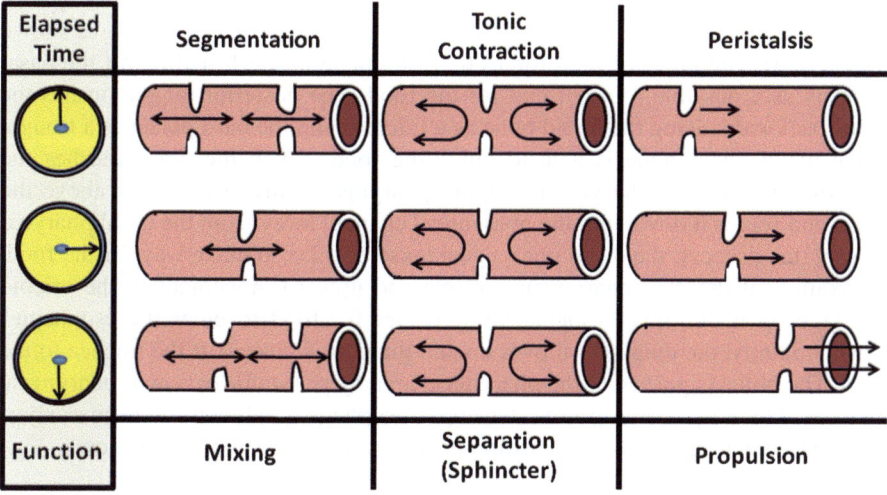

Fig. 2.9 Types of gastrointestinal movement. There are three major types of gut motility patterns observed in the gut, namely segmentation, tonic contraction and peristalsis. Each serves a specific function for the digestion and processing of luminal contents

simultaneously activates proximal excitatory neurons that stimulate contraction and inhibitory neurons in adjacent segments that induce receptive relaxation. Thus, the regulation of each segment of bowel must be dynamic and integrated with other parts of the gut, or the consequences could be potentially disastrous.

There are **three major types of motility pattern** in the gut (Fig. 2.9). (**1**) **Segmentation** is the non-propulsive annular contraction of the circular muscle layer that is predominantly found in the *small and large intestines*. Its major function is to *mix intestinal chyme* through its squeezing action; (**2**) **Tonic contractions** characterize certain regions of the gut that serve as *sphincters* for dividing the gut into functional segments. Important aspects of these regions are mechanisms that can inhibit tonic contraction to *allow luminal contents to pass at appropriate times*; (**3**) **Peristalsis** describes a highly integrated, complex motor pattern marked by sequential annular contractions of gut segments that produce a sweeping, propulsive wave forcefully *moving luminal contents distally*. Further discussion of these patterns in the context of specific bowel functions is provided below.

4 Chewing and Swallowing

Chewing of food is a *voluntary* act, though in some de-cerebrate patients chewing responses to food may occur reflexively. Chewing stimulates output from the salivary glands, stomach, pancreas and liver via neural and endocrine reflex pathways. **Swallowing** of food *begins voluntarily but then proceeds reflexively* through extrinsic nerves of the pharyngeal muscle.

An illustration of afferent and efferent systems involved in swallowing is shown in Fig. 2.10. The bolus of food forces the soft palate upward, sealing off the nasal pharynx as a possible route of exit for the food. The posterior tongue pushes the bolus backward, using the hyoid bone as a fulcrum and the hard palate as a trough, along which the tongue is able to guide the food. When the bolus reaches the posterior wall of the pharynx, the upper pharyngeal muscle contracts above the bolus and forces it down into the pharyngeal channel away from the nasal pharynx. During this process, the distal pharyngeal muscles relax to make way for the food. Relaxation of the cricopharyngeus muscle and upward movement of the cricoid cartilage open the upper esophageal sphincter for food to enter the esophagus. Concomitantly, the epiglottis moves up and the glottis closes off the trachea to the food. The bolus is pushed to the pharyngeal area by peristaltic action of pharyngeal muscles. Once the bolus passes the upper esophageal sphincter, the pharyngeal muscular structure returns to its resting position and awaits another swallow.

5 Esophagus Motility

The sole purpose of the esophagus, which is about 25 cm in length, is to convey food from the pharynx to the stomach. To achieve this, the esophagus is coordinated at both ends by a structure, called sphincter, the **cricopharyngeus** or the **upper esophageal sphincter (UES)** at the top and the **lower esophageal sphincter (LES)** at the bottom. The UES is easily identified anatomically with the thickening of

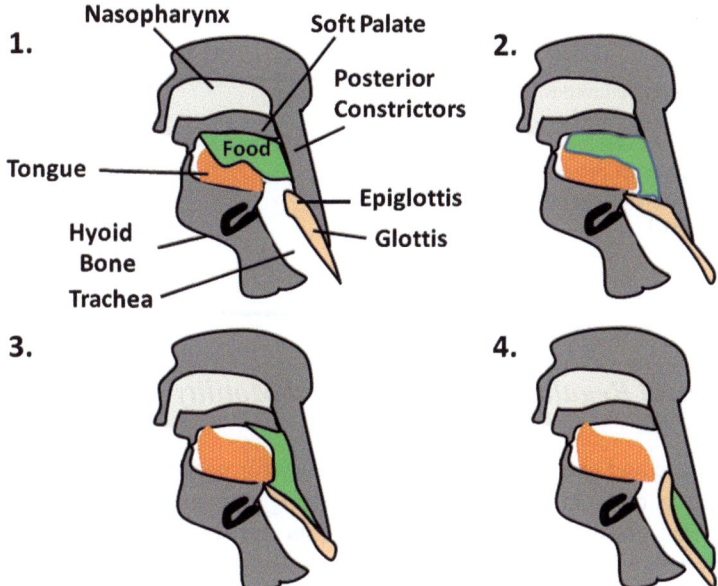

Fig. 2.10 Sequential events involved in chewing and swallowing food. (*1*) It is initiated voluntarily and swallowing of food thereafter proceeds reflexively. This is achieved by forcing the soft palate upward and the bolus of food seals off the nasal pharynx. (*2*) The posterior tongue pushes the bolus backward. (*3*) The upper pharyngeal muscle contracts above the bolus at the posterior wall of the pharynx, forcing it down into the pharyngeal channel away from the nasal pharynx. Distal pharyngeal muscles relax to receive the bolus, and the upper esophageal sphincter opens to allow passage into the esophagus. Concomitantly, the epiglottis moves up and the glottis closes off the trachea to prevent entry of food into the respiratory tract. (*4*) With passage of the bolus through the upper esophageal sphincter, the pharyngeal muscles return to their resting position and await another swallow

circular muscle and elevated pressure; while the LES is not evident as an anatomical entity, it exists functionally with elevated pressure.

In humans, the upper one third of the esophagus consists of striated muscle and the lower one third of smooth muscle, while the middle one third is mixed. The sequence of events following a swallow is termed ***primary peristalsis***. When swallowing is initiated, the UES opens, and food is pushed into the upper esophagus by a peristaltic wave that begins in the pharynx, closing the UES after it sweeps over it, and moves distally at about 2–4 cm/s through the body of the esophagus. When the peristaltic wave approaches the lower esophagus, the LES relaxes to let food enter the stomach. Contractions stop at the gastro-esophageal junction and do not proceed into the gastric musculature. ***Secondary peristalsis*** waves may also be initiated in the absence of swallowing. These are propulsive muscular contractions that are stimulated by distention of the body of the esophagus. They serve to clear retained food and fluid or remove refluxed gastric contents from the esophagus.

Fig. 2.11 Some factors that determine the muscle tone of the lower esophageal sphincter. Agents believed to have physiological roles in the regulation of lower esophageal sphincter pressure can be excitatory (*left*) or inhibitory (*right*)

Excitatory Agents	Inhibitory Agents
α-Adrenergic Agonists	Nitric Oxide
Bombesin	Acetylcholine
Met-enkephalin	VIP
Histamine	
	CGRP (?)
	CCK
	Somatostatin
	Glucagon

5.1 Neural Regulation of Esophageal Motility

Both intrinsic and extrinsic nerves are necessary for peristalsis to occur. However, their relative roles in initiating and controlling peristalsis remain uncertain. It is likely that **cholinergic fibers** of the **vagus nerve** play an important role in modulating the sequence of events. More recently, the role of **nonadrenergic, noncholinergic intrinsic neurons** in mediating peristalsis has been described, especially neurons that make *vasoactive intestinal peptide (VIP)* and *nitric oxide*.

The LES has a very important function in allowing food to enter the stomach and preventing the reflux of gastric contents back into the esophagus. At least two serious diseases namely *reflux esophagitis (gastro-esophageal reflux disease or GERD)* and *achalasia* (discussed below) can be attributed to **dysfunction of the LES**. Sphincter pressure is affected by many endogenous factors in the body, as well as by several food substances (Fig. 2.11). For example, coffee, fats, and ethanol sometimes decrease LES pressure, and many neural and hormonal agents have been shown to affect it, although only a few have been identified as physiologically important. Some appear to directly affect smooth muscle cells, and others modulate interneuron and/or motor activities. Those believed to be physiologically relevant inhibitory regulators include nitric oxide and VIP; they may be secreted from enteric neurons in the LES that receive input from the vagal fibers. Agents such as *calcitonin gene-related peptide (CGRP), CCK, somatostatin,* and *glucagon* appear to decrease sphincter tone experimentally, but neural fibers expressing these agents have not been found in the LES. Thus, these agents may work by an endocrine or possibly paracrine pathway and may have relatively minor or predominantly modulatory roles in LES regulation. Although stimulation of the sphincter by *prostaglandins* can also decrease tone, they are not believed to be important in normal circumstances. Of the agents that increase sphincter pressure, *acetylcholine, α-adrenergic agonists, bombesin,* and *met-enkephalin* have been implicated as having important physiological roles. *Histamine* can both increase and decrease sphincter tone, depending on the receptor subtype stimulated. *Gastrin* is now believed to have a role in regulating sphincter tone.

5.2 Esophageal Motility Dysfunctions

Given the complexity of the regulatory mechanisms of esophageal motility, it is not surprising that many clinical disorders have been described, some being quite common and other debilitating. *Achalasia* is a condition characterized by **dysphagia** (*i.e.* difficulty in swallowing) that results from the *failure of the LES to relax*, thereby causing a **functional obstruction**. Furthermore, there is a loss of peristalsis of the esophageal body. Although the etiology and pathogenesis of the disease are still uncertain, many patients demonstrate a loss of ganglion cells of the myenteric plexus. These findings have led us to believe that loss of critical inhibitory enteric neurons results in a LES that is tonically contracted. Alternatively, there is evidence of neural defects in the vagal dorsal nucleus of the brainstem, which could produce the motor abnormalities observed in achalasia. *Tertiary contractions* are often seen, particularly with age. These are non-peristaltic, spastic, uncoordinated contractions usually occurring in the distal esophagus. Although they are always abnormal, they usually do not cause symptoms. *Reflux esophagitis or GERD* is a disease that arises when *LES tone is reduced*, thereby allowing gastric contents to reflux into the lower esophagus. Since gastric juices are corrosive to the esophageal mucosa, the distal esophagus becomes inflamed and sometimes ulcerated. Patients experience extreme heartburn, and not infrequently severe scarring and stricture at the lower esophagus develop, which can sometimes cause an **anatomical obstruction** to the passage of food.

6 Gastric Motility

There are **three major important motor functions of the stomach**. First, its muscle relaxes to accommodate large volumes during a meal, *i.e. receptive relaxation*; second, its contraction mixes ingested material with gastric juice, thus facilitating digestion, solubilizing some constituents, and reducing the size of the food particles, *i.e. mixing and propulsion*; third, its contraction propels gastric content, called *gastric chyme*, into the duodenum at a regulated rate so as to provide optimal time for intestinal digestion and absorption, *i.e. gastric emptying*. The first of these functions is possible through a mechanism called *receptive relaxation*. After the LES opens, the fundus and upper body of the stomach relax to receive the bolus of food. This portion of the stomach has the ability to accommodate a wide range of gastric volumes (0.5–1.5 L) without significant increases in intragastric pressure. Relaxation of the proximal stomach associated with the act of swallowing is mediated by vagal inhibitory fibers because **vagotomy** (surgical interruption of the vagal nerve to the stomach) reduces distensibility of the proximal stomach. In contrast, distensibility of the proximal stomach is increased by release of GI hormones such as *secretin, gastric inhibitory polypeptide (GIP),* and *CCK*.

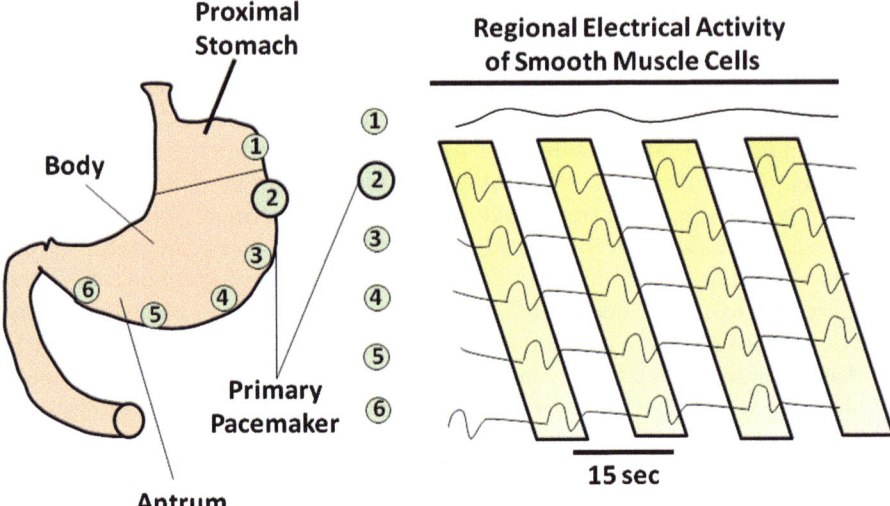

Fig. 2.12 Pacemaker cells in the regulation of gastric peristalsis. Peristalsis, which is the major motor activity of the distal stomach, is initiated and regulated by specialized pacemaker cells in the mid-portion of the greater curvature. Muscular contractions occur when spike potentials appear during the depolarizing phase of the BER. The probability of spike potentials is increased by vagal or gastrin stimulation

The distal stomach must mix gastric contents with gastric secretion and slowly empty the gastric chyme into the small intestine. The major motor activity in this part of the stomach is **peristalsis**, which is initiated and regulated by specialized pacemaker cells in the mid portion of the greater curvature of the stomach (Fig. 2.12). Despite the uniform BER frequency of the gastric motor cells from this area of the stomach to the pylorus, there is a **phase lag of distal segments**, essential to preventing simultaneous depolarization of all parts of the distal stomach. Muscular contractions occur when spike potentials appear during the depolarizing phase of the BER. The probability of contractions is *increased* by *vagal or gastrin stimulation*. With increasing appearance of spike potentials, peristaltic contractions approach and eventually equal the BER frequency of 3–4/min. Conversely, spike potentials are *decreased* by *vagotomy, sympathetic stimulation,* and *secretin*. Because of the phase lag from the midpoint of the greater curvature to the pylorus, peristaltic waves will proceed smoothly in a rostral-to-caudal direction. The motion, referred to as *antral systole*, can last up to 10 s. Pyloric sphincter function is closely coordinated with antral motor activity, augmenting the grinding process (see below) of gastric contents and slowing the emptying of gastric chyme into the small intestine. During antral systole, the pyloric sphincter partially closes, resulting in retropulsion of antral contents, as the antrum contracts. This process facilitates the breaking up of particulate matter and the mixing of food with gastric secretions (analogous to the agitation cycle of a washing machine). Small amounts of food (usually those that have been digested or are more liquid) pass through the pyloric sphincter and enter the small intestine. These events are summarized in Fig. 2.13.

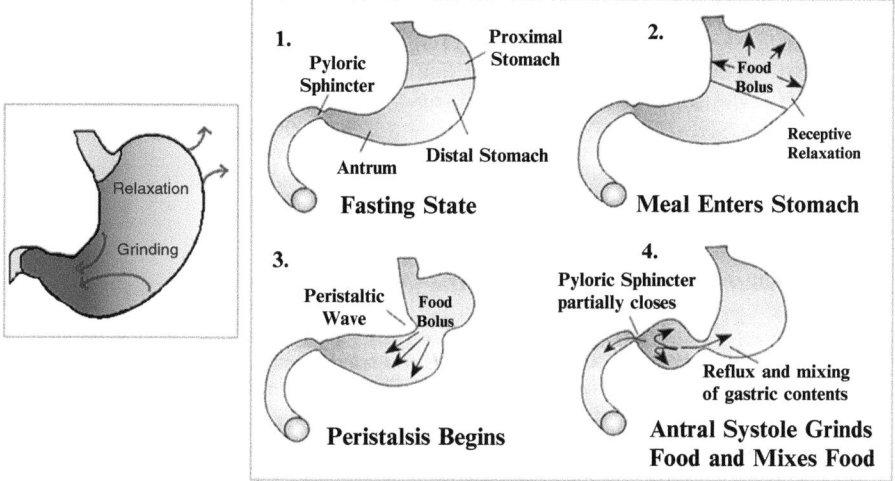

Fig. 2.13 Sequential events during gastric motility. (*1*) The stomach has distinct anatomical regions. (*2*) As the bolus of food enters the stomach, the fundus and upper body of the stomach relax to receive it. (*3*) Through peristalsis, the distal stomach mixes the gastric contents with gastric secretions and slowly empties the gastric chyme into the small intestine. (*4*) As peristalsis proceeds through the gastric antrum, the pyloric sphincter partially closes, resulting in retropulsion of antral contents as the antrum contracts. This process facilies the breakup of particulate matter and the mixing of food with gastric secretions. Small amounts of liquidized and digested food enter the small intestine through the narrow pyloric channel

6.1 Neural and Hormonal Regulation of Gastric Emptying

The small intestine has a limited capacity and can only accept small amounts of gastric chyme at any one period of time. The pyloric sphincter, therefore, plays a very important role in this respect. However, the small intestine also releases a number of hormonal factors that inhibit gastric emptying. These factors are stimulated by the presence of **unsaturated soaps** and *fatty acids of chain lengths of C_{12} or greater*. Denervation fails to abolish the effect of fat in the proximal small bowel on gastric motility. Gastric emptying is also inhibited by a number of **enterogastric reflexes**. For example, the presence in the duodenum and jejunum of *hyper- or hypotonic fluids, acid (pH < 3.5), polypeptides, oligosaccharides, and fatty acids* result in the inhibition of gastric peristaltic contractions. With the exception of the response to fat, inhibition by these substances in the upper small bowel is abolished by **vagotomy**, indicating a neurogenic regulatory pathway rather than a hormonal regulatory mechanism. Over-distension of the bowel also causes inhibition of gastric motor activity, as well as the inhibition of the rest of the GI tract. This reflex is called the **intestino-intestinal reflex**.

In general, receptor activation in the stomach and duodenum sensitive to distension, changes in osmolarity, pH, or lipid content can trigger pathways that regulate gastric emptying. In this regard, the excitatory influences that originate in the stomach are kept in check by inhibitory mechanisms that originate in the duodenum so as to have gastric chyme delivered in metered doses to the duodenum. These inhibitory mechanisms for gastric emptying, *i.e.* neural and hormonal pathways, prevent the upper small intestine from being overwhelmed by material from the stomach. In addition, various **emotional states**, such as *anger, fear* and *depression*, produce changes in gastric motor activity. Figure 2.14 summarizes the intestinal phase pathways that inhibit gastric emptying.

6.2 Gastric Motility Dysfunctions

As described, gastric emptying is regulated at a rate optimal for digestion and absorption of a meal. Abnormalities of gastric emptying thus lead to several clinical problems. Rapid emptying of gastric contents into the small intestine, for instance, can cause **"dumping syndrome,"** characterized by *nausea, pallor, sweating, vertigo*, and sometimes *fainting* within minutes after a meal or ingestion of a hypertonic solution. The symptoms are very similar to the *vasovagal syndrome*, in which massive systemic discharge of vagal fibers causes vasodilation and a fall in arterial blood pressure. The pathophysiological basis of this syndrome may be, in part, related to the exaggerated release of enteric hormones such as *GIP, neurotensin, VIP,* and *enteroglucagon*, which have significant vascular effects. On the other hand, **"gastroparesis"** is a condition characterized by impaired or absent ability of the stomach to empty. This condition is occasionally observed in **severely diabetic patients** who develop autonomic neuropathy. The loss of vagal stimulation to the stomach markedly impairs antral systole, preventing the proper digestion and emptying of gastric contents. These patients often complain of *early satiety (feeling of being full), abnormal bloating, and nausea*.

7 Motility of the Small Intestine

As the major site of digestion and absorption of nutrients, the small intestine has two major motor functions, namely *mixing* and *propulsion*. Multiple short (1–2 cm long) annular constrictions, called **segmentations**, frequently appear in the small bowel and cause an apparent to-and-fro movement of the intestinal chyme. The frequency of segmental contractions is dependent on the regional BER rate, which decreases from proximal to distal small intestine. Thus, the duodenum can contract with a basic rate up to approximately 11 cycles/min, whereas contractions in the ileum can only approach 8 cycles/min. The decreasing gradient of BER frequency promotes the distal movement of intestinal chyme (Fig. 2.15).

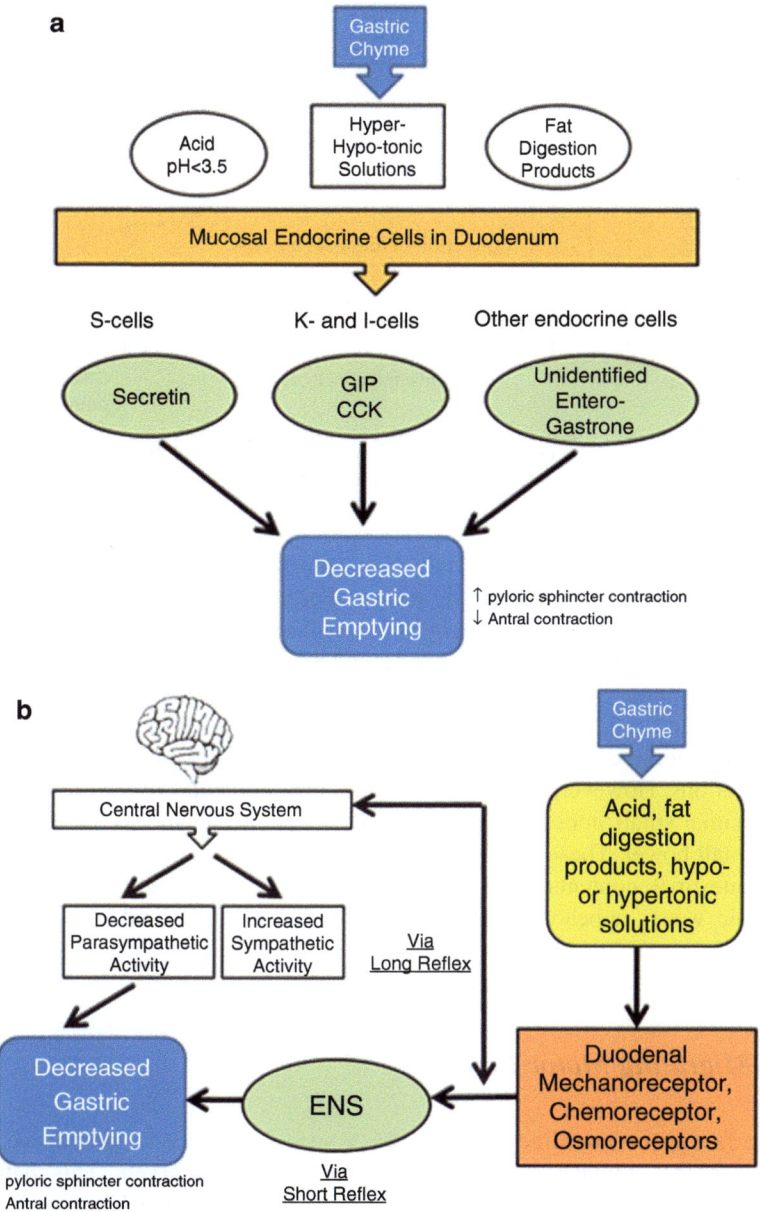

Fig. 2.14 Regulatory pathways for the inhibition of gastric emptying. (**a**) Hormonal pathway. (**b**) Neural pathway

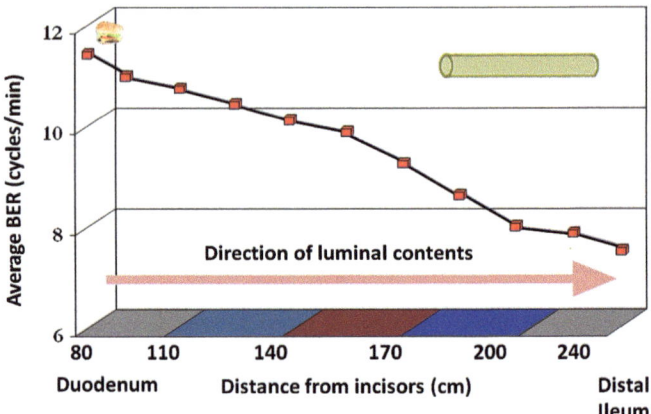

Fig. 2.15 Motility of the small intestine. The decreasing gradient of BER frequencies, which determines the maximal rates of contractions, promotes the distal movement of intestinal chyme

Like other parts of the bowel, muscular contractions of the small intestine are stimulated by extrinsic and intrinsic factors. For example, *CCK, bombesin, opioid peptides* (e.g. met-enkephalin), *tachykinins such as substance P*, and *acetylcholine* are stimulatory. *Sympathetic discharges, α- adrenergic agonists, CGRP, nitric oxide, VIP,* and *glucagon* are inhibitory. *Glucagon*, for example, is frequently used by radiologists and endoscopists to inhibit intestinal motility during diagnostic studies. Propulsive motor contractions of the small intestine are far less frequent than segmentations. However, after ingestion of a meal and entry of gastric chyme into the proximal small intestine, peristaltic-like waves occur, mostly confined to the upper small bowel. These waves travel relatively short distances but unequivocally move intestinal contents in an aboral direction. The stimulus for these short peristaltic waves appears to be the distention of the intestinal wall by luminal contents.

7.1 Migrating Motor Complex

During fasting or interdigestive periods, a specific pattern of propulsive motor activity in the stomach and small intestine can be observed. This pattern is characterized by cyclic motor activity migrating from stomach to distal ileum. Each cycle consists of a **quiescent period (phase I),** followed by a period of **irregular electrical and mechanical activity (phase II)**, and finally by a **short phase of regular activity (phase III)**, where electrical spike potentials peak and mechanical frequencies approximate the BER (Fig. 2.16). This specific peristalsis, called the ***migrating motor complex (MMC)**,* can best be followed by measuring

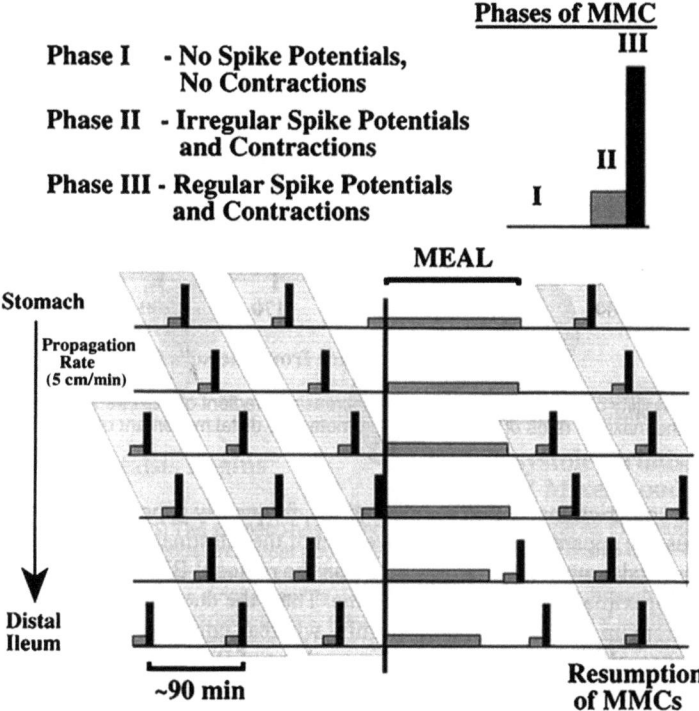

Fig. 2.16 Migrating motor complex of the intestine. Seen during fasting or interdigestive periods, this type of motor activity is characterized by cyclic motor activity migrating from the stomach to distal ileum. A quiescent (*phase I*) period is followed by a period of irregular electrical and mechanical activity (*phase II*) and finally by a short phase of regular activity (*phase III*). MMCs are believed to serve a "housekeeping" function, preparing the bowel for the next meal. MMCs are immediately interrupted when a meal is ingested. MMC indicates migrating motor complex

phase III events. Occurring about every 90 min, the MMC migrates from the stomach to terminal ileum at a rate of approximately 5 cm/m. Several hormones may be involved in the initiation of MMCs, including *motilin*, opioid peptides and somatostatin, albeit to varying degrees and in different regions of the bowel. Although their physiological role is still uncertain, MMCs are believed to have a "*housekeeping*" function, clearing the stomach and small bowel of luminal contents, thus preparing the bowel for the next meal. MMCs are immediately interrupted when a meal is ingested, an inhibitory effect probably mediated by extrinsic neural signals and, in part, by gut peptides such as gastrin and CCK. On the other hand, the orderly propagation of MMCs down the digestive tract is dependent on the regulation by enteric neurons. The enteric nervous system additionally coordinates MMC-motor activities with other intestinal functions such as gastric, biliary, and pancreatic secretions, as well as intestinal blood flow (Fig. 2.17).

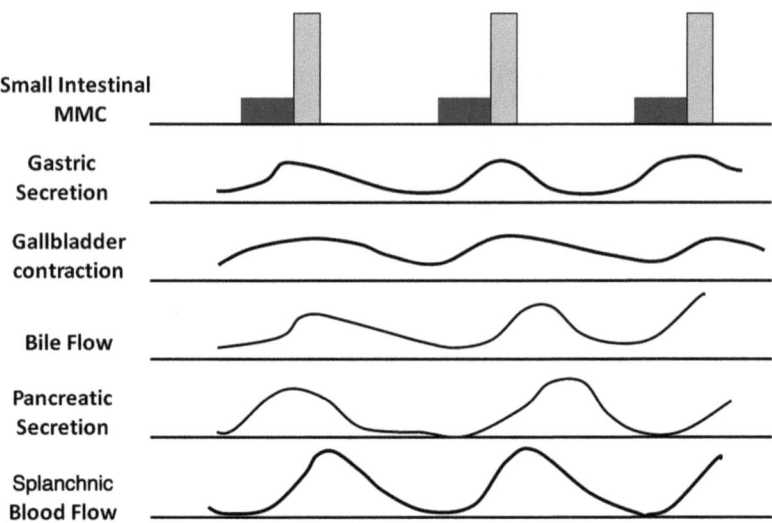

Fig. 2.17 MMC-coordinated digestive functions. The enteric nervous system coordinates MMC-motor activities with other intestinal functions such as gastric, biliary, and pancreatic secretions and intestinal blood flow (Adapted from Wood and Wingate, Gastrointestinal neurophysiology; American Gastroenterological Association Teaching Slide Collection 20, slide 94 ©, Bethesda, Maryland; Used with permission)

7.2 Small Intestinal Dysmotility

Two other motility patterns of the small intestine merit brief discussion, as they are clinically relevant and often occur as a consequence of other pathophysiological processes. The **first** is *"vomiting"*, which is frequently associated with nausea and can be evoked by many different stimuli. It is a stereotypic response which is dependent on both central and enteric nervous systems. When the vomiting center in the brain is stimulated, a diffuse autonomic neural discharge occurs that activates the enteric neural network responsible for the vomiting reflex. The reflex involves the closing of the glottis after inspiration, contraction of the abdominal muscles causing increased abdominal pressure, and retropulsion of intestinal and gastric contents by **retrograde power contractions (reverse peristalsis)** beginning in the upper small intestine. The **second** pattern is an *adynamic state of the small intestine called "ileus"* that is often initiated by noxious stimuli such as abdominal trauma or perforation and occasionally by medications such as opiates. During the state of ileus, bowel sounds are absent and air-fluid levels are observed on plain upright X-ray films of the abdomen. MMCs are absent and the bowel motility is characterized by prolongation of phase I, the quiescent phase of the motor activity.

8 Motility of Large Intestine

The colon of an adult human receives 0.5–2.5 L of chyme per day. This consists of undigested and unabsorbed residues of food, in addition to water and electrolytes. The colon must reduce the volume of this intestinal chyme to about 100–200 g of fecal material before excretion. Hence, the movements of the colon are slow and irregular and serve mainly to segment its content to increase contact with the absorbing surface.

The colon lacks a continuous layer of longitudinal muscle. Instead, the muscles are organized into three flat bands called the *teniae coli.* Contractions of teniae coli cause the colon to open and close in a manner similar to an accordion. Segmental contractions of the circular muscle layer divide the colon into segments called *haustrations* and represent the main motor activity in the colon. Like other parts of the bowels, the frequency of segmentations is dependent on regional BER rates. However, in contrast to the situation in the small intestine, there are increasing gradients of BER frequency from rostral to caudal (Fig. 2.18). This provides an effective means for the more distal part of the colon to mix the luminal contents, allowing the absorptive surface area to maximally extract water and electrolytes and solidify fecal contents.

Three to four times a day, a mass movement occurs wherein the segmental contractions of the left colon disappear and a simultaneous contraction of the right colon propels its content distally. This tends to occur after meals and has been referred to as the **gastric colic reflex**. True peristaltic waves are extremely rare in the colon. Most often adjacent segments of colon appear to contract independent of each other. These segmental contractions are generally increased in amplitude and stimulated by eating or by the presence of bulk in the colon.

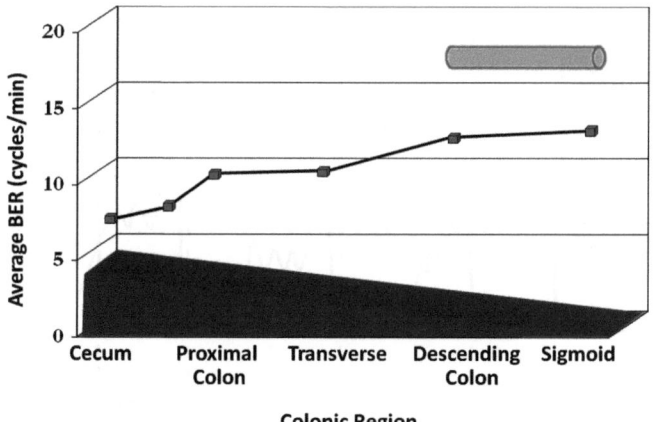

Fig. 2.18 Gradient of basic electrical rhythm (BER) in human colon. There appears to be an increasing gradient of BER frequency from rostral to caudal in the human colon. This provides an effective means for the mixing of luminal contents in the distal part, allowing the absorptive surface area to extract water and electrolytes efficiently and solidify fecal contents

Extrinsic cholinergic nerves can not only directly stimulate mechanical activity of the colon but they also modulate motility by their actions on interneurons and postganglionic enteric motor neurons. In contrast, sympathetic neurons inhibit colonic motility. However, the enteric nervous system plays a major role in regulating colonic motor, transport, and blood flow functions. Like other parts of the bowel, neurotransmitters of the enteric nervous system are inhibitory or excitatory.

9 Rectal Function and Defecation

Motility of the rectum and the anal sphincters are quite distinct from colonic motility. The rectum has **two primary functions: (1)** to serve as a storage site for feces and **(2)** to expel feces during defecation. Thus, the rectum must have the ability to accommodate a certain volume of stool. When this capacity is exceeded, intramural stretch receptors are activated, making the individual aware of the urgency of the forthcoming event and relax the internal anal sphincter. The individual then increases intra-abdominal pressure by forcing air against the closed glottis (**Valsalva maneuver**) and relaxes the external anal sphincter, which is made of skeletal muscle (Fig. 2.19). The individual can inhibit defecation by voluntarily

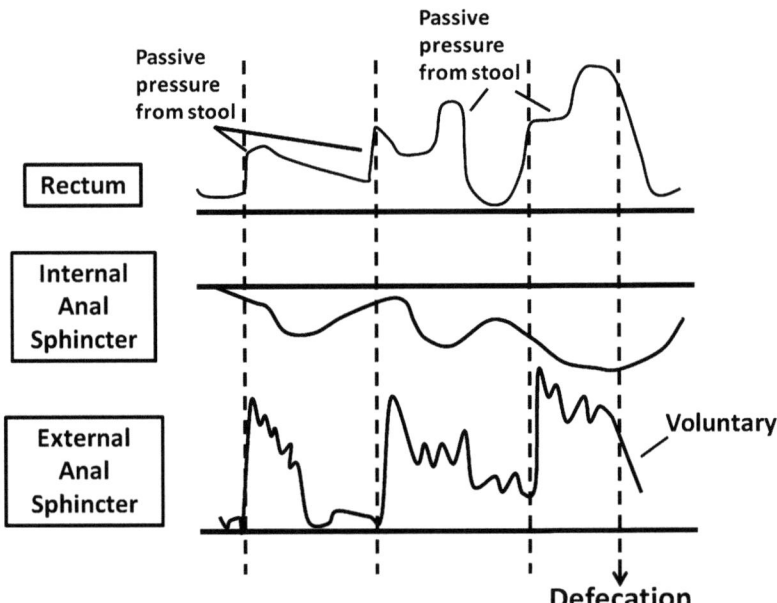

Fig. 2.19 Defecation involves a series of coordinated motor activities. The figure shows regional pressure measurements (Modified from Davenport [4])

increasing the tone of external anal sphincter, which transiently increases internal sphincter tone. However, as rectal pressure builds, the urge to defecate increases and the above processes are again initiated.

Summary

- The GI motility is well organized for optimal digestion and absorption functions.
- It is regulated by neural, hormonal and myogenic control systems.
- The rhythm of the GI tract is generated by myogenic system of the smooth muscle layer that undergoes spontaneous depolarization and repolarization cycles, known as slow-waves which contribute to the **basic electrical rhythm (BER)**.
- The frequency of BER is determined by the slow waves which is essentially constant.
- The force of contractions is determined by the number of spikes fired within each wave, which depends in turn on the neural and hormonal input.

Clinical Correlations

Case Study 1

A 1-week old infant boy presents with a history of vomiting, abdominal distention, and constipation. He has not been nursing well and appears irritable. Physical examination reveals a distended, tympanitic (drum-like) abdomen, and hyperactive bowel sounds, consistent with intestinal obstruction. A plain x-ray film of the abdomen shows a large ovoid mass mottled by small, irregular gas shadows in the right side of the colon, consistent with fecal retention and an area of obstruction is suspected. A barium enema shows narrowing in the distal segment of rectum. An open biopsy of the wall of the narrowed rectum is performed under general anesthesia.

Questions

1. **Based on the clinical observation, what is the diagnosis of the patient?**
   Answer: This child has **Hirschsprung's disease (Congenital megacolon)**.
2. **Explain briefly the pathogenesis of this disease.**
   Answer: Congenital megacolon is characterized by a *functional obstruction of the colon* resulting from **aganglionosis** (an absence of ganglion cells) of the bowel wall. The *rectum* and *sigmoid colon* are the most common sites of involvement. Full thickness biopsy (the diagnostic procedure of choice) shows the hallmark of the disease, which is the absence of ganglion cells in the myenteric and submucosal plexuses. The absence of enteric neurons, especially those secreting nitric oxide and VIP, results in a loss of inhibitory tone necessary to counteract excitatory motor neurons that contract muscle fibers in the anal sphincter and other distal segments of the colon. Consequently, colonic contents cannot pass through the involved segment and the patient experiences symptoms of intestinal obstruction.

3. **How do you treat the patient?**
   Answer: Treatment of the disease involves **surgical removal** of the involved segment.

Case Study 2

A 20-year-old student has a history of type 1 diabetes mellitus (T1DM: juvenile onset and insulin deficient) since age 15 who presents with severe constipation, abdominal distention, and severe heartburn, especially in the supine position. Occasionally, she experiences dysphagia (difficulty in swallowing), particularly when solid foods are ingested. In the past several months, she has also had abdominal discomfort and fullness immediately after eating small amounts of food. On physical examination, she has diabetic retinopathy, orthostatic hypotension (drop in blood pressure after standing), and moderately severe peripheral neuropathy. Upper GI and small bowel barium X-ray studies reveal a dilated stomach, retained food material, and retention of over half of the barium after 30 min. The small intestine is noted to have some nonspecific dilatations. The colon is filled with feces.

From this case presentation, it is a type of T1DM-induced gut dysmotility. Based on your knowledge of diabetes and on the findings of the physical examination, how would you explain the patient's symptoms and radiologic findings at different levels of GI tract dysfunction?

Questions

1. **Esophageal dysfunction?**
   Answer: A major consequence of severe and long-standing diabetes mellitus is the development of **peripheral and autonomic neuropathy**. The latter has numerous effects on bowel function, a major manifestation being the development of abnormal bowel motility. Because the extent and severity of autonomic dysfunction vary considerably from patient to patient, the clinical presentation of patients are quite protean. Many patients experience *esophageal and swallowing symptoms* as a consequence of altered or aberrant sphincter functions and/or peristalsis. **Tertiary contractions** may become more pronounced and frequent. *Loss of lower esophageal sphincter tone* can result in symptoms **of reflux esophagitis** *(heartburn, chest pain)* because of acid-pepsin injury of the esophageal lining. The barium X-ray study shows reflux of the barium column into the esophagus.

2. **Gastric dysfunction?**
   Answer: Autonomic neuropathy can also affect gastric emptying. The stomach becomes flaccid and antral peristalsis is defective or lost. This condition, called **diabetic gastroparesis**, is not uncommon. Barium remains in the stomach for 10 hours after ingestion, indicative of delayed emptying. Gastroparesis can result in the formation of a *bezoar,* a conglomerated mass of undigested vegetable and fruit fiber.

2 Gastrointestinal Motility

3. **Small intestinal dysfunction?**
   Answer: Autonomic neuropathy of the small intestine can cause **intestinal stasis**, a condition that favors luminal colonization and overgrowth by bacteria. These organisms compete for luminal nutrients, thus causing nutrient malabsorption and diarrhea.
4. **Colonic dysfunction?**
   Answer: Finally, the normal function of the colon and rectum is frequently affected by diabetic autonomic neuropathy. Patients can develop *severe constipation or diarrhea.* In addition, the *loss of anal sphincter tone* and the afferent nerves required for sensing the presence of stool in the rectal vault may develop. These patients may lose normal defecation function and experience *fecal incontinence*. The treatment of intestinal complications associated with diabetes is particularly problematic, as the neuropathy is usually irreversible.

Case Study 3

A 40-year-old woman presents with a progressive history of dysphagia. She has difficulty swallowing both solids and liquids and finds that it is worse when she experiences periods of emotional stress or eats too rapidly. She has found that alcohol is of some benefit in helping the food pass into the stomach. Occasionally, when she is supine or is exercising, she regurgitates food particles from meals eaten hours before. Several weeks ago, she was treated with antibiotics for bronchopneumonia. On physical examination, she appears to be normal. The only significant finding is that she has halitosis (bad breath). On chest x-ray film, an air-fluid level is noted in the posterior mediastinal area. A barium swallow is taken which reveals a dilated lower esophagus. Esophageal manometry of the patient is subsequently ordered.

Questions

1. **What would you expect these examinations to show? Explain the pathophysiology of this disease.**
   Answer: This patient has **achalasia,** a condition in many ways similar to Hirschsprung's disease. It is a neuromuscular disorder that affects the ***LES*** and is characterized by the *failure of the sphincter to relax* during the process of swallowing. Normally, the LES is richly innervated by VIP- and nitric oxide-secreting enteric neurons of the myenteric plexus that provide inhibitory tone during the terminal phase of esophageal peristalsis, allowing the food bolus to pass into the stomach. The loss of inhibitory tone and impairment of sphincter relaxation causes a *functional obstruction of the esophagus*. Alcohol can result in transient symptomatic relief, as it causes proximal dilation of the esophagus and retention of luminal contents, which explain the patient's *halitosis* (fermentation of retained esophageal contents) and the air-fluid level observed radiographically. On occasion, especially at night when the patient is supine, the luminal contents of the esophagus are regurgitated and aspirated. This can cause ***bronchopneumonia***, as was the case with this patient.

2. **What would you recommend for the treatment of this patient?**
   Answer: The treatment of this disease is either **esophageal dilation** or **surgical myotomy of the lower esophageal sphincter**. Esophageal dilation is performed with an inflatable balloon or luminal dilator, which splits the sphincter. Patients often require repeated dilation over time, as the condition recurs. Recently, **botulinum toxin injection** is an alternative approach to treating the disease; it is achievable based on the rationale that the toxin is to decrease cholinergic input, thus reducing the LES pressure.

Further Reading

1. Bharucha AE, Brookes SJH (2012) Neurophysiologic mechanisms of human large intestinal motility. In: Johnson LR (ed) Physiology of the gastrointestinal tract, 5th edn. Academic, New York, pp 977–1022
2. Bitar KN, Gilmont RR, Raghavan S, Somara S (2012) Cellular physiology of gastrointestinal smooth muscle. In: Johnson LR (ed) Physiology of the gastrointestinal tract, 5th edn. Academic, New York, pp 489–509
3. Brierley SM, Hughes P, Harrington A, Blackshaw LA (2012) Innervation of the gastrointestinal tract by spinal and vagal afferent nerves. In: Johnson LR (ed) Physiology of the gastrointestinal tract, 5th edn. Academic, New York, pp 703–731
4. Davenport HW (1982) Physiology of the digestive tract. Year Book Medical Publishers, Chicago, p 93
5. Goyal RK, Chaudhury A (2008) Physiology of normal esophageal motility. J Clin Gastroenterol 42:610–619
6. Koeppen BM, Stanton BA (2008) Gastrointestinal physiology. In: Berne & Levy's physiology, 6th edn. Mosby-Elsevier, Philadelphia, pp 487–553
7. Mittal RK (2012) Motor function of the pharynx, the esophagus, and its sphincters. In: Johnson LR (ed) Physiology of the gastrointestinal tract, 5th edn. Academic, New York, pp 919–950
8. Poole DP, Furness JB (2012) Enteric nervous system structure and neurochemistry related to function and neuropathology. In: Johnson LR (ed) Physiology of the gastrointestinal tract, 5th edn. Academic, New York, pp 557–581
9. Sanders KM, Koh SD, Ordög T, Ward SM (2004) Ionic conductances involved in generation and propagation of electrical slow waves in phasic gastrointestinal muscles. Neurogastroenterol Motil 16:100–105
10. Tack J, Janssen P (2010) Gastroduodenal motility. Curr Opin Gastroenterol 26:647–655

Chapter 3
Gastric Physiology

Eugene B. Chang and Po Sing Leung

1 Introduction to the Gastric Functions

Our **stomach** produces about 2L/day of gastric acid secretion/juice. Its high H^+ ions (pH=1–2) kill most of the ingested germs; it catalyzes the conversion of inactive pepsinogen to pepsin. The presence of *acid/pepsin* begins the digestion of dietary protein. However, pancreatic proteases can hydrolyze all ingested protein in the absence of pepsin. An important component of gastric juice is *intrinsic factor* (IF), which binds vitamin B_{12} in the duodenum, allowing it to be eventually absorbed in the distal ileum (see B_{12} absorption of Chap. 9); it is the only indispensable substance of gastric juice. Its deficiency, following gastric surgery or in pernicious anemia, must take injections of B_{12} or oral B_{12} with IF. The surface epithelia cells of stomach secrete mucus and bicarbonate that protect the mucosa from acid/pepsin erosion.

Understanding gastric physiology is important for all physicians in light of the numerous patients who present with stomach diseases. The pharmaceutical industry has taken advantage of the knowledge gained through basic research to develop therapeutic compounds that have revolutionized the treatment of acid peptic-related diseases. In fact, medications for the treatment of dyspepsia and peptic ulcer diseases are currently the most frequently prescribed on the market (see Sect. 5.2).

E.B. Chang, M.D. (✉)
Department of Medicine, University of Chicago, Chicago, IL, USA
e-mail: echang@medicine.bsd.uchicago.edu

P.S. Leung, Ph.D. (✉)
School of Biomedical Sciences, Faculty of Medicine, The Chinese University of Hong Kong, Hong Kong, People's Republic of China
e-mail: psleung@cuhk.edu.hk

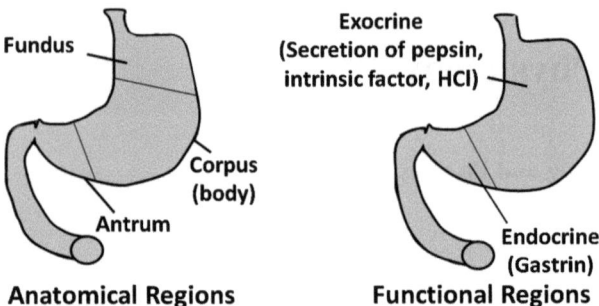

Fig. 3.1 Anatomical and functional regions of the stomach. The anatomical division of the stomach consists of the fundus, antrum, and corpus (body) that reflects differences in motor function. Functionally speaking, the stomach has two major regions: an exocrine or glandular portion that consists of the fundus and body or acid-secreting area, and an endocrine portion that is located in the antrum or gastrin-secreting area

1.1 Functional and Anatomical Organization of the Stomach

As shown in Fig. 3.1, the **anatomical subdivision** of the stomach into the *fundus, antrum* and *body (corpus)* mirrors differences in motor function (Chap. 2). The fundus functions as a reservoir for ingested meals, the body as the initial site of peristalsis, and the antral-pyloric region as the site of the greatest mechanical agitation and mixing of food with gastric secretions. In terms of **gastric mucosal function,** however, the stomach can be divided into **two major regions:** the *"exocrine" or glandular portion (acid-secreting area)* found in the mucosa of the fundus and body and the *"endocrine" part (hormone-secreting area)* located in the antral mucosa. Discussion in this chapter will focus on the gastric mucosal functions.

Acid-Secreting (Exocrine) Area

The mucosa of the exocrine portion of the stomach consists of **simple columnar epithelial cells** that line the luminal surface. These cells secrete mucus and an alkaline fluid, both necessary for protecting the stomach against its own potentially harmful juices. Opening into the mucosal surface are numerous gastric pits that serve as conduits for the secretions of 3–7 oxyntic gastric glands (Fig. 3.2a) into the gastric lumen. The area occupied by the gastric pits is at least 50 % of the total luminal surface area. As shown in Fig. 3.2b, the remainder of the gastric or oxyntic gland can be further divided into **two regions.** The *neck* of the oxyntic gland contains *parietal* and *mucous neck cells*, the latter resembling intestinal goblet cells and secreting mucus. The neck region is also the site of germinal cell proliferation and differentiation, giving rise to surface cells that migrate up to the surface or glandular cells that move downward toward the base. Cells progressively mature as

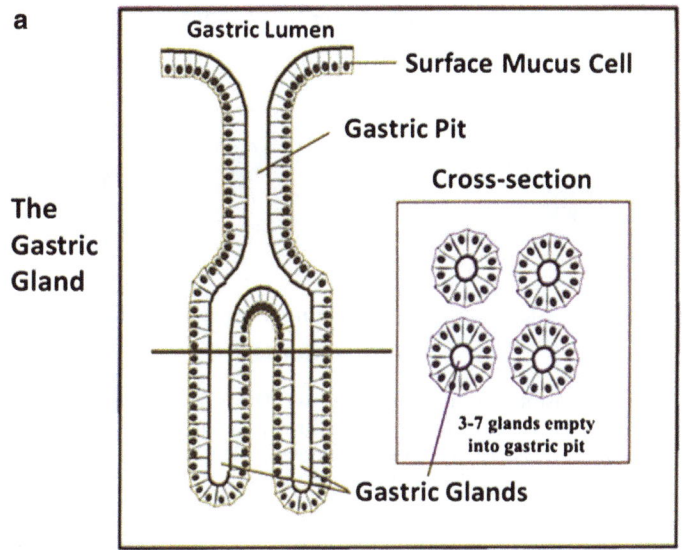

Fig. 3.2 The structure of gastric gland. (**a**) Side and cross-sectional views of gastric pits. (**b**) Various types of cells that are found in the gastric exocrine mucosa

they reach their final destinations and then are continuously replaced by new cells, the turnover of various cell types ranging from days to weeks. The *base* of the gland contains *chief cells* in addition to some parietal and mucous neck cells. The oxyntic gland also contains a number of endocrine-type cells dispersed among chief and parietal cells that play a role regulating their secretory functions.

Each cell type has different functions. **Parietal cells** are the acid-secreting cells, and their cellular physiology will be covered in greater detail later in the chapter. **Chief cells** make and secrete pepsinogen, which is converted by gastric acid to

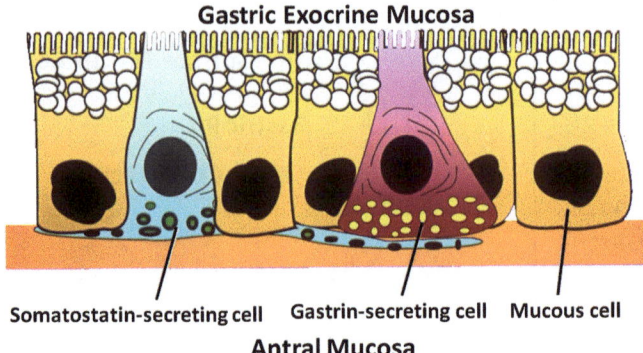

Fig. 3.3 Cells of the gastric endocrine antral mucosa. The two major endocrine cells of the antral mucosa are somatostatin-secreting D cells and gastrin-secreting G cells. The D cells are believed to have dendritic-like extensions of their cell membranes that span to adjacent G cells, which are consistent with the paracrine regulation of mucosal function

the active form pepsin. **Mucous cells** make mucus, an essential component that lubricates and protects the gastric mucosa, and those at surface and in gastric pits also secrete bicarbonate. Interspersed among these cells of the oxyntic gland are *enterochromaffin-like (ECL)* **cells** that regulate their functions by secretion of bioactive amines and peptides. Finally, a number of **mast-like cells** found in close apposition to epithelial cells of the gastric gland secrete histamine. In humans, they are believed to be important physiological regulators of gastric acid and pepsin secretion, an effect mediated through a paracrine action.

Hormone-Secreting (Endocrine) Area

The **antral-pyloric or hormone-secreting area** contains **mucus-secreting cells** similar to those seen in oxyntic glands, as well as a limited number of **parietal and pepsinogen-secreting cells.** However, this region of the stomach is the major source of *gastrin* secreted by the *endocrine G cells* of the mucosa. These cells are pyramidal in shape and have a narrow apical surface with long microvilli (Fig. 3.3). Secretory granules are located at the base of the cell, and the release of gastrin occurs by fusion and exocytosis with the basolateral membrane. *Endocrine D cells*, also found in the antral-pyloric region, elaborate *somatostatin*, an important physiological regulator of gastrin release and gastric acid secretion. D cells appear to have dendritic-like extensions of their cell membranes that span to adjacent G cells, consistent with a paracrine regulation of mucosal function. These processes make the delivery of somatostatin to G cells rapid and highly selective. At least seven additional types of *enteroendocrine cells* can be found in the gastric mucosa; however, their physiological role in the regulation of gastric function is not as well-defined.

2 Gastric Acid Secretion

2.1 Neural and Hormonal Regulation of Gastric Acid Secretion

Acid secretion is regulated by nerves and GI hormones. Although the understanding of their mechanisms of actions is still incomplete, accumulated evidence has been learned about the regulation and cell biology of acid secretion to have had a major impact on the treatment of patients with peptic ulcer disease.

There are **three major stimulators of acid secretion: (1)** *gastrin* (predominantly secreted by antral G cells), **(2)** *histamine* (in humans probably arising from mast-like cells in the lamina propria subjacent to the gastric epithelium), and **(3)** *acetylcholine* (secreted from postsynaptic vagal fibers innervating the gastric mucosa). On the other hand *Somatostatin*, secreted from *antral* and *oxyntic gland D cells* as well as *pancreatic islet cells*, and *prostaglandins* from *mucosal cells*, are the **major paracrine inhibitors of gastric acid secretion.**

The interaction of these regulators is extremely complex, and several issues remain controversial. For instance, it is still not clear at what level certain agents regulate acid secretion. Specific receptors on parietal cells for histamine, gastrin, and acetylcholine have been demonstrated, and their stimulation results in acid secretion (Fig. 3.4). **Histamine** stimulates H_2-*type receptors* and activates adenylate cyclase to increase cellular content of cyclic adenosine monophosphate (cAMP). The increase in cAMP, in turn, activates cAMP-dependent protein kinase, which appears to be an important mediator in initiating acid secretion. However, there is evidence that histamine may also increase cytosolic Ca^{2+} concentrations, through coupling with another signal transduction pathway such as phospholipase C (mediating phosphatidylinositol metabolism). Gastrin and acetylcholine appear to activate the *phosphatidylinositol (PI) pathway*, **gastrin** through the stimulation of a specific *gastrin receptor (CCK$_B$)* and **acetylcholine** through activation of a M_3-*type muscarinic receptor*. These agonists both increase intracellular free calcium concentrations. However, there are some differences in the activation pathways. Acid secretion stimulated by acetylcholine is dependent on the presence of extracellular calcium, whereas the effects of gastrin are only partially affected by this manipulation. Thus, the cellular mechanisms mediating these responses are clearly more diverse and complex than previously thought. The physiological role of *protein kinase C*, which is activated by the generation of diacylglycerol from phosphatidylinositol metabolism, is unclear. Some evidence suggests that it has a minor role in stimulation of acid secretion and may be more important in feedback inhibition of parietal cell function. Finally, cAMP and calcium may interact in a variety of ways to modulate cell activation. For example, potentiation can be seen between agents that increase cytosolic calcium (gastrin and acetylcholine) and cAMP (histamine), an observation that may have considerable clinical and therapeutic relevance. The nature of this interaction remains undetermined.

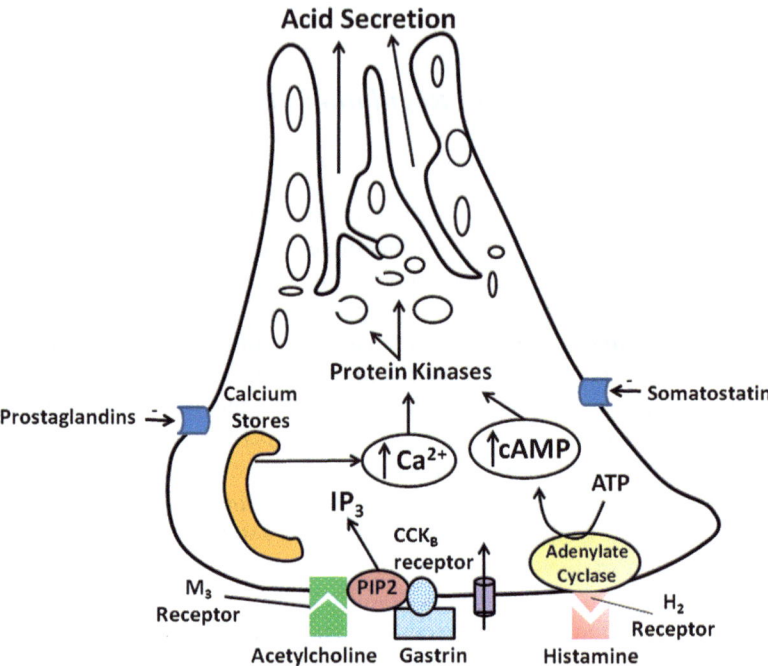

Fig. 3.4 Cellular mechanism of acid secretion by the parietal cell. There are three basic acid stimulants involved in the regulation of acid secretion, i.e. the neurocrine, acetylcholine; the endocrine, gastrin; the paracrine, histamine. *PIP$_2$* phosphatidyl-4,5-bisphosphate, *IP$_3$* inositol 1,4,5-triphosphate

In addition to their direct effects on parietal cell function, these agents appear to affect other target cells involved in the regulation of acid secretion (Fig. 3.5). For example, *gastrin* from antral G cells and acetylcholine from cholinergic neurons stimulate specific receptors on mast or ECL cells, promoting the secretion of histamine. *Histamine* potentiates the effects of each of these agents on parietal cell secretion. Noradrenergic nerves inhibit histamine release from mast cells through activation of β receptors, the net effect being the inhibition of acid secretion.

The role of **somatostatin-secreting D cells** in both the antral and acid-secreting regions of the stomach also appears to be a major component of the regulatory systems. In the ***antrum***, somatostatin inhibits G-cell secretion of gastrin and is released by the presence of luminal acid and gastrin (serving as negative feedback mechanisms) and by stimulation from enteric neurons containing acetylcholine, vasoactive intestinal peptide (VIP), and gastrin-releasing peptide (GRP). In the ***exocrine portion of the stomach,*** D-cell release of somatostatin directly inhibits parietal cell acid secretion. Alternatively, somatostatin also inhibits acid secretion by decreasing histamine release by **mast-like and ECL cells**. *Gastrin and cholecystokinin (CCK)* stimulate somatostatin secretion, whereas cholinergic neurons inhibit it.

Fig. 3.5 Interaction of regulatory pathways in gastric acid secretion. Apart from the direct effects of the three basic acid stimulants of histamine, gastrin and acetylcholine on parietal cell functions, various regulatory agents also affect other target cells for modulating acid secretion such as the inhibitory action of somatostatin

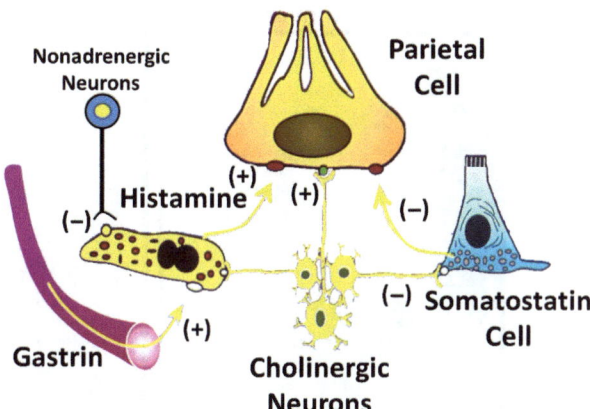

The relative contribution of each of the aforementioned pathways in regulating gastric acid secretion is highly variable among mammalian species and not entirely elucidated in humans. However, it does appear that histamine plays a major role in stimulating gastric acid secretion. Specific inhibitors of the histamine H_2 receptor, such as *cimetidine* and *ranitidine,* are clinically and experimentally very effective in reducing acid secretion, in spite of the fact that the stimulatory effects of gastrin and acetylcholine are still present. How can this be accounted for? A possible explanation is that in the absence of histamine stimulation of parietal cells, there is no potentiation of the gastrin or acetylcholine effects, thus diminishing acid secretion. In addition, gastrin stimulation of somatostatin release would continue unabated, further rendering a decrease in acid secretion.

2.2 Cellular Mechanisms Mediating Gastric Acid Secretion

Cellular Events

When the **parietal cell** is stimulated, it undergoes rapid, dramatic morphologic changes that are now known to be important in setting up the acid-secretory apparatus. The resting, or non-secreting, parietal cell has a specialized network of narrow channels that extend from the luminal surface through much of the cell body (Fig. 3.6a). These are **secretory canaliculi**, lined by short, stubby microvilli. There is also an elaborate system of tubular and vesicular membranes (**tubulovesicles**) in the luminal aspect of the cell. Large mitochondria, characteristically seen in these cells, are essential for providing a high oxidative capacity during acid secretion.

Within a few minutes of stimulation, microvilli lining the canaliculi become longer and more elaborate. It is estimated that the luminal surface area increases

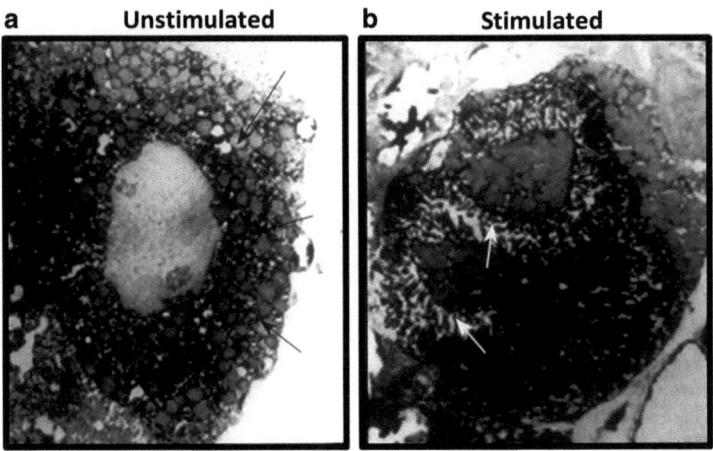

Fig. 3.6 Morphological changes associated with parietal cell activation and secretion. (**a**) The resting, or nonsecreting, parietal cell has many mitochondria and an elaborate system of tubular and vesicular membranes near the luminal aspect of the cell. (**b**) With stimulation, the luminal surface area increases six to tenfold, with a concomitant disappearance of tubulovesicle membranes. These changes are caused by the fusion of the tubulovesicle membranes with the plasma membrane, increasing the number and density of proton pumps in the apical surface (Courtesy of Dr. George Sachs)

to 6–10 times that of the resting state (Fig. 3.6b). Concomitant with these changes is the disappearance of tubulovesicle membranes. Withdrawal of the stimulating agents leads to a collapse of the canaliculi spaces and a return of the tubulovesicle forms in the cytoplasm. These observations are consistent with the membrane recycling that occurs during the stimulated and resting states of the parietal cell. Tubulovesicles contain large amounts of the gastric proton-pump enzyme ***H^+/K^+-ATPase***. Following stimulation of the parietal cell, tubulovesicle membranes fuse with the plasma membrane, increasing the number and density of proton pumps in the apical surface. Conversely, withdrawal of the stimulus leads to endocytosis of these membranes, a decrease in the density of proton pumps in the apical membrane, and decreased acid secretion. The proton pump is only active when inserted in the luminal membrane; it is inactive when re-sequestered into tubulovesicle membranes.

Electrolyte Transport

The mechanism for proton secretion into the lumen is fairly well understood and summarized in Fig. 3.7. The essential element here is a membrane-bound enzyme

Fig. 3.7 Electrolyte transport and acid secretion at the luminal membrane of a stimulated parietal cell. The major component of acid secretion is the H^+/K^+-ATPase, or proton pump, which actively exchanges H^+ for K^+. For vectorial H^+ secretion to occur, conductance pathways for K^+ and Cl^- must exit for recycling of K^+ ions and extrusion of Cl^-, respectively

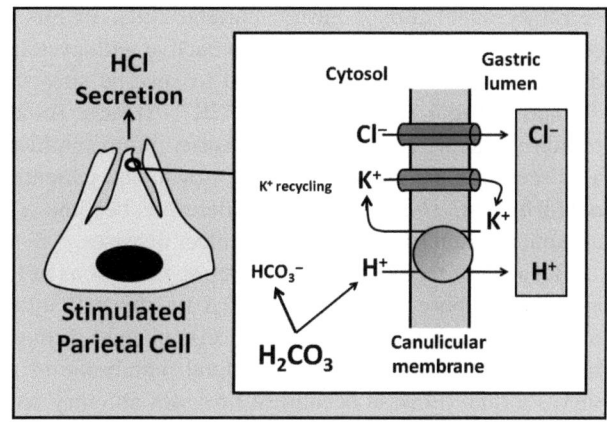

that actively extrudes protons into the lumen in exchange for potassium ions. This enzyme requires adenosine triphosphate (ATP) and hence is referred to as the **H^+/K^+-ATPase, or proton pump**. Gastric H^+/K^+-ATPase is a member of the phosphorylating, ion-motive ATPase gene family, which includes Na^+/K^+-ATPase and Ca^{2+}-ATPase. It is a heterodimer consisting of a *larger catalytic subunit (α)* and a *smaller (β) glycosylated subunit*. The role of the β subunit is unknown, but it appears to be required for the proton to work as a functional unit. The proton pump is electrically neural; that is, there is a one-for-one exchange of H^+ and K^+, with no net transfer of charge. Essential to the proton pump and acid secretion is the presence of an exit pathway for K^+, which is then recycled in exchange for H^+. This occurs through the coupled transport of K^+ and Cl^- out of the cell via respective conductive channels. The Cl^- channel is distinct from the cystic fibrosis transmembrane regulator (CFTR) channel. These exit pathways operate in tandem with the proton pump, the net effect being the rapid secretion of hydrochloride into the canalicular lumen. The **K^+ and Cl^- exit pathways** may be the major rate-determining factors in acid secretion, as both are regulated by cAMP-dependent protein kinase. In addition, being voltage-gated channels, they are active only when inserted into the luminal membrane where a favorable membrane potential exists. To date, there is no evidence that supports direct kinase regulation of the proton pump.

The **proton pump** is somewhat unique in its actions. It has an enormous active secretory capacity, capable of secreting protons against large electrochemical gradients. As a consequence, luminal pH can approach the 1–2 range, representing a 5–6 log unit difference from intracellular pH values of approximately 7. No other part of the body can approach this level of acidification. Recently, several highly specific and clinically useful inhibitors of the proton pimp have been developed, *specifically substituted-benzimidazole compounds* such as *omeprazole* and *lansoprazole*. Being weak acids with pK_a values of approximately 4, these agents are selectively (if not exclusively) taken up and concentrated in the acidified secretory canaliculi of gastric parietal cells. In this acid environment, they are converted to sulfonamides that are reactive with cysteine SH-thiol, but not cysteine disulfides. Thus, these agents bind to specific sites in the extracellular (luminal) domain of the α subunit of the H^+/K^+-ATPase. As a result, ATPase activity and proton transport cease. Because proton pump inhibitors block the final pathway, acid secretion can be virtually shut down, and patients may become a state, called *achlorhydria*. *Omerprazole* has therefore become a major part of the medical armamentarium for treating peptic ulcer diseases.

Because of the voluminous secretion of protons at the apical surface, equivalent amounts of base must be delivered by the cell into blood plasma to maintain normal intracellular pH. This is accomplished through a **Cl^-/HCO_3^- exchange mechanism** located in the basolateral membrane of parietal cells. This pathway serves a dual purpose because it provides an entry pathway for chloride ions that are then vectorially secreted with protons. Coupled with Cl^-/HCO_3^- exchange is **Na^+/H^+ exchanger**, which is important for providing intracellular sodium for recycling out of the cells via the sodium pump. This allows the sodium pump to operate efficiently at high capacity to maintain the electrochemical gradients required for proton secretion and to provide an entry pathway for potassium ions (since gastric juices typically contain potassium concentrations that exceed plasma potassium concentrations by twofold or more).

2.3   *Organismal Regulation of Gastric Functions*

Meal-Stimulated Gastric Acid Secretion

At the organismal level, **meal-associated stimulation of gastric secretion** can be divided into **three phases** (Fig. 3.8a). The *cephalic phase* includes responses evoked by stimuli from the central nervous system, which can be initiated by the sight, smell, or taste of food, as well as conditioned reflexes such as the sound of a bell, and hypoglycemia induced by insulin injections or experimental administration of a GABA-like agonist. This can best be demonstrated experimentally by "sham feeding," in which food is chewed but not allowed to enter the stomach. As the duration of sham feeding lengthens, the magnitude of the gastric acid and motor response increases. **Three areas of the brainstem** appear to be important in **relaying information to the stomach:** the area *postrema, nucleus tractus solitarii (NTS),* and the *dorsal motor nucleus (DMN).* A variety of *neural peptides, such as pancreatic polypeptide, neuropeptide Y,* and *peptide YY,* may mediate the cephalic-phase response, since administration of these peptides into the brain can initiate the response. When these areas receive input from other parts of the brain (sensory processing areas), and from peripheral hormonal and metabolic signals, they activate gastric secretions through vagal nerve stimulation. Preganglionic vagal fibers synapse with enteric neurons containing gastrin-releasing peptide (GRP) and

3 Gastric Physiology

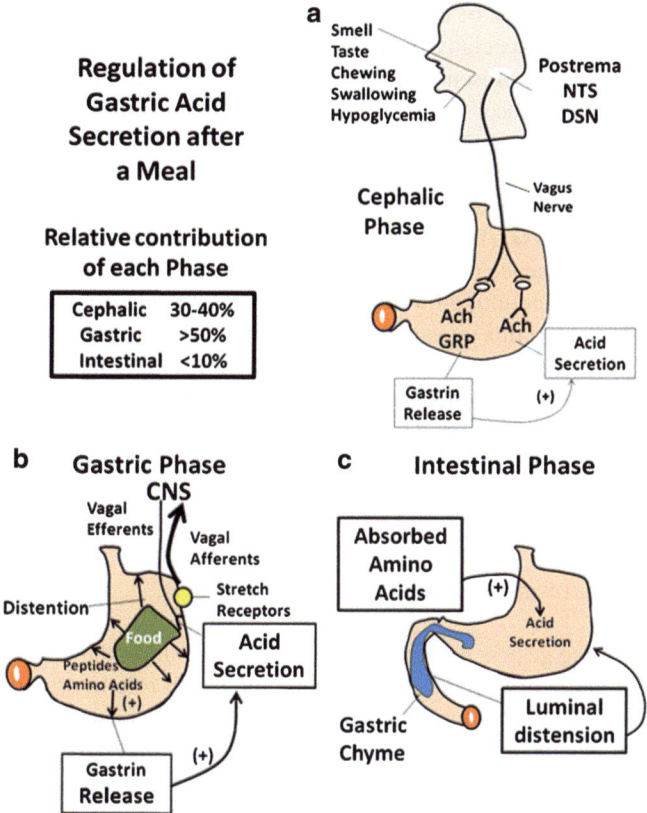

Fig. 3.8 Meal-stimulated gastric acid secretion. There are three phases that are important for regulation of meal-stimulated gastric acid secretion; they are cephalic phase (**a**), gastric phase (**b**), and intestinal phase (**c**). Of these, the gastric phase is believed to be the most important. *NTS* nucleus tractus solitarii, *DMN* dorsal motor nucleus, *Ach* acetylcholine, *GRP* gastrin releasing polypeptide

acetylcholine, which stimulate gastrin release in the antrum and acid secretion in the body and fundus. Surgical secretion of the vagal nerve will completely abolish the cephalic phase of gastric acid stimulation.

The **gastric phase** (Fig. 3.8b) occurs when fluid or food is present in the gastric lumen. This phase accounts for approximately *60 % of the total acid secretion stimulated by a meal*. A major component of this response is **gastric distention** produced by luminal contents, particularly in the antrum. The greater the distention is, the greater the gastric acid output. This response is mediated by both vagal nerve fibers and local intrinsic factors, the latter being gastrin in particular. When the stomach is minimally distended, vagal innervation appears to predominate, since gastrin release is minimal. However, when the antrum is further distended, large amounts of gastrin are promptly released, and acid secretion increases accordingly.

Another major component of the gastric phase is the stimulation of gastrin release by luminal nutrients, especially peptides and aromatic amino acids (phenylalanine and tryptophan). Their presence is somehow sensed by antral G cells, which are then stimulated to release gastrin. In addition, these **luminal agents** appear to activate afferent enteric neurons that initiate reflex neural networks stimulating G-cell secretion. In contrast to protein hydrolysates, carbohydrates and fats are not potent stimuli of gastric acid secretion.

The *intestinal phase* (Fig. 3.8c) is that component of gastric acid secretion stimulated by the presence of food in the small intestine. It is the least important of the various phases, accounting for less than *10 % of total gastric acid secretion*. The recognized initiators of this intestinal phase are distention and the digestion products of protein. Part of the humoral mechanism of the intestinal phase is probably due to absorbed amino acids, since intravenous administration of amino acids does stimulate gastric acid secretion. The intestinal phase by itself is a weak stimulus of gastric acid secretion but does appear to strongly potentiate the effects of histamine and gastrin.

Inhibitory Mechanisms of Gastric Acid Secretion

In each phase of the gastric response to meals, factors inhibit the acid secretion that accompanies stimulatory events. They play an important role in providing negative feedback to control the magnitude of the secretory response and in eventually bringing the stomach back to its resting state during interdigestive periods.

Inhibition in the cephalic phase can be demonstrated by injection of *bombesin, calcitonin, neurotensin, interleukin-1 (IL-1), prostaglandins,* or *corticotropin-releasing factor (CRF)* into the brain. Their inhibitory effects appear to be mediated by vagal and sympathetic fibers to the stomach. During the gastric phase, acidification in the antral mucosa inhibits gastrin release stimulated by sham feeding or by distention of the antrum. This effect is probably mediated by enteric neurons, as administration of atropine inhibits it. Increased luminal acid concentration also appears to stimulate the release of somatostatin from antral D cells, which inhibits gastrin release via a paracrine action. In the intestinal phase, three agents are known to inhibit acid secretion when instilled in the small intestine; they are acid, hyperosmolar solutions, and fat (the latter being most potent). The inhibition appears to be mediated by humoral substances collectively called **"*enterogastrones.*"** Although no agent has been definitively identified, the candidate hormones for enterogastrones include *gastric inhibitory peptide (GIP), neurotensin, somatostatin, secretin, vasoactive intestinal peptide (VIP), enteroglucagon,* and *peptide YY*. Of these, peptide YY is the most likely agent. It is released by fat from the distal small intestine and inhibits pentagastrin-stimulated acid secretion. In addition to these mechanisms, gastric secretion is also inhibited by *secretin, gastrin releasing polypeptide (GRP),* and *cholecystokinin (CCK)* (see Chap. 1).

3 Pepsinogen Secretion

Pepsin is an important digestive enzyme secreted predominantly from gastric chief cells in the form of **pepsinogen**, its precursor **zymogen**. These peptic cells are located on the walls of the *oxyntic glands* and appear to be regulated by many of the same agents involved in the regulation of gastric acid secretion. Not surprisingly, therefore, pepsinogen secretion parallels that of acid secretion. Agents such as *gastrin, acetylcholine, CCK,* and *GIP* **stimulate pepsinogen release** by increasing cytosolic Ca^{2+} through receptor-mediated phosphatidylinositol metabolism (Fig. 3.9). Agents such as *secretin, VIP, E-series of prostaglandins,* and *β-adrenergic receptor agonists* **stimulate peptic cells** by activating adenylate cyclase, resulting in the generation of cAMP and activation of a cAMP-dependent protein kinase. Both signal transduction pathways appear to stimulate the release of pepsinogen via exocytosis of secretory granules, but cAMP-mediated stimuli may also directly stimulate *de novo* synthesis of pepsinogen.

On release into the lumen, pepsinogen is immediately activated by acid, with optimum pH being about 2. The formation of pepsin has a **positive feedback effect**, leading to a more rapid and complete conversion of pepsinogen into pepsin (Fig. 3.10).

Pepsin belongs to the general class of **aspartic protease**, so named because of the presence of two aspartic acid residues that are part of the catalytic site. It is a very good proteolytic enzyme and, because of its high activity on collagen, is more important for the digestion of meat than for vegetable protein. However, pepsin digestion of proteins is usually incomplete, since large peptides called peptones are

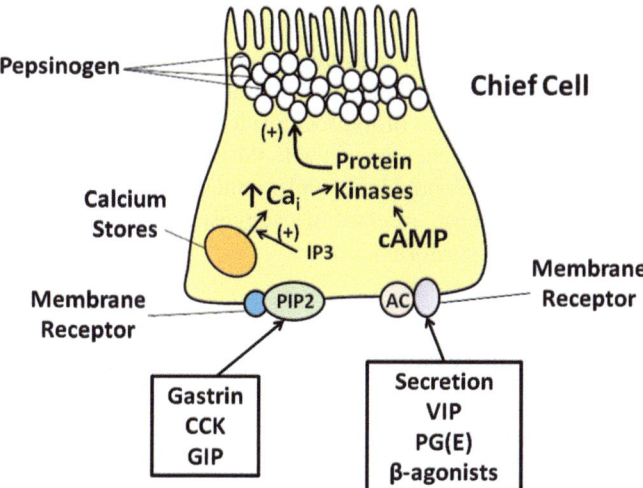

Fig. 3.9 Regulation of pepsinogen secretion by gastric chief cells. Cellular mechanisms and mediators involved in the action of regulatory agents of pepsinogen secretion illustrated above

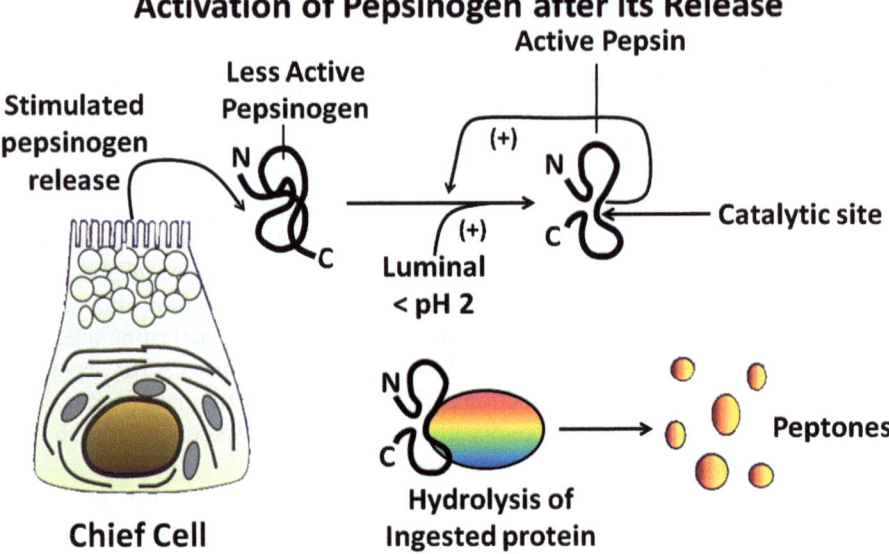

Fig. 3.10 Activation of pepsinogen. After its release into the lumen, pepsinogen is immediately activated by acid, at an optimum pH of 2. Activated pepsin has a positive feedback effect, leading to more rapid and complete conversion of pepsinogen into pepsin

frequently found in gastric chyme entering the small intestine. Peptones serve as potent signals for the release of various hormones, including gastrin and CCK.

There are **two immunologically distinct classes of pepsinogens**. *Group I pepsinogens* are secreted by the peptic and mucous neck cells of the *oxyntic gland*, whereas *group II pepsinogens* are made in *pyloric and Brunner's glands*. There has been considerable interest in pepsinogen as a possible etiologic agent in the formation of gastric and duodenal ulcers. For example, several studies have shown that instillation into the stomach of hydrochloride alone does not cause ulceration, but the inclusion of gastric juice or pepsin with the acid results in ulcer formation. In many duodenal ulcer patients, both basal and stimulated pepsin secretion are greater than in normal controls. These observations suggest a possible contributory, albeit minor, role for pepsin in ulcer formation.

4 Mucus Bicarbonate Secretion

4.1 Mucus Secretion

Mucus is made up of glycoproteins that have very interesting physiochemical properties. They are extremely hydrophilic and can form gels that contain up to 95 % water. They consist of a **polypeptide core** and are **highly glycosylated at serine and threonine residues** (Fig. 3.11). The *glycosidic portion* of the molecule makes

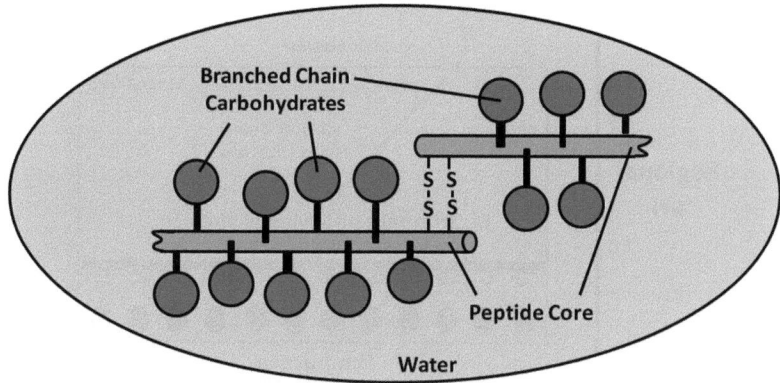

Fig. 3.11 Mucin structure and polymerization. Gastric mucin glycopeptides are extremely hydrophilic and can form gels that contain 95 % water. They consist of a polypeptide core that is highly glycosylated at serine and threonine residues

up 80 % the total molecular weight, which usually exceeds 500 kD. It also confers resistance to proteolytic digestion, an important property since mucus appears to protect the stomach against acid-pepsin damage. The ***non-glycosylated regions*** of mucin proteins serve as sites for polymerization, which occurs via disulfide linkages. Proteolysis or reduction of these linkages by N-acetyl-L-cysteine (Mucomyst) or dihydrothiothreitol (DTT) depolymerizes mucus molecules, destroying their gel-like properties and making the glycoproteins soluble.

Two forms of mucus are found in the stomach, namely a ***soluble*** and an ***insoluble*** form. Insoluble or adherent mucus is a viscous, slippery gel that covers most of the mucosal surface of the stomach and is continuously secreted by surface and gastric pit cells. It has an important role in providing protection against acid-peptic injury (discussed below). Soluble mucin results from the degradation of insoluble mucus by peptic action. The **soluble fraction** mixes with and lubricates gastric chyme during gastric motility. The **insoluble fraction** forms a semi-impermeant layer that protects underlying cells from damage by gastric acid (Fig. 3.12). Thus, the gel slows the permeation of acid from the lumen and bicarbonate secreted from underlying epithelial cells, establishing a buffer zone where the pH near the surface of the epithelial cells is in fact alkaline.

Factors That Affect Mucus Secretion

Mucus secretion and hence the thickness of the mucus layer can be affected by a number of neuro-hormonal agents. **Cholinergic stimulation**, for example, results in copious release of mucus in the stomach, whereas noradrenergic and adrenergic neurotransmitters do not appear to influence the function of these cells. Other agents that stimulate mucus secretion include *serotonin and prostaglandins E and F*. A possible contributory factor in **aspirin-induced gastric ulceration** is the inhibition

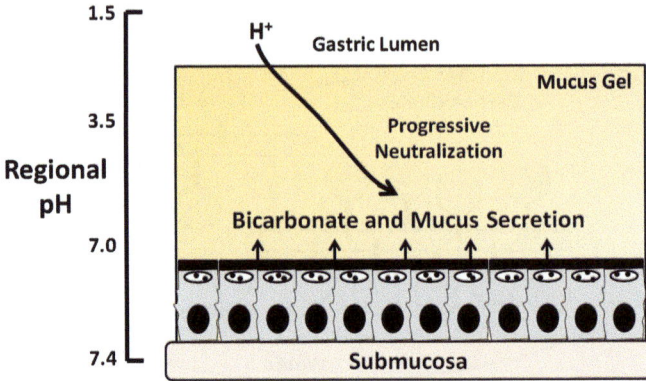

Fig. 3.12 Essential role of mucus in gastric mucosa protection. The gastric mucus gel slows the permeation of acid from the lumen and of bicarbonate secreted from underlying epithelial cells. This establishes a buffer zone, where the pH near the surface of the epithelial cells is in fact alkaline

of prostaglandin formation resulting in a decrease in the mucus protective barrier. In addition, **peptic ulcers** and **gastritis** associated with *Helicobacter pylori* infection (see below) may also be caused by compromise of the gastric mucus layer.

4.2 Bicarbonate Secretion

A major function of cells at the surface and in gastric pits is the secretion of bicarbonate, which in conjunction with the mucus layer serves to protect the gastric epithelium against acid-pepsin damage or called *gastric mucosal barrier*. Bicarbonate secretion alone is insufficient to counteract the magnitude of proton secretion. The mechanism of bicarbonate secretion remains unknown, although some studies have implicated the presence of a bicarbonate channel or Cl^-/HCO_3^- exchange. Bicarbonate production is dependent on **carbonic anhydrase** activity. Its transport appears to be active and stimulated by *vagal nerves* and *E-series prostaglandins*.

5 Pathogenesis of Peptic Ulcer Diseases

5.1 Pathophysiology

Over the past several decades, the understanding and treatment of acid-peptic diseases of the stomach have significantly advanced. The original concept of no acid/no ulcer still appears correct, but is also recognized that the balance between

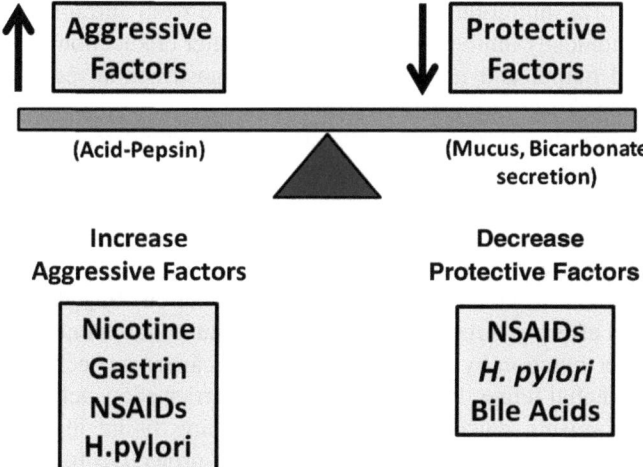

Fig. 3.13 Some factors that are believed to be important for the development of peptic ulcer disease. The well-being of the gastric mucosa is dependent on protective factors that counteract the effects of aggressive factors such as acid in pepsin. Agents that are involved in increasing aggressive factors are shown in the lower *left* and those that decrease protective factors are shown on the *right*

aggressive and protective factors is critical for maintaining the well-being of the stomach. **Peptic ulcers** and **gastritis (inflammation of the stomach lining)** result if there is an increase in aggressive factors or a decrease in protective factors (Fig. 3.13). *Aggressive factors* include increased acid-pepsin activity stimulated by *smoking (nicotine), excessive gastrin secretion (gastrinoma),* and *non-steroidal anti-inflammatory drugs (NSAIDs)*. **NSAIDs** have a variety of effects that can promote ulcer formation, including the inhibition of prostaglandin especially PGE_2, which suppresses acid secretion. Thus, NSAIDs usage may augment the acid-secretory response to histamine and gastrin. Another important aggressive factor recently recognized is *H. pylori,* a small, gram-negative bacterial organism that is the likely cause of most *non-NASID-associated* peptic ulcer diseases and gastritis. In fact, *H. pylori* now appear to be the most common infection worldwide, found in both ulcer and non-ulcer subjects. Unlike most bacterial organisms, *H. pylori* can survive the harsh acid milieu of the stomach, colonizing the mucus layer of the gastric antrum. Its ability to survive this environment is due, in part, to the production of *urease*, an enzyme that converts urea to NH_3, which is used to buffer H^+ by forming NH_4. Though being not invasive, the organism can elicit an inflammatory response of the underlying mucosa. How *H. pylori* elicits this response is not entirely clear, but elaboration of lipopolysaccharide (endotoxin), toxins, adhesions, and chemotactic substances by the organism has been implicated.

H. pylori do appear to cause **increased basal acid output** and **meal-related hypergastrinemia** in some subjects, especially patients with duodenal ulcer disease. The mechanism for increased gastrin release from antral G cells is not clear, but it may involve inhibition of somatostatin release from antral D cells, which is important for tonically inhibiting gastrin release. After eradication of the organism, the basal acid output and meal-stimulated hypergastrinemia response appear to normalize, suggestive of a causative relationship.

It is intriguing that *H. pylori* do not cause acid-peptic diseases in everyone. In fact, a large proportion of carriers appear who have no symptoms or gastric mucosal abnormalities. This observation underscores the importance of other contributing factors in the pathogenesis of peptic ulcer disease. Among these, **decreases in mucosal barrier function** are thought to be the most important. Several agents and factors can alter or compromise mucosal-barrier function. *H. pylori*, for instance, either by direct effect or through its stimulation of mucosal inflammation, decreases mucosal-barrier function by altering mucus and bicarbonate secretion and breaks the gastric epithelial lining. Breaches in mucosal barrier allow acid-pepsin to enter, further compromising mucosal function and causing tissue injury. Aside from *H. pylori,* other factors can also decrease gastric-protective functions. **NSAIDs** inhibit arachidonate metabolism by blocking cyclooxygenase activity, thus resulting in decreased prostanoid formation, which is believed to be important for maintaining mucus and bicarbonate secretion (see above). ***Smoking*** and ***increased nicotine exposure*** may also inhibit mucus and bicarbonate secretion, accounting for the increased incidence of gastric acid-peptic disorders in smokers.

5.2 *Pharmacological Approaches to the Treatment of Peptic Ulcer Diseases*

The management of most peptic ulcer diseases has now moved into the arena of medical therapy because of the enormous advances made in the development of effective and safe pharmacological agents (Fig. 3.14). Detailed discussion of these agents is beyond the scope of this text. The following therefore represents a brief overview of the classes of compounds that have been used to decrease acid secretion and/or increase mucosal cyto-protective functions. It should be noted that the medical treatment of *H. pylori*-associated peptic ulcer disease now includes the concomitant use of antibiotics. In this regard, the most frequent and effective regimen called **"triple therapy"** is the use of either a ***proton pump inhibitor (omeprazole)*** plus ***two to three antibodies*** (e.g. amoxicillin and clarithromycin). Below are some treatments and the proposed mechanisms of action for peptic ulcer diseases.

Antacids. Antacids, compounds that **neutralize acid**, have been the mainstay of medical therapy. These agents include preparations containing *aluminum hydroxide, calcium carbonate*, and *magnesium hydroxide.* Although these agents are effective in treating patients with peptic ulcers, they must be taken frequently and in large quantities, making for poor patient compliance.

Treatment	Mechanism of action
Antacids e.g. Al(OH)$_3$, Mg(OH)$_2$ Mg-induced diarrhea & Al-induced constipation	Neutralization of gastric acid - Nonspecific with many disadvantages
H$_2$ receptor antagonists e.g. Cimetidine & ranitidine	Inhibition of histamine-dependent acid secretion
H$^+$/K$^+$-ATPase inhibitors e.g. Omeprazole & Lansoprazole (Losec)	Inhibition of proton pump - More potent & longer lasting, e.g. DU caused by gastrinoma
Muscarinic receptor antagonist (Anticholinergic) e.g. Pirenzipine & atropine	Inhibition of Ach-stimulated acid secretion
Prostaglandin agonist e.g. Misoprostol	Inhibition of acid secretion; "Direct cytoprotection", such as mucus and HCO$_3^-$ stimulation
Sucralfate	Stimulation of prostaglandin synthesis
Bismuth salts	Eradication of H. pylori
Carbenoxolone	Stimulation of prostaglandin synthesis
Vagotomy/Antrectomy	Removal of Ach- or gastrin-mediated acid secretion
Antimicrobials e.g. Amoxycillin, clarithromycin	Eradication of H. pylori *Triple therapy = H$^+$/K$^+$-ATPase blocker + Two antibiotics for 2 weeks ↑ Healing rate ↓ Recurrence (relapse) rate

Fig. 3.14 Some treatments for peptic ulcer disease and the mechanisms of action

Anticholinergics. In theory, anticholinergics should be included in the medical treatment of peptic ulcer diseases because they can **reduce acid secretion**. However, these agents are companied by a number of *side effects* that patients find intolerable. In addition, these agents *slow gastric motility,* promoting *increased contact time between gastric acid and the gastric mucosa*, potentially a confounding factor to their efficacy.

H$_2$-receptor antagonists. Histamine appears to be a major regulator of gastric acid secretion. In fact, in some mammalian species, but to a lesser extent in humans, histamine mediates the actions of gastrin. Thus, **inhibition of actions of histamine by blocking H$_2$ receptors** is an extremely effective way to suppress gastric acid secretion. Thus, agents such as *cimetidine, ranitidine, and famotidine* work as competitive antagonists and have been clinically effective. Although they do not completely inhibit gastric acid secretion, their effects on total acid output is sufficient in most cases to allow the stomach to heal.

Proton pump inhibitors. **Substituted benzimidazoles**, such as *omeprazole*, are extremely potent **inhibitors of gastric acid secretion**. They are weak bases that are selectively taken up and concentrated by gastric parietal cells. In an acid milieu, these agents are converted to **sulfonamides**, which *irreversibly* bind to α subunit of the H$^+$/K$^+$-ATPase (proton pump). Omeprazole can completely inhibit all acid secretion. In view of this, the use omeprazole as an acid suppressant is potent (it binds to the final pathway of acid secretion) and long-lasting (it binds irreversibly to the H$^+$/K$^+$-ATPase).

Prostaglandin agonists. Prostaglandins, especially of the E and I series, have a number of actions that make them useful in the treatment of peptic ulcer diseases. First, these agents **inhibit parietal cell secretion,** possibly through the activation of receptors linked to inhibitory G proteins of adenylate cyclase; these results in the inhibition of effects of histamine on acid secretion. In addition to anti-acid secretory actions, these agents also appear to have **cytoprotective actions**. Though the factors mediating this effect are not well understood, they may involve the stimulation of mucus and bicarbonate secretion by gastric surface cells.

Cytoprotective agents. These agents include compounds such as *sucralfate* and *colloidal bismuth subcitrate* that avidly bind to the base of ulcers, physically providing a **protective layer against acid injury.** *Sucralfate* is an octasulfate of sucrose that is highly, negatively charged, allowing it to interact with positively charged protein molecules. ***Colloidal bismuth subcitrate*** is a complex bismuth salt of citric acid that has a similar mode of action, especially in an acid milieu. Both agents have been effective in treating peptic ulcer disease.

Triple therapy. It is the most frequent and effective regimen that contains either a **proton pump inhibitor** (e.g. omeprazole) or H_2 **receptor antagonist** plus **two to three antibiotics** (e.g. amoxycillin and clarithromycin) for 2 weeks. This sort of treatment not only heals the ulcers but also prevents the recurrence of ulcers.

Summary

- Acid is secreted by the parietal cells which contain the enzyme H^+/K^+**-ATPase** on their apical secretory membranes.
- The three major stimulants of acid secretion are the **hormone gastrin**, the **neurocrine Ach,** and the **paracrine histamine**, the latter being released from ECL cells in response to gastrin and Ach.
- The gastric mucosal barrier of a normal individual can protect the stomach even in the elevation of HCl and pepsin levels.
- When HCO_3^- and/or mucus are suppressed, the barrier is compromised and ulceration may occur.

Clinical Correlations

Case Study 1

A 40-year-old man presents to the emergency room with history of episodic, sharp, epigastric abdominal pain frequently accompanied by nausea and vomiting. He noted that his vomitus often looked like coffee grounds and that he had been passing dark-colored, tarry-like stools. He had endured his pain by drinking milk and taking two aspirins four times a day, but frequently wakes up in the middle of the night with abdominal pain. His past history is essentially negative, except that he is a heavy smoker. Because he is anemic and orthostatic (blood pressure falls when he stands up, indicating vascular volume depletion), he is admitted to the hospital for treatment and further evaluation. After receiving several units of blood, he undergoes an endoscopic examination, which reveals large duodenal and gastric ulcers. A mucosal biopsy is performed and subjected to a *Campylobacter-*

like organisms (CLO) test, a colorimetric assay for the presence of bacterial urease, a marker of *H. pylori* (formerly known as *Campylobacter pylori*). The CLO test is markedly positive.

Questions

1. **What caused the gastric lesions in this patient?**
   Answer: This patient has **peptic ulcer disease** that could have been caused by several factors. Since the patient's symptoms preceded his use of aspirin, it is likely that his duodenal ulcer was caused by *H. pylori* infection and perhaps increased acid secretion stimulated by smoking. Although the pathogenic mechanisms mediating *H. pylori*-associated peptic ulcer disease are not entirely clear, increased acid secretion and decreased mucosal-barrier function are likely to be involved. Duodenal and gastric ulcers are, however, also caused by aspirin. **Aspirin and other NSAIDs** inhibit the synthesis of prostaglandins, which are important for maintaining mucus and bicarbonate secretion and barrier function by surface epithelium. As a consequence, mucosal-barrier function is compromised, and the gastric mucosa becomes vulnerable to acid-peptic injury. Decreased endogenous production of prostaglandins also results in increased acid secretion, as prostaglandins may stimulate somatostatin release. It is also possible that NSAIDs decrease mucosal blood flow by inhibition of prostaglandin production, further compromising mucosal integrity.

2. **What would you recommend as treatment for this patient?**
   Answer: The treatment would involve the use of **antibiotics**, possibly **bismuth-containing compounds** (e.g., Pepto-Bismol), and **acid-suppressing medications**. Antibiotics are necessary to eradicate *H. pylori* infection; otherwise, the rate of ulcer recurrence can be substantial. Bismuth compounds are effective, possibly by counteracting some of the deleterious actions of *H. pylori*; acid-suppression by H_2-receptor blockers or proton-pump inhibitors (e.g. Omeprazole) can accelerate ulcer healing. Currently, triple therapy with the use of one proton pump inhibitor (e.g. omeprazole) and antibiotics (e.g. amoxicillin and clarithromycin) is the most effective treatment option. The patient should also be strongly advised to quit smoking, a known risk factor in the development of peptic ulcer disease, and to stop taking aspirin or other NSAIDs.

Case Study 2

A medical student presents to the emergency room with a 2-day history of probable viral gastroenteritis, characterized by severe nausea and vomiting. He has been unable to keep down any fluids and says that he is weak and dizzy (especially when he stands up). Serum electrolytes are drawn for further assessments.

Questions

1. **What kind of metabolic abnormalities would you expect?**
   Answer: Serum electrolytes showed **hypokalemia (low K^+)**, **hypochloremia**, and the presence of **metabolic alkalosis**. These abnormalities arise from **two sources**. First, the *loss of gastric juices*, rich in H^+, K^+, and Cl^-, accounts

for a major part of the metabolic disturbance, especially since the student is unable to replace them orally. These are ions secreted by parietal cells via the transport processes present in the luminal or canalicular membrane. The metabolic abnormalities are further exacerbated by the ***dehydration*** of the student. The contraction of vascular volume activates renal mechanisms, i.e. activation of the renin-angiotensin-aldosterone system, important for preserving volume. As a consequence, water and Na^+-bicarbonate are reabsorbed at the cost of urinary excretion of K^+ and H^+.

Case Study 3

A 32-year-old woman presents with severe abdominal pain, weight loss, and occasional nausea and vomiting. She was diagnosed as having duodenal ulcer disease on two previous occasions when she presented with identical symptoms. Since her last episode 2 months ago, she has been religiously taking her medications, which include a histamine H_2 blocker and antacids. She denies any history of smoking or use of NSAIDs. On upper endoscopic examination, several severe duodenal ulcers are found, some post-bulbar (past the first portion of the duodenum) and in the jejunum. Stool examination reveals some fat malabsorption and occult blood.

Questions

1. **Because of the severity of the symptoms, what important blood test should be drawn?**
   Answer: This patient has a rare cause of duodenal ulcer disease. **Serum gastrin levels** should be obtained, and in this patient they are found to be profoundly elevated. With her history of ulcer disease refractory to conventional medical management, together with increased serum gastrin, a **diagnosis of gastrinoma** or **Zollinger-Ellison syndrome** must be entertained. After an extensive work-up, this diagnosis is confirmed. Gastrinomas are characteristically slow growing, and the major immediate morbidity to this patient is related to complications of severe peptic ulcers, including *bleeding, perforation,* and *obstruction from duodenal scarring.*

   The **pathophysiology of this disease** involves excessive gastrin secretion (**hypergastrinemia**) by a tumor, causing gastric acid hypersecretion. The increased acid load to the duodenum and jejunum can often exceed the capacity of the pancreas to secrete bicarbonate. Thus, luminal pH of the proximal small intestine can be very low and injurious to the mucosa. In addition, the abnormal luminal pH prevents activation of pancreatic-proenzyme activation and proper micellar formation. This causes maldigestion of luminal nutrients and malabsorption, explaining this patient's loss of weight.

2. **How would you medically treat this patient if she refuses surgical intervention?**
   Answer: The mainstay of medical management of this condition is acid secretion suppression. Use of H_2-receptor blockers is often inadequate because they are overwhelmed by the potency of gastrin-stimulated acid secretion. By

blocking at the final pathway of acid secretion, i.e. using *omeprazole* to block the proton pump, acid secretion can often be significantly diminished or blocked. Alternatively, patients have been effectively treated with ***Octreotide**®*, a somatostatin analog that suppresses the synthesis and release of gastrin by tumor cells. In addition, Octreotide may inhibit acid secretion by inhibiting histamine release and parietal cell acid secretion. Finally, it should be noted that omeprazole is so effective in inhibiting acid secretion in many patients that it actually causes **achlorhydria** (complete absence of gastric acid). Because luminal acid is a negative feedback mechanism for gastrin secretion by antral endocrine cells, the use of omeprazole can cause increases in serum gastrin levels **(hypergastrinemia)** by itself.

Further Reading

1. Atherton JC (2006) The pathogenesis of Helicobacter pylori-induced gastro-duodenal diseases. Ann Rev Pathol 1:63–96
2. Bornstein JC, Gwynne RM, Sjövall H (2012) Enteric neural regulation of mucosal secretion. In: Johnson LR (ed) Physiology of the gastrointestinal tract, 5th edn. Academic, New York, pp 769–790
3. Chu S, Schubert ML (2012) Gastric secretion. Curr Opin Gastroenterol 28:587–593
4. Fock KM, Graham DY, Malfertheiner P (2013) *Helicobacter* pylori research: historical insights and future directions. Nat Rev Gastroenterol Hepatol 10:495–500
5. Kopic S, Murek M, Geibel JP (2010) Revisiting the parietal cell. Am J Physiol 298:C1–C10
6. Kozyraki R, Cases O (2013) Vitamin B12 absorption: mammalian physiology and acquired and inherited disorders. Biochimie 95:1002–1007
7. Li H, Meng L, Liu F, Wei JF, Wang YQ (2013) H^+/K^+-ATPase inhibitors: a patient review. Expert Opin Ther Patients 23:99–111
8. Malfertheiner P, Chan FK, McColl KE (2009) Peptic ulcer disease. Lancet 374:1449–1461
9. Okamoto C, Karvar S, Forte JG, Yao X (2012) The cell biology of gastric acid secretion. In: Johnson LR (ed) Physiology of the gastrointestinal tract, 5th edn. Academic, New York, pp 1251–1279
10. Schubert ML (2012) Regulation of gastric acid secretion. In: Johnson LR (ed) Physiology of the gastrointestinal tract, 5th edn. Academic, New York, pp 1281–1309

Chapter 4
Pancreatic Physiology

Eugene B. Chang and Po Sing Leung

1 Introduction to the Pancreatic Functions

The human **pancreas** consists of **two organs in one structure**: the **exocrine gland** made up of *pancreatic acinar cells* and *duct cells* that produce digestive enzymes and sodium bicarbonate, respectively; the **endocrine gland** made up of *four islet cells*, namely *alpha-, beta-, delta-, PP-,* and *ipsilon- cells* that produce glucagon, insulin, somatostatin, pancreatic polypeptide, and ghrelin respectively. While the physiological role of **exocrine pancreas** (>80 % by volume) is to secrete digestive enzymes responsible for our normal digestion, absorption and assimilation of nutrients, the **endocrine pancreas** (<2 % by volume) is to secrete islet peptide hormones for the maintenance of our glucose homeostasis. The pancreatic functions are finely regulated by neurocrine, endocrine, paracrine and/or intracrine mechanisms. Thus, dysregulation of these pathways should have significant impacts on our health and disease. Nevertheless, the underlying mechanisms by which pancreatic functions are regulated remain poorly understood.

Embryologically, the human pancreas originates from **two separate outgrowths**, designated as the *dorsal* and *ventral* buds, from the foregut endoderm directly posterior to the stomach; it is similar to the pancreas development in murine. The **dorsal bud** arises from evagination of the dorsal side of the primitive duodenum at around 3.75th week of gestation while the **ventral bud** arises from the base of the hepatic diverticulum at around 4.5th week of gestation. After undergoing the rotation of the ventral bud to the right of and then behind the developing

---

E.B. Chang, M.D. (✉)
Department of Medicine, University of Chicago, Chicago, IL, USA
e-mail: echang@medicine.bsd.uchicago.edu

P.S. Leung, Ph.D. (✉)
School of Biomedical Sciences, Faculty of Medicine, The Chinese University of Hong Kong, Hong Kong, People's Republic of China
e-mail: psleung@cuhk.edu.hk

duodenal loop, the dorsal and ventral buds come into contact with one another and fusion of the two buds occurs at the end of 6th week of gestation. The ventral bud gives rise to the head and uncinate process of the pancreas while the dorsal bud forms the remaining portion of the organ. Meanwhile, the ventral bud duct is also fused with the distal portion of the dorsal bud duct and thus forms the subsequent **duct of Wirsung**, the main pancreatic duct which runs through the entire pancreas. The proximal portion of the dorsal bud duct becomes the future **duct of Santorini**, the accessory duct. During the fusion of the two pancreatic buds at 6th–7th week of gestation, the pancreatic architecture is observed with tubular structures surrounded by dense mesenchymal tissues next to the duodenal structure. The **mesenchymal layer** probably provides signals to the invading epithelium that regulates the balanced development of the future endocrine and exocrine portions of the pancreas. The dual origin of the organ accounts for the regional differences in the islet cell distribution in adult pancreas. In addition, the **arterial blood supply** of the pancreas arises from branches of the *splenic, gastroduodenal* and *superior mesenteric arteries*. **Extrinsic neural innervation** comes from both parasympathetic and sympathetic fibers through the splenic subdivisions of the celiac plexus. These nerves innervate all the major components of the pancreas, including blood vessels, pancreatic acinar cells, and duct and islet cells.

2   Digestive Enzyme Secretion of the Exocrine Pancreas

2.1   *Synthesis and Exocytosis of Protein by the Pancreas*

Acinus as the Functional Unit of Exocrine Pancreas

The major functional unit of the exocrine pancreas is the **acinus** or acini (in pleural), composed of contiguous, pyramid-shaped glandular cells with their apex facing the lumen of the acinus (Fig. 4.1). These cells have many noteworthy specialized features. First, they are **highly polarized**, having distinct functional and structural differences in the apical and basolateral plasma-membrane domains. Second, acinar cells have **well developed Golgi and rough endoplasmic reticulum complexes**, essential for the synthesis and storage of secretory proteins. **Zymogen or storage granules** can be also found in the apical (luminal side) cytoplasm of the cell, and they vary in number depending on the stage of development and state of stimulation by neuronal and hormonal agents. **Nuclei** are located at the very *base* of the cell.

Mechanism of Protein Secretion by the Acinus

The primary function of the pancreatic acinar cell is to produce large amounts of digestive enzyme proteins that are eventually transported through the ductal system into the duodenum to be mixed with intestinal chyme. The cellular events

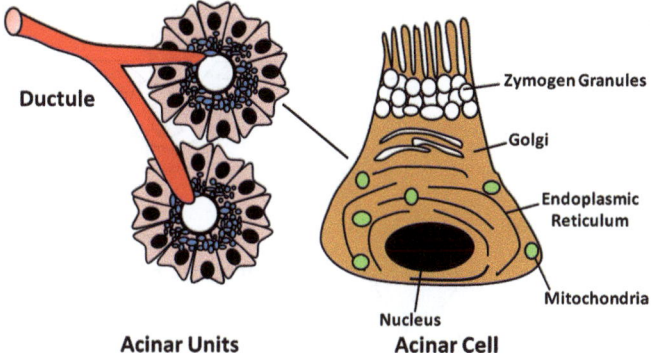

Fig. 4.1 A schematic diagram showing the structure and functional unit of the exocrine pancreas. The acinus is the functional unit of the exocrine pancreas, which is composed of contiguous pyramid-shaped glandular cells with their apex toward the lumen of the acinus. Acinar cells have well-developed Golgi and rough endoplasmic reticulum complexes, essential for synthesizing and storing large amounts of secretory proteins. Zymogens or storage granules can be found in the apical or luminal side of the cell

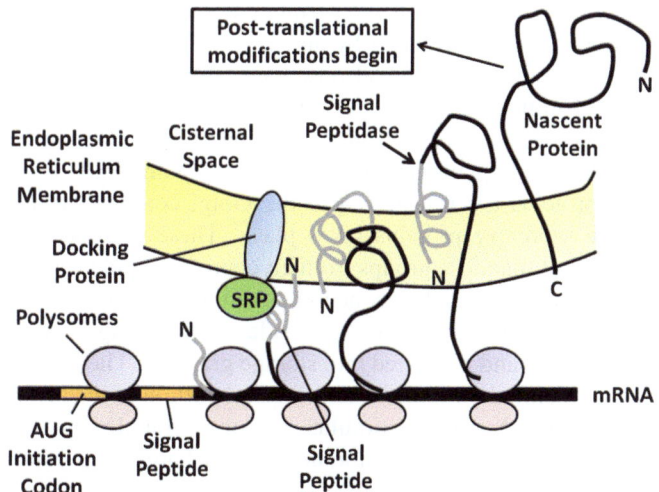

Fig. 4.2 The processing and synthesis of pancreatic digestive enzymes. Proteins for export are first synthesized on polysomes attached to the outer or cytosolic aspect of the rough endoplasmic reticulum at the base of the acinar cell. As translation continues, the nascent protein transverses the endoplasmic reticulum membrane and enters the cisternal space

involved in the synthesis and export of these proteins have been well-characterized. Proteins for export are synthesized on polysomes attached to the outer or cytosolic aspect of the rough endoplasmic reticulum located at the base of the acinar cell. A special signal sequence after the AUG initiation codon is translated into an amino-terminal extension called the **signal peptide** (Fig. 4.2). The signal peptide is avidly

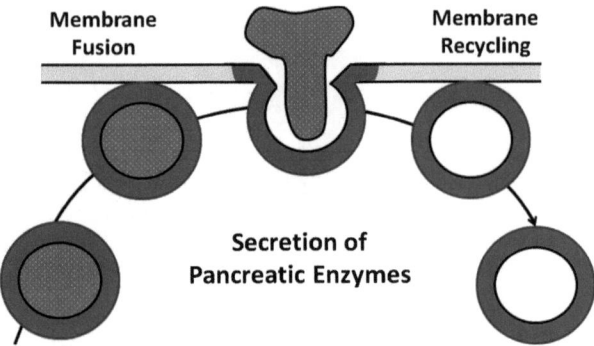

Fig. 4.3 The process of exocytosis of pancreatic enzyme secretion. After an appropriate neural or hormonal stimulus, zymogen granules move to apical membrane, fuse with plasma membrane, and discharge their contents into the luminal space by the process of exocytosis

bound by a cytosolic protein called the **signal-peptide recognition particle (SRP)**, which facilitates the binding of the mRNA-ribosomal complex to the endoplasmic reticulum (ER) membrane. It does so by recognizing and binding to a specific ER-membrane receptor or docking protein. Because of the hydrophobicity of the signal peptide, it enters the internal ER compartment. As translation continues, the rest of the protein traverses the ER membrane and enters the cisternal space. The signal peptide is then cleaved off by an enzyme called **signal peptidase**. The nascent protein then undergoes several post-translational modifications, including the formation of disulfide bridges, glycosylation, sulfation, and phosphorylation. Post-translational processing of secreting proteins is important in folding them into proper tertiary and quaternary configurations. Within 20–30 min of their synthesis, these proteins are transferred to the Golgi complex, where additional processing of the secretory proteins takes place. These modifications generally involve the removal of mannose groups from glycoproteins and progressive buildup to a complex glycosylated form by sequential additions of monosaccharides. These glycoproteins move from the cis to trans side of the Golgi complex and are eventually concentrated and packaged into storage granules. The secretory granules then move by an undefined mechanism to the apical portion of the acinar cell. Upon an appropriate neural of hormonal stimulus, zymogen granules move to the apical membrane, fuse with the plasma membrane, and discharge their contents into the luminal space by the process of exocytosis (Fig. 4.3).

Regulation of Digestive Enzyme Secretion

Several intracellular messengers appear to play a role in regulating the secretion of digestive enzymes. Increases in intracellular calcium are stimulated by agents such as *cholecytokinin (CCK), acetylcholine, bombesin* (gastrin-releasing peptide or GRP is the mammalian equivalent), and *substance P*. Although receptors on acinar cells for all these agents have been identified, it is likely that **cholinergic muscarinic receptors** are the major pathway for regulation (discussed below). These agents

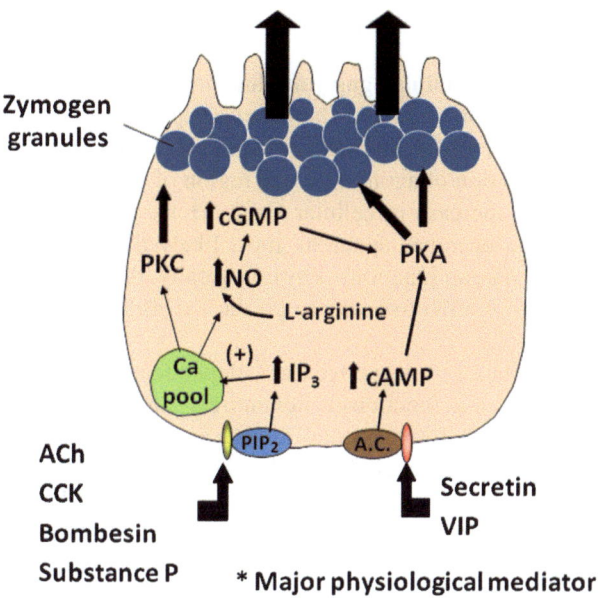

Fig. 4.4 A cellular model showing the regulation of protein secretion by a pancreatic acinar cell. Upon stimulation by agonists such as cholecystokinin (CCK), secretin, vasoactive intestinal peptide (VIP), and acetylcholine (ACh), signal transduction pathways are evoked in the pancreatic acinar cell. ACh and CCK stimulate acinar cell secretion by activating inositol triphosphate (IP$_3$)/diacyl glycerol (DAG) signaling pathways, thus leading to increased cytosolic Ca^{2+} and protein kinase C (PKC). Secretin and VIP stimulate secretion by elevating intracellular cAMP, thereby activating protein kinase A (PKA)

stimulate phosphatidylinositol (PI) metabolism, leading to the formation of inositol 1,4,5-triphosphate (IP$_3$) and 1,2-diacylglycerol (DAG) (Fig. 4.4). IP$_3$ activates IP$_3$ receptors of calcium storage organelles, releasing Ca^{2+} and increasing cytosolic Ca^{2+}. Though being transient, this stimulation serves to trigger a number of more sustained biochemical and functional events. Increased Ca^{2+} activates calcium-dependent, constitutively expressed **nitric oxide synthase** (**NOS**), which produces nitric oxide (NO) from L-arginine. The activity of this enzyme also appears to be sensitive to the level of intracellular calcium stores, as depletion of these stores can independently activate NOS activity. NO then appears to stimulate soluble guanylate cyclase activity to increase cellular cyclic guanosine monophosphate (cGMP) levels. cGMP stimulates increased plasma-membrane permeability to Ca^{2+} to sustain increased cytosolic Ca^{2+}, possibly through activation of a cGMP-dependent cation channel or through the action of cGMP-dependent protein kinases. Increases in cytosolic Ca^{2+} stimulate exocytosis of zymogen granules, probably through calcium-calmodulin-dependent protein kinases. The **activation of protein kinase C** by DAG may also play a role in stimulating acinar cell secretion, particularly in sustaining the effect after receptor activation.

Synchronization of Secretory Responses

Recently, the role of **intercellular gap junctions** in "synchronizing" acinar cells of a functional unit has become evident. Gap junctions provide points of low electrical resistance and for intercellular permeation by small-charge molecules such as IP_3 or Ca^{2+}. Local application of agonists to one region of an acinar unit, for example, stimulates a regional increase in cellular Ca^{2+}. However, the activation of other cells of the unit soon becomes apparent, most likely through the propagation of an activating signal via gap junctions. Physiologically, this **intercellular coupling mechanism** provides an effective way to produce a unified secretory response by the acinar unit.

The relative importance of the above Ca^{2+}-dependent agents for the physiological regulation of pancreatic functions is not entirely clear (see discussion at greater length later in the chapter). Experimentally, these agents do appear to have different effects on acinar cell function, even though they all appear to activate PI metabolism. This could be attributed to differences in their effects on PI metabolism or on receptor coupling with other transduction pathways. For instance, *acetylcholine* and *CCK* vigorously stimulate IP_3 and DAG, the latter eventually desensitizing both receptors. *Bombesin*, on the other hand, has far less effect on DAG formation and does not cross-desensitize receptor activation by CCK or acetylcholine. As another example, *CCK*, but not acetylcholine, has trophic effect on pancreatic acinar cells. This may arise from CCK-receptor coupling with the phospholipase D pathway, which causes the formation of phosphatidic acid and choline, important for mediating CCK trophism.

Agents such as *secretin* and *vasoactive intestinal peptide (VIP)* also stimulate acinar cell secretion, apparently through the activation of adenylate cyclase and increased cyclic adenosine monophosphate (cAMP). Although the exact mechanisms that result in enzyme secretion are not well understood, cAMP-dependent protein kinases are probably involved. The physiological role of cAMP-mediated agonists in regulating acinar cell functions is also not well characterized, although it is believed to be less important than calcium-mediated agonists. However, these agents do appear to augment or potentiate the actions of Ca^{2+}-mediated agonists (Discussion below).

Finally, it should be noted that **microtubules** or **microfilaments** of the **actin-myosin system** play an important role in the movement of secretory granules to the site of exocytosis. For example, *cytochalasin B,* an *inhibitor of the microfilament system*, inhibits the exocytic process. After discharge of the contents of the zymogen granules, excess membrane can be retrieved by the cell and reutilized. Several lines of evidence suggest that this reutilization process involves endocytosis and relocation of these internalized membranes to the trans side of the Golgi complex and to lysosomes.

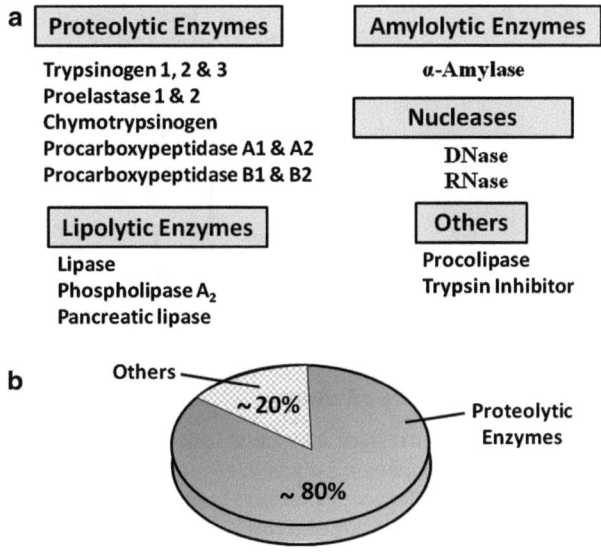

Fig. 4.5 Common proteins of human pancreatic juice. (**a**) These secreted proteins are essential for digestion and absorption of ingested nutrients. (**b**) Proteolytic enzymes make up the majority of proteins that are secreted by pancreatic acinar cells

2.2 Protein Secretion Essential for Digestion

The exocrine pancreas makes and secretes a variety of proteins, most of which are essential for digestion and absorption of ingested nutrients (Fig. 4.5). However, if these proteins were secreted as active enzymes within the parenchyma of the pancreas, the consequences would be potentially disastrous, as extensive tissue destruction would result (**i.e.** *autodigestion*). To **prevent autodigestion**, the pancreas protects itself in several ways. First, all potentially *harmful enzymes are made in an inactive or proenzyme form* and are packaged in zymogen granules within acinar cells. Secreted enzymes remain inactive until their reach the duodenal lumen (Fig. 4.6a). Second, in the duodenum, *trypsinogen, a major proteolytic enzyme, is converted to active trypsin* by an enzyme, called *enterokinase*, a brush-border enzyme expressed by duodenal mucosa. Trypsin catalyzes the activation of more trypsin through hydrolysis of the N-terminal hexapeptide of the trypsinogen molecule, thus rapidly accelerating the entire process. Active trypsin is also essential for activation of several other proteolytic and lipolytic pancreatic enzymes. Thus, the activation site for potentially destructive enzymes is geographically removed from the pancreas and compartmentalized within the duodenal lumen (Fig. 4.6b). Finally, acinar cells make *trypsin inhibitor*, which is packaged with trypsinogen in zymogen granules. Its role is to activate small amount of trypsin that may form within cells or the body of the pancreas.

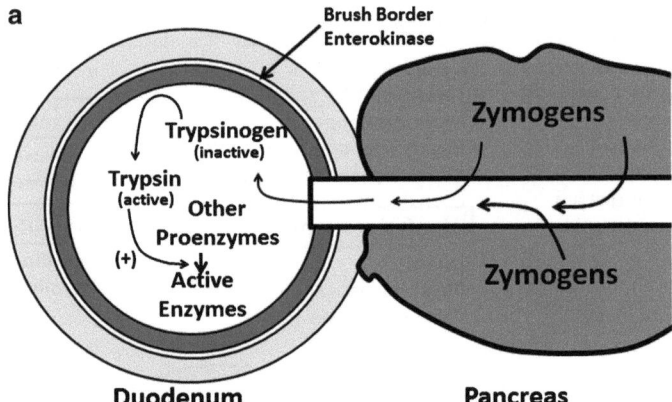

Fig. 4.6 Mechanisms and sites of activation of pancreatic zymogens. (**a**) Secreted enzymes remain inactive until they reach the duodenal lumen. Trypsinogen, a major proteolytic enzyme, is converted to active trypsin by brush-border enterokinase. Trypsin catalyzes the activation of more trypsin, thus accelerating the entire process. Active trypsin is also essential for activation of other proteolytic and lipolytic pancreatic enzymes. (**b**) Mechanisms for preventing autodigestion are shown

Pancreatic juice contains several kinds of hydrolytic enzymes capable of digestive micronutrients (proteins, peptides, fats, carbohydrates, and nucleic acids) into their basic subunits. Proteolytic enzymes make up almost 80 % of all protein found in pancreatic juice (Fig. 4.5b). These include **endopeptidases** (cleaving internal peptide linkages) such as *trypsin* and **exopeptidases** (cleaving from ends of proteins) such as *carboxypeptidases*. Other enzymes include *nucleases*, *amylase* (digestion of carbohydrates), and those involved in lipid digestion such as *lipase, phospholipase A_2*, and *carboxylesterase*. The functional aspects of these enzymes will be discussed in greater depth in the sections on nutritional physiology.

3 Fluid and Electrolyte Transport of the Pancreas

The *basal volume of pancreatic secretion* is estimated to be 0.2–0.3 mL/min, although, when stimulated, pancreatic secretion can reach 4.0–4.5 mL/min. The *daily output of pancreatic juices* into the duodenal lumen in humans is approximately 2.5 L/day. The fluid secreted by the pancreas represents the combined secretions of both duct and acinar cells, but the compositions of these secretions differ significantly. **Acinar cells** secrete a plasma-like fluid that is predominantly

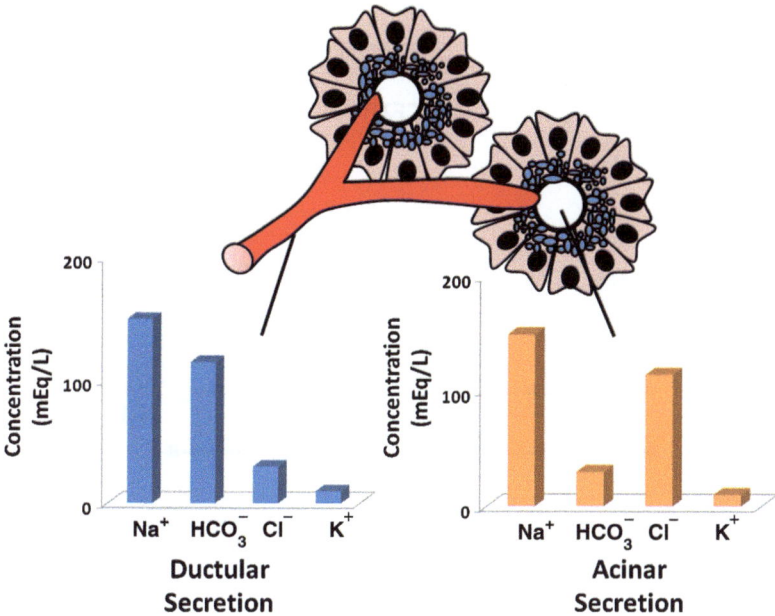

Fig. 4.7 Difference in ionic composition of ductular and acinar fluid. Acinar secretion is a plasma-like fluid that is predominantly sodium chloride. Ductular secretion is predominantly sodium bicarbonate

sodium chloride (Fig. 4.7). The rate of acinar cell fluid secretion is dependent on the rate of enzyme secretion, and the maximal flow rate is much less than that seen in ducts. The primary function of acinar cell fluid secretion therefore is to transport secreted enzymes into the duct system.

Duct cells, on the other hand, secrete a *bicarbonate-rich* fluid, at a considerably variable flow rate of 0–4.0 mL/min depending on the state of pancreatic stimulation. The purpose of the alkaline secretion is to neutralize gastric acid that enters the duodenum, a process essential for achieving optimal conditions for pancreatic enzyme activity. Inadequate bicarbonate secretion with failure to reach a neutral liminal pH, as occurs in **chronic pancreatitis**, contributes to maldigestion of ingested nutrients seen in this condition.

3.1 Cellular Mechanism for Ductal Bicarbonate Secretion

The cellular mechanisms for bicarbonate secretion by duct cells have been partially elucidated (Fig. 4.8). Briefly, bicarbonate is derived from carbonic acid, which is formed from carbon dioxide and water diffusing in from the interstitial side. Carbon dioxide derived from metabolism is believed to account for less than 5 % of bicarbonate in pancreatic juice. Carbonic anhydrase catalyzes the production of

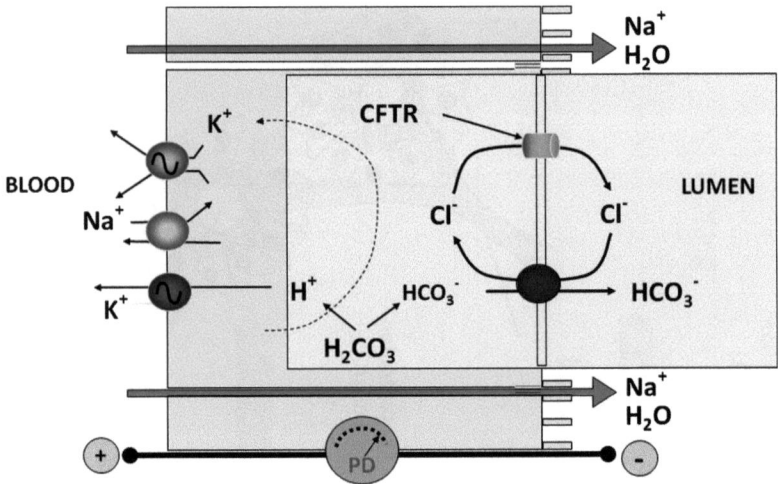

Fig. 4.8 Mechanism of active bicarbonate secretion by pancreatic duct cells. Note that the recycling of Cl$^-$ at the luminal membrane through the cyclic fibrosis transmembrane regulator (CFTR) transporter appears to be important for sustained bicarbonate secretion

HCO_3^- and H^+ from carbonic acid. HCO_3^- is then transported across the luminal plasma membrane by a HCO_3^-/Cl^- exchanger. The major source of luminal Cl$^-$ is now believed to be from the concomitant secretion of the anion via a luminal-membrane Cl$^-$ channel. This channel is regulated by **cAMP-dependent protein kinase** or **cystic fibrosis trans-membrane regulator (CFTR) protein**, which is defective in cystic fibrosis. The recycling of Cl$^-$ is, therefore, a major factor in determining HCO_3^- secretion. Inhibition of Cl$^-$ channel activity will decrease HCO_3^- secretion. This may explain why *pancreatic insufficiency develops in some cystic fibrosis patients*, as it results from defective ductular secretions. In such a condition, proteinaceous acinar secretions become concentrated and their precipitation can potentially cause blockage and destruction of pancreatic ducts.

Protons generated during the production of HCO_3^- must be rapidly transported out of the cells or cell pH would drop precipitously. This occurs at the basolateral membrane through **two different mechanisms**. One involves *Na^+/H^+ exchange*, although it is estimated that the capacity of this pathway is limited especially during maximal HCO_3^- secretion. Therefore, the presence of an *H^+/K^+-ATPase (proton pump)* in the basolateral membrane may provide an alternative and perhaps primary mechanism for rapid proton extrusion. This proton pump is different from the one found in parietal cells of the stomach, being functionally more analogous to proton pumps found in the kidney and distal colon. Na^+/K^+-ATPase is also present in the basolateral membrane, necessary for producing favorable electrochemical gradients for Cl$^-$ secretion. Na^+, some K^+, and water accompany HCO_3^- secretion, mostly entering the duct lumen by passive paracellular diffusion, their rate of transport determined by prevailing electrochemical and osmotic forces.

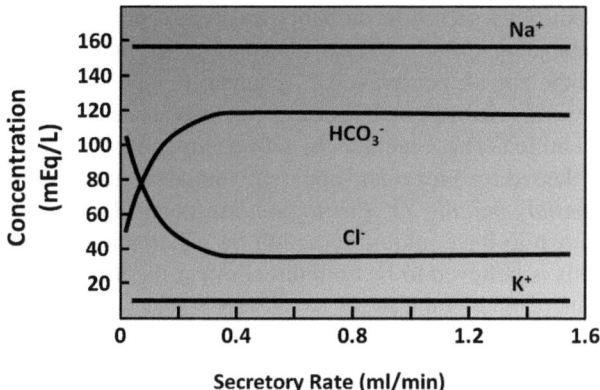

Fig. 4.9 Changes in ionic composition and concentration of pancreatic juice with secretory rate

3.2 Modifications of Electrolyte Composition

Considerable modifications of the pancreatic-juice electrolyte composition occur as secretory rates change. At **low flow rates**, pancreatic fluid is mostly sodium chloride, with small amounts of K^+ and HCO_3^- (Fig. 4.9). However, as **flow rates increase**, the concentration of HCO_3^- increases and a reciprocal decrease in those of Cl^- is observed. It is important to note that Na^+ and K^+ concentrations, the two major cations in pancreatic juice, are not affected by flow rate. Although the mechanisms underlying the rate-dependent alterations in anion composition are not fully understood, several possible explanations have been offered. First, the final concentration of HCO_3^- in pancreatic juices depends on the net HCO_3^-/Cl^- exchange in pancreatic ducts. At low rates, pancreatic juice has greater contact time with ducts, sufficient for ample HCO_3^-/Cl^- exchanger to occur. Pancreatic fluid becomes more Cl^--rich. At faster flow rates, contact time is insufficient for passive HCO_3^-/Cl^- exchanger, and secretions remain HCO_3^--rich. An alternative mechanism involves the proportions of duct and acinar fluids that are admixed. Acinar fluid is mostly sodium chloride and the maximal flow rate is substantially lower than the maximal flow rate observed in duct cells. At low flow rates, HCO_3^- concentration of pancreatic juice approaches that of duct cell fluid as it becomes the contributor to the total volume.

3.3 Regulation of Electrolyte and Water Secretion

Electrolyte and water secretion by the pancreas is regulated by **neural** and **hormonal** agents. The major **stimulant** for sodium bicarbonate and water secretion is *secretin*. This 27-amino-acid peptide is structurally similar to vasoactive intestinal peptide (VIP), and both hormones are known to stimulate adenylate cyclase in duct cells isolated from the pancreas. The mode whereby cAMP induces electrolyte secretion probably involves critical phosphorylation events by cAMP-dependent protein kinase (A-kinase). As mentioned above, A-kinase regulation of the apical membrane Cl^- channel (CFTR) may regulate bicarbonate secretion. Several **inhibitors** of fluid and electrolyte secretion by the pancreas have also been identified. These include the *tetradecapeptide somatostatin, pancreatic polypeptide* (released by a meal and apparently under vagal cholinergic control), *glucagon, and possibly peptide YY. Prostaglandins,* particularly those of the E series, have been shown to have inhibitory action on pancreatic bicarbonate fluid secretion in vivo. This is believed to be an indirect effect mediated by the effects of these agents on pancreatic blood flow.

4 Organismal Regulation of Pancreatic Exocrine Secretion

The amount of pancreatic enzymes in the fluid and the timing of their release are critical for the efficient digestion of micronutrients as they enter the duodenum. The pH of gastric chyme must also be quickly neutralized for several reasons. First, acid-pepsin damage to the duodenal mucosa must be prevented. Second, neutral pH is optimal for pancreatic-enzyme activation and function. Finally, neutral pH increases the solubility of bile acids and fatty acids. The mechanisms for maintaining intraluminal pH at or near neutral values are so efficient that only the very proximal duodenum is normally exposed to pH values below 6. To achieve this level of fine control, the timing and extent of pancreatic secretions must be closely integrated with luminal events and digestive demands. The **regulation of pancreatic function** can be divided into **three phases**: *cephalic, gastric,* and *intestinal*. These phases are defined on the basis of where the stimulant acts.

4.1 Cephalic-Phase Pancreatic Secretion

The cephalic phase of pancreatic secretion results from **central integration of stimuli** such as *sight* and *smell of food or eating food*. Sham feeding, for instance, can stimulate up to 50 % maximal secretion. These stimuli activate efferent vagal impulses that stimulate the secretion of enzymes and bicarbonate (Fig. 4.10a). This action is partially mediated by cholinergic fibers but may also involve peptidergic neurons.

4.2 Gastric-Phase Pancreatic Secretion

This phase is initiated by the **distention of the stomach by food** and by the **presence of amino acids and peptides** in the lumen (Fig. 4.10b). These stimuli

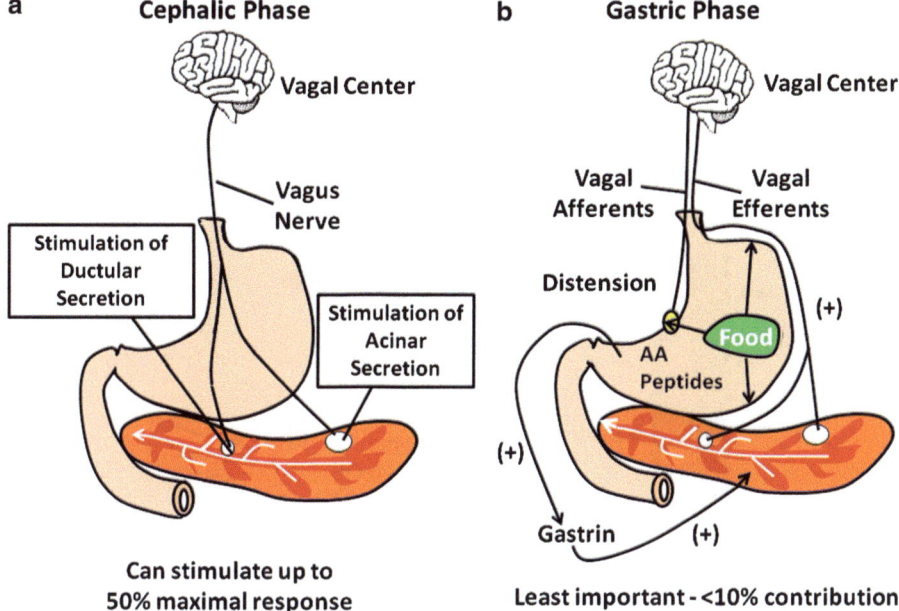

Fig. 4.10 Cephalic and gastric phases of pancreatic exocrine regulation. Meal-stimulated pancreatic secretions are illustrated. (**a**) Cephalic phase and (**b**) gastric phase

activate vagovagal reflexes and gastrin release to stimulate predominantly pancreatic enzyme secretion. **Vagotomy** will abolish most of these effects. This phase is probably the *least important*, accounting for less than 10 % of meal-stimulated pancreatic secretions. Pancreatic bicarbonate secretion is also relatively unaffected by stimuli evoked during the gastric phase.

4.3 Intestinal-Phase Pancreatic Secretion

The intestinal phase of pancreatic secretion is physiologically and quantitatively the *most important* (Fig. 4.11). It is initiated by the **entry of gastric chyme** into the intestinal lumen and is primarily mediated by cholinergic reflexes and the release of cholecystokinin (CCK) and secretin. *CCK* and *secretin* are made by **endocrine cells** of the upper small bowel and their release stimulated by both the composition and quantity of food. Luminal fatty acids, for instance, stimulate CCK release, saturated fatty acids being more potent than unsaturated ones. Additionally, relative potency of fatty acids correlates with chain length, i.e. $C_{18} > C_{12} > C_8$. Neutral triglycerides do not stimulate pancreatic secretion unless lipolysis occurs. It is also believed that there are specific receptors for amino acids or oligopeptides on the small intestinal cells capable of eliciting a neural or hormonal response. Stimulation of the

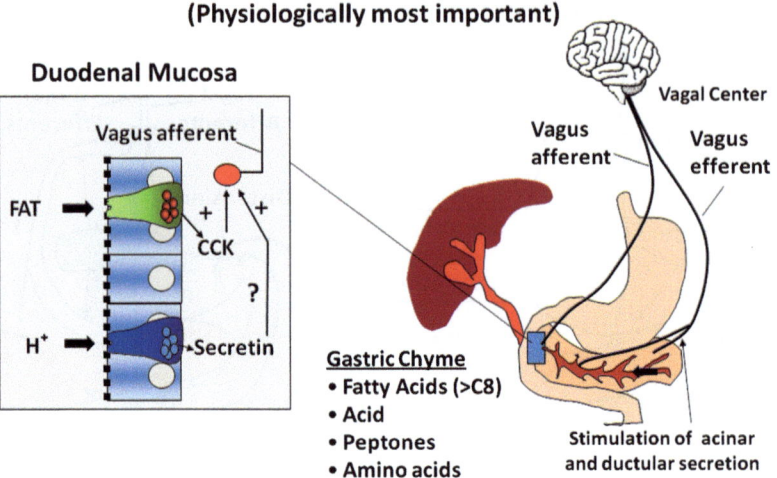

Fig. 4.11 Intestinal phase of pancreatic exocrine regulation. This phase of pancreatic secretion is physiologically and quantitatively the most important. It begins with entry of gastric chyme into the intestinal lumen and is primarily mediated by cholinergic enteric neurons. Ultimate stimulation of acinar secretion is mediated by vagal efferents from the vagal center of the central nervous system

sensory receptors causes the release of peptides such as *CCK, secretin*, and *gastric inhibitory polypeptide (GIP)*, which are crucially involved in activating the second phase of digestion involving coordination of hepatic, biliary, intestinal, gastric, and pancreatic functions.

Regulation of Release of CCK

CCK release from duodenal endocrine cells is stimulated by the presence and composition of luminal contents. Until recently, it was believed that CCK stimulated acinar cell secretion via an *endocrine* action, i.e. entry into the blood circulation and stimulation of distinct CCK receptors expressed by pancreatic acinar cells. However, this notion now appears to be *incorrect*. In fact, the physiological actions of CCK are **predominantly paracrine**, involving afferent cholinergic enteric neurons. Administration of the muscarinic receptor antagonist *atropine* or the perivagal or mucosal application of the sensory neurotoxin *capsaicin* significantly decreases CCK-stimulated pancreatic acinar cell secretion. These findings place into question on the physiological significance of CCK receptors expressed by pancreatic acinar cells. Although it is possible that these represent a second line of regulation, the issue remains unresolved.

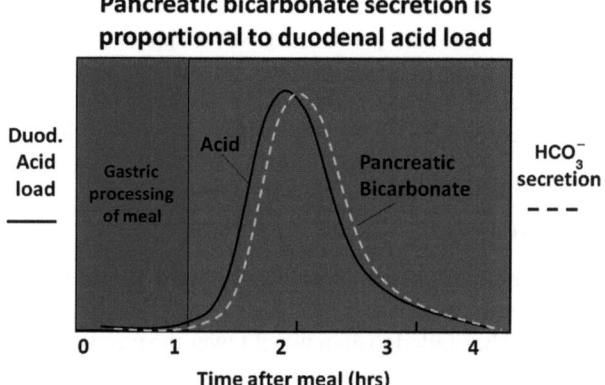

Fig. 4.12 Pancreatic bicarbonate secretion is proportional to the duodenal acid load (Adapted from American Gastroenterological Association Teaching Slide Collection 16 ©, Bethesda, Maryland. Used with permission)

Regulation of Release of Secretin

Increases in secretin stimulated by gastric acid entering the duodenum are tightly controlled and proportional to luminal pH (Fig. 4.12). As a consequence, pancreatic bicarbonate secretion is always proportional to the luminal acid load. At present, the mechanism for this process is believed to involve secretion of secretin by **duodenal endocrine cells** into the blood circulation, although an intermediary role of enteric neurons has not been fully explored. Through an endocrine action, secretin stimulates specific receptors expressed by pancreatic duct cells, initiating bicarbonate secretion. In addition, secretin appears to inhibit gastric acid secretion and motility, allowing time for intestinal digestion before the next bolus of gastric chyme is delivered to the small intestine.

Feedback Mechanism for Pancreatic Secretion

The concept of **feedback regulatory mechanisms** for pancreatic secretions has long been based primarily on studies in rats. Removal of pancreatic proteases from the intestinal lumen or blocking intraluminal proteolytic activity results in stimulation of pancreatic enzyme secretion. The strongest evidence in support of this hypothesis comes from rats in which bile and pancreatic juices are diverted from the intestine. Intestinal perfusion with *trypsin, chymotrypsin,* or *elastase* suppresses CCK release and the pancreatic-enzyme secretory response. Conversely, perfusions with *amylase* or *lipase* have no effect. Inactivated proteolytic enzymes are ineffective, indicating that the inhibitory action of proteases is due to their proteolytic activity. The precise mechanism by which these proteases inhibit pancreatic secretion is still unknown. Recent studies have suggested that a protease-sensitive protein called **CCK-releasing peptide** is secreted by the proximal small intestine in response to meal-stimulated CCK release from duodenal endocrine cells. During a meal, luminal proteins are the dominant targets of pancreatic protease activity, making more CCK-releasing factor available for stimulating mucosal endocrine cells. As protein digestion approaches completion, progressively more CCK-releasing peptides become substrates for hydrolysis, reducing the stimulus for more CCK release. During the fasting state, stimulated secretion of the CCK-releasing peptide ceases, and residual peptide in the lumen is rapidly degraded.

Clinical Correlations

Case Study 1

The parents of a small infant bring their child in for evaluation of abnormal bowel movements characterized by loose, bulky, extremely malodorous, and oily stool. The child has failed to gain weight over the past several months and was recently discharged from the hospital for a pulmonary infection. A sweat test is consistent with the diagnosis of cystic fibrosis (CF). Stool examination reveals the presence of fat.

Questions

1. **What GI complication of CF does this child have?**
   Answer: CF is the most common lethal genetic disorder of Caucasians. The genetic basis is now well understood and involves the inheritance of a defective gene called the **cystic fibrosis transmembrane regulator (CFTR)**. The CFTR gene encodes for a membrane protein that appears to be a cAMP-regulated Cl^- channel, although it may have other functions as well. The **clinical manifestations** mainly involve *pulmonary infections* and *complications of the GI tract*. Patients typically have ***abnormal sweat electrolytes***, characterized by abnormally elevated levels of Na^+ and Cl^-. Although several mutations of CFTR have been described, the most common involved is the **deletion of three nucleotides** encoding a **phenylalanine at position 508**. This causes failure of CFTR to be inserted in the plasma membrane and defective regulation by cAMP-dependent protein kinase.

   CFTR is now recognized to be important in the pathways that mediate pancreatic bicarbonate secretion as discussed in this chapter. The failure of its membrane insertion and activation by cAMP-dependent protein kinase prevents recycling of Cl^- required for vectorial bicarbonate secretion. As a consequence, CF patients have **diminished ductal secretion** that impairs the ability of the pancreas to move pancreatic zymogens to the duodenal lumen. Although the precise pathogenic mechanisms causing pancreatic insufficiency in 85–90 % of CF patients are not known, it is possible that the **ductal inspissation of pancreatic juices** and **activation of zymogens** may lead to *ductal obstruction*, *inappropriate activation of pancreatic enzymes*, and *tissue injury*.

2. **How does pancreatic exocrine insufficiency affect the patient?**
   Answer: CF can be accompanied with *pancreatic insufficiency* that causes **maldigestion** (see why below) and **malabsorption** of nutrients (discussed in greater detail in the nutritional physiology chapters). The pancreas has a large functional reserve, as demonstrated by the fact that significant maldigestion in humans with chronic pancreatitis does not occur until the maximal pancreatic secretory capacity for enzyme drops to less than 10 % of normal. In addition to

the **loss of pancreatic enzyme** output (due to tissue destruction), **inadequate bicarbonate secretion** by the pancreas results in failure to adequately neutralize the pH of gastric chyme. Thus, luminal conditions for activation of pancreatic zymogens (best at neutral pH) and micellar formation (necessary for fat digestion and absorption) are suboptimal. This explains this child's *fatty stools* and *failure to gain weight*.

Case Study 2

A 5-year-old boy presents with a history of growth retardation and bulky, oily, malodorous stools. Although he has no history of pulmonary infections, a sweat test for CF is ordered; it is found to be normal. Peroral biopsy of the duodenal mucosa is also normal, and bacterial overgrowth of the bowel is ruled out by a number of diagnostic tests. Because of the extremely high content of stool fat (20 g/day), the possibility of pancreatic insufficiency is entertained. The serum amylase level and imaging studies of the pancreas, however, are normal. Finally, studies to determine the exocrine function of the pancreas are performed. Bicarbonate secretion stimulated by administration of intravenous secretin is found to be normal. However, examination of duodenal contents reveals very low trypsin activity.

Questions

1. **What might be the underlying problem of this patient?**
   Answer: When the pancreatic juice is further analyzed, it appears to have normal protein content, arguing against a problem with pancreatic enzyme production or secretion. However, activities of all pancreatic enzymes are low, suggesting a failure in their activation. This is confirmed, as most *zymogens* are still found to be in their *inactive, proenzyme form*. A mucosal biopsy is performed again and tested for the presence of enterokinase, which is found to be absent. Thus, this child has a **congenital absence of brush-border enterokinase**, which is required for activation of trypsinogen to trypsin (see below). Trypsin in turn activates all other proenzymes of pancreatic juice.
2. **How can you treat this patient?**
   Answers: With **oral replacements of enterokinase**, the child rapidly gains weight and no longer has intestinal symptoms.

Case Study 3

A 58-year-old woman is found to have a slightly elevated 24 hour stool fat content (7 g/day). She is otherwise asymptomatic. She had a history of having a truncal vagotomy 20 years ago for peptic ulcer disease.

Questions

1. **In a patient who has had a vagotomy (surgical interruption of the vagus nerve to the gut), what would be the mechanisms involved in decreased pancreatic secretion thus fat maldigestion and malabsorption?**
   Answer 1: Traditionally, the actions of CCK and possibly secretin on pancreatic function were thought to be mediated by a hormonal pathway. However, recent

studies of humans and animals have dispelled this notion. It is now believed that the **vagal nerve** is essential for mediating CCK-stimulated acinar-cell zymogen secretions (and possibly secretin-stimulated bicarbonate secretion from ducts). CCK is released by endocrine cells of the duodenal mucosa in response to luminal fat and proteolytic products. CCK then stimulates afferent vagal fibers that in turn initiate an effector response emanating from the vagal nuclei of the brainstem. Efferent vagal stimulation thus appears to account for the majority of stimulated acinar cell secretion.

Critical thinking:
Small amounts of pancreatic enzymes normally escape from the gland into the plasma, and those that are absorbed across the intestinal epithelia likewise enter the plasma. Because they are of low molecular weight, they appear in the urine. *Amylase*, for example, will enter the plasma when there is ductal obstruction or rupture or pancreatic destruction. The enzyme is filtered through the glomerular filtration apparatus and along with other proteins is partially reabsorbed by the renal tubules. **A rise in urine amylase** will result from increased liberation of the enzyme from the pancreas or from reduced reabsorption by the renal tubule. Hence the enzyme has been useful in **making the diagnosis of pancreatitis**. Elevated urine amylase is particularly significant in a diagnosis of pancreatitis when the *renal clearance of amylase* **far exceeds** the *renal clearance of creatine* (a crude measure of glomerular filtration rate).

Further Reading

1. Argent BE, Gray MA, Steward MC, Case RM (2012) Cell physiology of pancreatic ducts. In: Johnson LR (ed) Physiology of the gastrointestinal tract, 5th edn. Academic, New York, pp 1399–1423
2. Brelian D, Tenner S (2012) Diarrhoea due to pancreatic diseases. Best Pract Res Clin Gastroenterol 26:623–631
3. Forsmark CE (2013) Management of chronic pancreatitis. Gastroenterology 144:1282–1291
4. Gittes GK (2009) Developmental biology of the pancreas: a comprehensive review. Dev Biol 326:4–35
5. Gorelick FS, Jamieson JD (2012) Structure-function relationships in the pancreatic acinar cell. In: Johnson LR (ed) Physiology of the gastrointestinal tract, 5th edn. Academic, New York, pp 1341–1360
6. Lai KC, Cheng CHK, Leung PS (2007) The ghrelin system in acinar cells: localization, expression and regulation in the exocrine pancreas. Pancreas 35:e1–e8
7. Leung PS (2010) The renin-angiotensin system: current research progress in the pancreas. In: Advances in experimental medicine and biology book series, vol 690. Springer, Dordrecht, pp 1–207
8. Leung PS, Chan YC (2009) Role of oxidative stress in pancreatic inflammation. Antioxid Redox Signal 11:135–165
9. Leung PS, Ip SP (2006) Cell in focus – pancreatic acinar cell: its role in acute pancreatitis. Int J Biochem Cell Biol 38:1024–1030
10. Liddle RA (2012) Regulation of pancreatic secretion. In: Johnson LR (ed) Physiology of the gastrointestinal tract, 5th edn. Academic, New York, pp 1425–1460

11. Sah RP, Dawra RK, Saluja AK (2013) New insights into the pathogenesis of pancreatitis. Curr Opin Gastroenterol 29:523–530
12. Shih HP, Wang A, Sander M (2013) Pancreas organogenesis: from lineage determination to morphogenesis. Ann Rev Cell Dev Biol 29:81–105
13. Williams JA (2010) Regulation of acinar cell function in the pancreas. Curr Opin Gastroenterol 26:478–483
14. Williams JA, Yule DI (2012) Stimulus-secretion coupling in pancreatic acinar cells. In: Johnson LR (ed) Physiology of the gastrointestinal tract, 5th edn. Academic, New York, pp 1361–1398

Chapter 5
Intestinal Water and Electrolyte Transport

Eugene B. Chang and Po Sing Leung

1 Introduction to Gut Absorption of Fluid and Electrolyte

In addition to *digestion* and *absorption* of nutrients, the **intestinal tract** has several essential functions, including a *barrier* to the outside environment, *synthesis* of secreted proteins, such as those required for fat absorption and immunoglobulin secretion, and elimination of waste products, as well as *transport* of salt and water. Failure to efficiently absorbing water and electrolyte, it will lead to *dehydration* and *electrolyte imbalance*. Most of these processes are dependent on the specialized mucosal functions and structural requirements discussed in this chapter. Particular emphasis will be placed on aspects of intestinal epithelial biology and the transport of fluid and electrolytes. The discussion of other mucosal functions such as digestion and absorption of nutrients will be covered in subsequent chapters in Part Two of this book.

The *cells relevant to intestinal water and electrolyte transport* can be roughly divided into **two major groups:** those involved in **electrolyte transport** and those having predominantly a **regulatory role,** i.e. integrating the functional responses of the first group. This organization of the intestinal mucosa is extremely important considering the type of work it must perform. The absorptive epithelium of the intestine receives a luminal load averaging 9 L/day (Fig. 5.1). Approximately 2 L come from oral ingestion and 7 L from endogenous secretions from a variety of

---

E.B. Chang, M.D. (✉)
Department of Medicine, University of Chicago, Chicago, IL, USA
e-mail: echang@medicine.bsd.uchicago.edu

P.S. Leung, Ph.D. (✉)
School of Biomedical Sciences, Faculty of Medicine, The Chinese University of Hong Kong, Hong Kong, People's Republic of China
e-mail: psleung@cuhk.edu.hk

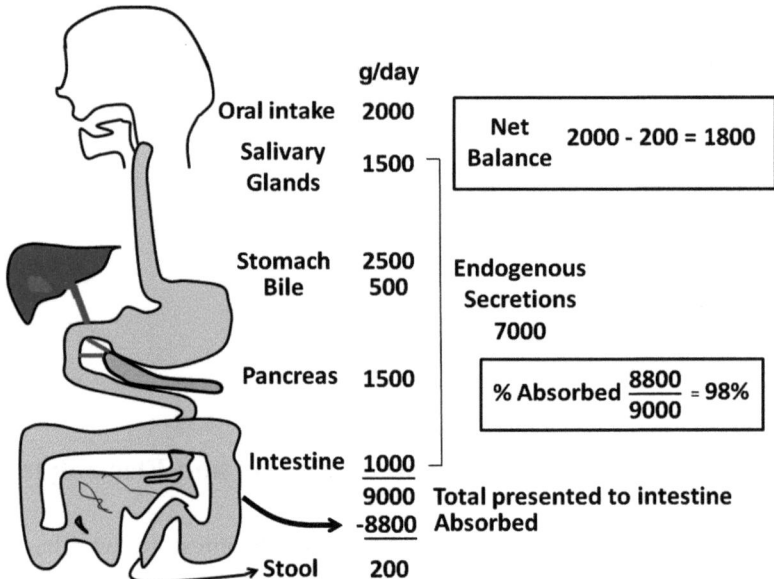

Fig. 5.1 Daily volumes of fluid entering and being absorbed by the gastrointestinal system, and excreted into feces. Fully 98 % of the fluid load is absorbed in the intestine, with only 200 g/day or less excreted via the stool. Compromise of absorption efficiency results in diarrhea (Adapted from American Gastroenterological Association Teaching Slide Collection 25 ©, Bethesda, Maryland; Used with permission)

sources including the salivary glands, gastric juices, bile, pancreas, and enteric secretions. The endogenous secretions provide the necessary conditions, such as *pH, aqueous medium, salts,* and *osmolality,* for rapid and efficient digestion and absorption of nutrients and electrolytes. Of the 9 L presented to the intestinal mucosa each day, approximately 8.8 L or more are absorbed, resulting in less than 200 g/day of stool output (assuming the subject is eating a typical Western, low-roughage diet). Therefore, the gut is capable of absorbing greater than 98 % of the fluid load it is presented, making it a highly efficient organ for absorption of water and electrolytes. However, the gut also has the ability to adjust to large variations in luminal composition and volumes, especially increases. Diarrhea, which results when the absorptive capacity of the gut is exceeded, infrequently occurs under physiological circumstances. This can be attributed to the gut's ability to fine tune its absorptive and secretory processes, achieved through its intricate and often redundant regulatory mechanisms and by input from extra-intestinal sources. If any aberrations occur in the homeostatic regulation of ion-transport functions, stool output could exceed 200 g/day and diarrhea would result.

2 Organization of the Intestinal Mucosa

2.1 Intestinal Cells and Factors Involved in Water and Electrolyte Transport

The transport of water and electrolytes by the intestinal mucosa involves several types of cells and structural relationships. Important components of this process are discussed below.

Epithelial Cells

Epithelial cells represent the largest population of cells of the intestinal mucosa, of which there are **four major types:** (**1**) columnar, polarized epithelial cells, capable of vectorial transport of nutrients and electrolytes; (**2**) mucosal endocrine cells; (**3**) mucus producing and secreting goblet cells; and (**4**) defense-producing Paneth cells located at the base of intestinal crypts (Fig. 5.2). The latter two cell types will not be discussed, as their role in intestinal water and electrolyte transport is questionable or unknown. Gut epithelial cells emanate from a stable stem-cell population located near the base of the crypt and, with the exception of Paneth cells, differentiating as they migrate up the crypt-villus axis. As the cells reach the villus tip, they undergo **apoptosis** (physiologically programmed cell death) and are sloughed off into the lumen, with the entire sojourn taking 3–5 days. As cells undergo crypt-to-villus differentiation, significant *changes in cellular morphology* are evident (Fig. 5.3a), characterized by increasing polarity of the cells, differences in cellular organelle components, development of the microvillus membrane and terminal web, and alterations in other cytoskeletal and tight-junctional structural features. The latter are accompanied by decreases in junctional permeability, a property that may arise out of necessity to achieve efficient absorption of nutrients and electrolytes with a minimum of back flux. The regional differences in tight-junction permeability are reflected by the number of intercellular strands that make up the tight junctional complex. The number of strands in this anastomosing network in villus cells is much greater than in the crypt cells, making the villus regions more impervious to passive diffusion of water and electrolytes.

Alterations in functional characteristics along the crypt-villus axis also occur (Fig. 5.3b). Absorption and secretion of water and electrolytes are two distinct processes in the gut that take place in the villus and crypt cells, respectively. **Cystic fibrosis transmembrane regulator (CFTR),** a cyclic adenosine monophosphate (cAMP) regulated Cl^- channel involved in anion secretion, is predominantly found in **crypt cells**, consistent with a secretory role. As these cells become **villus cells**, the protein synthetic and secretory capabilities increase, reflected by increased development of the Golgi-endoplasmic reticulum complex and numbers of secondary endosomal vesicles. Properties of nutrient absorption and digestion appear, exemplified by increases in brush-border hydrolase activities, glucose

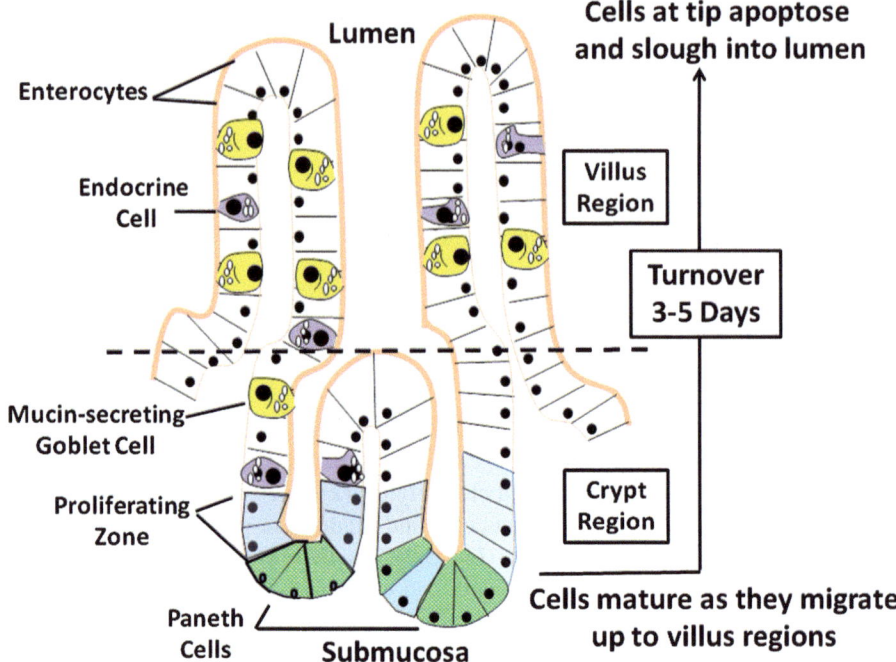

Fig. 5.2 Absorptive and secretory flows that determine the next fluid movement of the intestinal epithelial cells. There are four major types of epithelial cells making up the intestinal mucosa: enterocytes, endocrine cells, goblet cells, and Paneths cell. With the exception of Paneths cells, cells originating from the proliferative zone migrate up the villus axis and mature during the process, which eventually have a turnover rate of 3–5 days

transport and increased surface area of the apical membrane. Thus, the observed morphological, phenotypical and functional changes assumed by differentiating villus cells are consistent with their enhanced capacity for absorption of nutrients and electrolytes.

Blood Capillaries and Lymphatics

Blood capillaries and lymphatics have a major role in **intestinal water** and **electrolyte transport**. During absorption, for instance, they rapidly remove absorbed nutrients, water and electrolytes from the interstitium, thereby allowing vectorial transport to proceed. Similarly, active secretion of water and electrolytes is accompanied by increased mucosal blood flow and capillary filtration and by decreased villus lymph pressure and total lymphatic flow. In each case, in order for efficient absorption and secretion to occur, the events of mucosal transport, capillary flow, and lymphatic function must be coordinated. **Neural and hormonal** agents play a major role in the integration of these events. During secretion, for example, the net

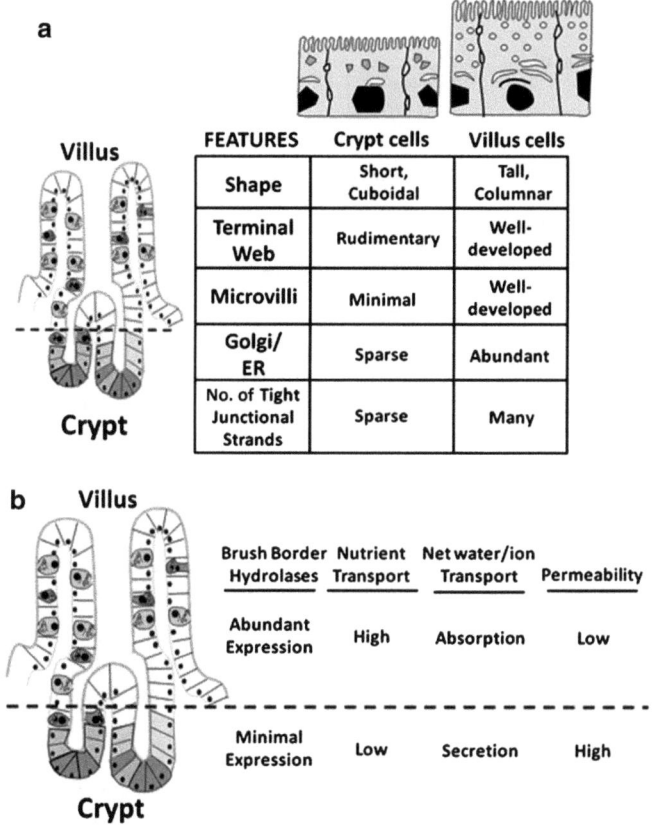

Fig. 5.3 Changes in structures and functions of villus cells and crypt cells. (**a**) As cells undergo crypt-to-villus differentiation, significant changes in cellular morphology become evident. (**b**) Alterations of functional characteristics of enterocytes occur along the crypts-villus axis (Adapted from American Gastroenterological Association Teaching Slide Collection 25 ©, Bethesda, Maryland; Used with permission)

effect of these agents is to increase delivery of plasma fluid and electrolytes to match the demands of active intestinal secretion. Another purpose of increased blood flow during active secretion is to increase tissue delivery of oxygen to meet metabolic demands.

Intestinal Motility

The relationship between intestinal motility and mucosal water and electrolyte transport is extremely complex and incompletely understood. Although the functional properties of motility and ion transport can be readily studied independently, their physiological roles in intestinal water nutrient and electrolyte transport in vivo are

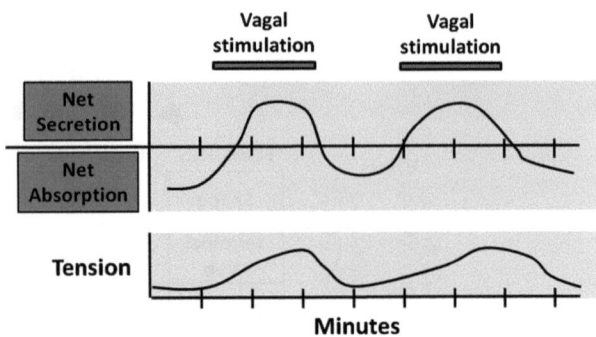

Fig. 5.4 Neural integration of intestinal motility and secretion. Coordination of water and electrolyte transport with intestinal motor function. Enteric reflexes coordinate intestinal water and electrolyte secretion with smooth muscle contractions

very much interdependent. This was recognized as early as 1912 by Babkin and Ishikawi, who noted that the periodicity of intestinal secretion coincided with that of intestinal motor activity. Consistent with this notion, there is ample evidence that **enteric reflexes** coordinate intestinal water and electrolyte secretion with smooth muscle contractions (Fig. 5.4). The integration of these responses serves several important purposes. First, it provides an immediate response to begin the process of digestion and absorption of luminal contents. Increased net secretion is essential for providing the aqueous milieu to reach isotonicity and for digestive enzymes to function properly. Coordinated patterns of intestinal motility ensure that mixing and propulsive activity of luminal contents proceed appropriately. This mechanical activity can significantly enhance the absorption of nutrients, electrolytes, and water through several potential mechanisms. Segmentation of the gut helps mix luminal contents with secretions and digestive juices and increases contact time between the luminal phase and the absorptive mucosal surface. **Increased intestinal motor contractility** is believed to alter the unstirred water layer, present as the result of the overlying layer of mucus gel. This facilitates diffusion of nutrients and electrolytes to the transporters of the brush-border membranes.

Decreased propulsive motor activity may enhance contact time between the absorptive surface area and luminal contents and is probably the major mechanism of action of commonly used antidiarrheal agents. In clinical studies of *loperamide*, *codeine*, and the α_2-receptor agonist *clonidine*, the predominant proabsorptive effect of these agents appeared to be due to their ability to increase the "enteropooling" capacity of the gut. The net effect of this action is to increase contact time between luminal fluid and the absorptive surface area.

2.2 Mucosal Cells Involved in the Regulation of Gut Water and Electrolyte Transport

Mucosal Endocrine Cells

Endocrine cells are widely distributed in the intestinal mucosa and represent a *major source* of **active amines** and **polypeptide hormones** important in the

Regulatory Agents of Intestinal Water and Electrolyte Transport

Source	Stimulate Net Secretion	Stimulate Net Absorption
Mucosal Epithelial Cells	Serotonin Gastrin/CCK Neurotensin Guanylin	Somatostatin
Lamina Propria Cells	Arachidonate metabolites Active Oxidants Nitric Oxide Some cytokines Bradykinin	?
Enteric Neurons	Acetylcholine Serotonin VIP Substance P Purinergic Agonists	Norepinephrine Neuropeptide Y
Blood	VIP Calcitonin Prostaglandins Atrial Natriuretic Peptides	Epinephrine Corticosteroids Mineralocorticoids Angiotensins

Fig. 5.5 Common regulators for the control of gut water and electrolyte transport. There are peptides, active amines, and other agents from different gut layers that can modulate intestinal fluid and electrolyte transport

regulation of fluid and electrolyte transport. Although many of the contents of their secretory granules are also found in nerve cells of the enteric nervous system, it is likely these cells have unique and independent role in modulating various mucosal functions. Like their transporting epithelial counterparts, mucosal endocrine cells are **polarized**. Their *apical or luminal pole* is characterized by tufts of microvilli and coated vesicles, whereas their *secretory granules and nuclei* are located at the basal domain of the cell. These structural characteristics most likely represent the functional requirements of the cell to sense alterations in luminal content such as pH, osmolality, and chemical content and initiate an integrated mucosal response through the release of secretory-granule contents at the basal surface of the cells. Several peptides, active amines, and other agents appear to modulate intestinal fluid and electrolyte transport (Fig. 5.5). The regional distribution of these agents throughout the gastrointestinal (GI) tract differs markedly. For example, cells containing *gastrin, cholecystokinin (CCK)*, and *secretin* are more prominently found in the **stomach** and **proximal small intestine,** whereas *neurotensin-containing cells* are largely restricted to the **ileum**. *Serotonin-containing enterochromaffin cells* are found throughout the mucosa but predominantly in the **crypt regions**, where they may project basal processes that run subjacent to neighboring epithelial cells and nerves containing other peptides. Recently, *guanylin,* the natural endogenous peptide agonist of the *E. coli* heat-stable enterotoxin (ST_a) receptor, has been localized to epithelial (nonclassic endocrine) cells of the colonic mucosa (and possibly Paneth cells of the human small intestine). Guanylin is released into the crypt lumen and stimulates luminal receptors on intestinal epithelial cells by a paracrine action to stimulate net secretion.

Mucosal endocrine cells most likely regulate intestinal ion-transport functions through a paracrine action. They can be activated by a variety of luminal and serosal stimuli. Thus, mucosal endocrine cells probably serve **two important functions** relevant to the control of intestinal water electrolyte transport. First, they provide a means for the mucosa to sense and rapidly respond to alterations in luminal content and milieu. The stimulated release of hormonal peptides and active amines causes appropriate changes in ion-transport functions, blood and lymphatic flow, and intestinal motility. Secondly, they may function as a fine tuning mechanism or amplifier for modulatory signals received from neural sources.

Enteric Neurons

The GI tract is one of the most richly innervated organs of the body and has **two major categories of nerves:** the *intrinsic or enteric nervous system* and the *extrinsic autonomic nervous system* consisting of parasympathetic and sympathetic nerve pathways. Although most **postganglionic sympathetic fibers** terminate in enteric ganglia, a few have been reported in close proximity to intestinal epithelial cells, where they may form actual synapses. Release of norepinephrine from these neurons stimulates α_2-adrenergic receptors on the basolateral membranes of enterocytes, causing increased electroneutral absorption of sodium chloride and inhibition of anion secretion. **Parasympathetic neurons** are also believed to be important in the regulation of intestinal salt and water transport. However, the nature of vagal postganglionic fibers to the intestine is less well understood. Some fibers make up interneurons that modulate enteric system tone or responses.

The **enteric nervous system** is a dense plexus of efferent, afferent, and interneurons, exceeding in number the neurons of the spinal cord. It is composed of *cholinergic* and *non-cholinergic* neurons that regulate numerous mucosal and motor functions. The number of neurotransmitters found in these nerves is quite large and includes *active amines* (such as *serotonin* and *acetylcholine*), *neuropeptides* (such as *substance P, neurotensin, CCK, neuropeptide Y, somatostatin, calcitonin gene-related peptide, vasoactive intestinal peptide* and *galanin*), and *purinergic neurotransmitters* (such as *adenosine* and *adenosine triphosphate, ATP*). This tremendous diversity of neurotransmitters is probably required to regulate and coordinate the numerous mucosal and motor functions involved in salt and water transport. Many of the enteric nervous system neurons are part of programmed reflexive circuits that can immediately respond to various stimuli. Mucosal sensory fibers respond to a number of stimuli, including mechanical factor (touch, pressure, and tension), changes in luminal content and composition (pH, osmolality, and amino acids), temperature, and pain. These sensory signals are then relayed to interneurons that rapidly process and sort the signals so that efferent (motor) neurons (regulating smooth muscle, blood vessel, absorptive and secretory cells, and other cells of the lamina propria) can produce patterned and coordinated responses for efficient transport of water and electrolytes. The enteric nervous system plays a major role in regulating intestinal water and electrolyte transport, as evidenced by the fact that mucosal and motor functions can proceed independently of extrinsic neural input.

Mesenchymal Cells of the Lamina Propria

Mesenchymal cells of the lamina propria and submucosa play a **juxtacrine** role and modulate intestinal mucosal transport, blood flow, and motor functions. They include many cell types such as sub-epithelial fibroblasts, endothelial cells, mast cells, neutrophils, macrophages, and eosinophils. Their numbers within the mucosa vary considerably depending on the species, the region of the intestine, and the prevailing physiological or pathophysiological circumstances. Although their relative roles in the physiological regulation of intestinal water and electrolyte transport have not been established, it is likely that many of these cells play a major role in causing net secretion, altered motor function, and changes in blood flow under pathophysiological situations such as *mucosal inflammation.*

Many of these cells appear to have important **interactions with neural and epithelial elements** within the intestinal mucosa (Fig. 5.6). For example, *mast cells*, found throughout the intestinal mucosa and often in close proximity or juxtaposed to enteric nerve fibers, activate enteric neurons. This serves to amplify and extend the effects of mast-cell mediators such as *histamine*. These cells have an important immunoregulatory role and are involved in the allergic and anaphylactic reactions to food antigens and helminth parasites, as well as in diseases such as *systemic mastocytosis*. The number of inflammatory mediators released by the mast cell is large and includes agent such as *histamine, adenosine, platelet-activating factor (PAF), serotonin,* and *arachidonic acid metabolites.* Agents such as *prostaglandin E_2 (PGE$_2$), adenosine,* and *serotonin* stimulate net intestinal secretion in part by directly activating epithelial receptors, causing decreased absorption of Na^+ and Cl^- and activate secretion of anions. However, these agents and others such as *histamine* and *PGD$_2$* also stimulate net secretion by promoting the release of various neurotransmitters, thus amplifying or augmenting the overall secretory response.

Resident tissue macrophages are also prevalent in the **intestinal mucosa**, constituting 10–20 % of the total cell number in the lamina propria. This makes

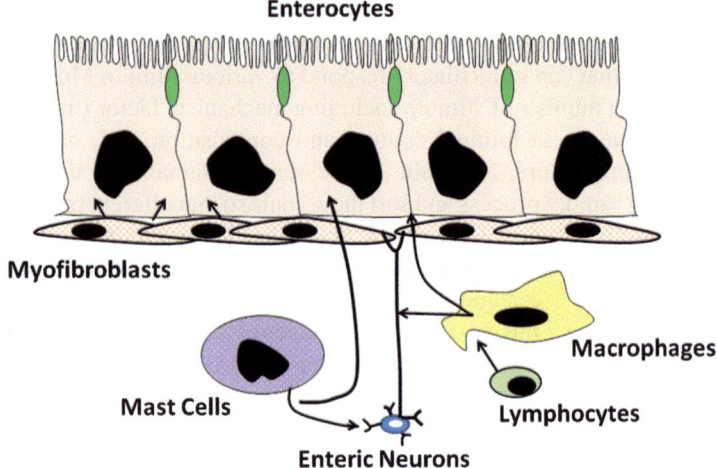

Fig. 5.6 Interaction of enteric cells and other mucoal cells for intestinal water and electrolyte transport. Mesenchymal cells of the lamina propria play an important role in regulating intestinal water and electrolyte transport (Adapted from American Gastroenterological Association Teaching Slide Collection 25 ©, Bethesda, Maryland; Used with permission)

the small and large intestine one of the largest repositories of macrophages in the body. Under steady-state and physiological conditions, there appears to be a constant turnover of macrophages (on the order of days to weeks), mostly from the replacement of existing tissue macrophages with incoming monocytes. In the intestine, the macrophages may be conditioned by tissue-specific influences, such as *bacterial products from the lumen* (including endotoxin), *extracellular matrix,* and *cytokines.* The continuous exposure to foreign antigens and pathogens, particularly in the colon, may be important in sensitizing macrophages, allowing them to rapidly react to various stimuli. Thus, macrophages function as a *first-line defense* against pathogens and antigens and are capable of orchestrating and amplifying an appropriate immune and inflammatory response. The intestinal macrophages are veritable factories for synthesis and release of numerous immune and inflammatory mediators. They are a major source of **carbon monoxide** and **5-lipoxygenase (5-LO) metabolites** and may mediate the secretory effects of many of the secretagogues that are known to activate the arachidonic acid cascade. In response to numerous immune and inflammatory stimuli, they also elaborate **cytokines** such as *interleukin-1 (IL-1), IL-6, granulocyte-macrophage colony-stimulating factor (GM-CSF),* and other inflammatory mediators such as *purinergic agents*. Most of these products have now been shown to affect intestinal water and electrolyte transport and have potent effects on modifying intestinal motor functions and capillary blood flow and permeability. Like mast cell products, they have numerous sites and mechanisms of actions that affect ion transport. These will be discussed later in the chapter.

3 Mucosal Electrolyte Transport Processes

3.1 Absorptive Pathways for Water and Electrolytes

Water transport by the intestine is closely coupled with solute movement and is passive. In theory, water flow could occur by transcellular and paracellular routes, but the prevailing evidence indicates that water transport by the intestine occurs through the **paracellular pathway.** Like other tissues that transport electrolytes, the intestine has a variety of **specialized transport proteins**, which can be divided into **three major types** (Fig. 5.7).

Pumps such as sodium pump (Na^+/K^+-ATPase) and the proton pump (H^+/K^+-ATPase) are energy-driven and capable of transporting ions against large electrochemical gradients. In intestinal epithelium, for example, the sodium pump is essential for establishing and maintaining electrochemical gradients (low intracellular Na^+ and electronegative membrane potential) that are required for other types of passive and facilitated transport processes.

Channel proteins are selective membrane "pores" for ions such as Na^+ and Cl^-. Channel transport is dependent on favorable electrochemical gradients, is often membrane-potential sensitive, and is generally electrogenic, i.e. causing a potential difference across the epithelial layer that promotes passive diffusion of a counter ion. For example, electrogenic Cl^- secretion in the gut causes a potential difference (serosa is positive relative to lumen) across the mucosa that promotes passive transport of Na^+, resulting in net sodium chloride secretion.

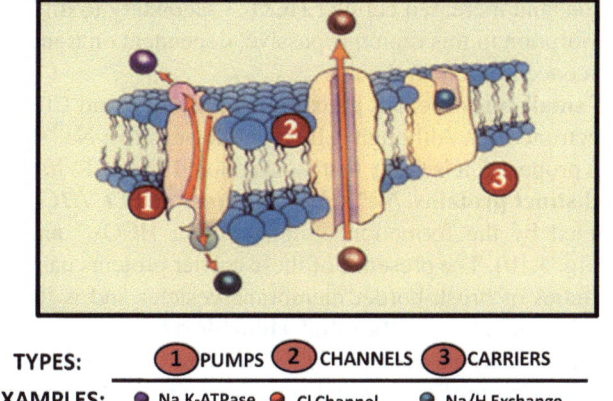

Fig. 5.7 Intestinal transport proteins. There are three types of transport proteins that are found in the plasma membranes of intestinal cells: pumps, channels and carriers (Adapted from American Gastroenterological Association Teaching Slide Collection 25 ©, Bethesda, Maryland; Used with permission)

Carrier transport proteins facilitate transport of ions and nutrients across the cell membrane, and their transport activity is dependent on existing electrochemical gradients. *Several types of carrier proteins* exist in gut epithelium. **Uniport** carrier proteins, such as the *facilitated glucose transporter GLUT2*, mediate the transport of a single ion or nutrient molecule; **symport** carrier proteins, such as *sodium/glucose cotransport-1 (SGLT-1) protein*, are carriers that simultaneously transport two or more molecules, often taking advantage of favorable electrochemical gradients for one molecule to actively transport others. Transport by these carriers occurs only if all solutes are present. In addition, glucose transport will not occur if an inwardly directed Na^+ gradient is absent; **antiport** carriers, such as the Cl^-/HCO_3^- and Na^+/H^+ *exchangers,* exchange one molecule for another.

Absorptive Pathways for Electrolytes

Na^+ **and** Cl^- are avidly absorbed by the intestinal mucosa, albeit differing in amount and by region-specific transport mechanisms of the gut. Several of these pathways and their relative distribution along the horizontal axis of the intestine are illustrated in Fig. 5.8.

In the **small intestine**, for example, Na^+ is in part absorbed by **solute-dependent Na^+-cotransport processes** that are quantitatively greater in the proximal than in the distal small bowel. In addition, the luminal bioavailability of nutrients and digestive enzymes is also largest in the proximal small intestine. **Non-nutrient dependent Na^+ absorption** in humans occurs predominantly via luminal membrane-located, amiloride-sensitive Na^+/H^+ exchangers that are not coupled to Cl^-/HCO_3^- exchangers in the jejunum, as is the case in the more distal regions of the bowel (Fig. 5.9). Consequently, Na^+ absorption is accompanied by an apparent HCO_3^- absorption, resulting from extrusion of protons, carbon dioxide formation, and increased cellular HCO_3^- secondary to diffusion of carbon dioxide. Cl^- absorption in this region is passive, dependent on transmural potential differences and concentration gradients.

In the **distal small intestine** and **proximal colon**, Na^+ and Cl^- absorption are coupled and electroneutral. Although a furosemide-sensitive Na^+-Cl^- cotransport protein has been proposed, it is likely that most coupled Na^+-Cl^- transport is carried out by the **two distinct proteins,** *Na^+/H^+ exchanger* and *Cl^-/HCO_3^- exchanger,* which are coupled by the formation of intracellular HCO_3^- and protons from carbonic acid (Fig. 5.10). The presence of these carrier proteins has been confirmed by numerous studies of brush-border membrane vesicles and is further supported by the findings in patients with **familial chloridorrhea**, a rare inborn error of transport manufactured by an *absence of Cl^-/HCO_3^- exchanger activity*. These patients develop *moderate diarrhea* and may have *metabolic alkalosis* and *stool pH in the acidic range*. Luminal perfusion studies demonstrate abnormality in the ileum and colon, where HCO_3^- secretion appears to be replaced by H^+ secretion. In the ileum and colon, Cl^- is also absorbed by a *HCO_3^--dependent pathways*, probably involving a luminal membrane Cl^-/HCO_3^- exchanger not coupled to Na^+/H^+ exchanger, as well as by **potential difference-dependent diffusion.**

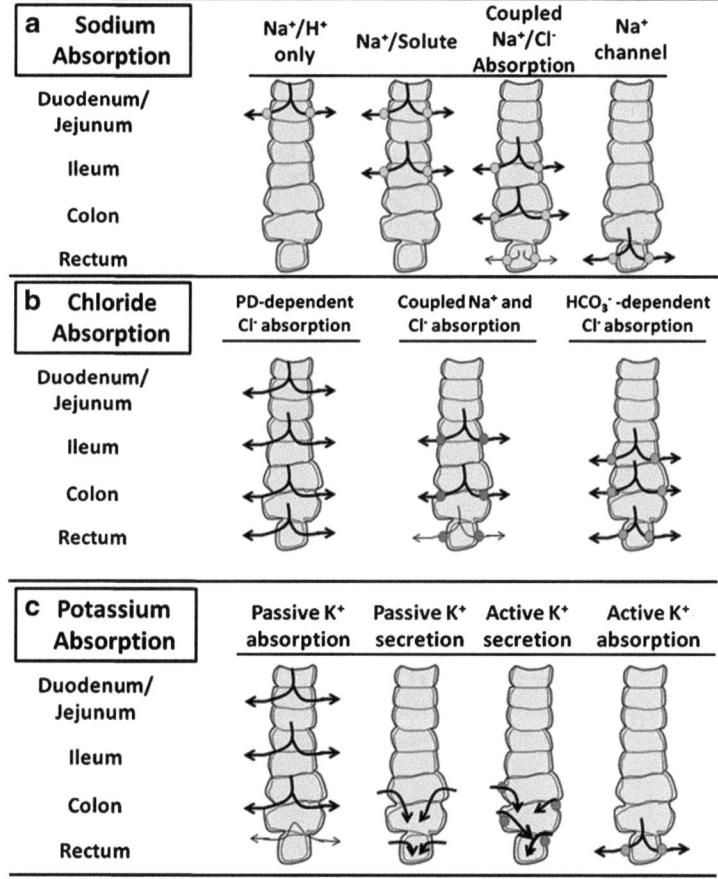

Fig. 5.8 Intestinal transport pathways for absorption. There are three types: sodium absorption (**a**), chloride absorption (**b**), and potassium absorption (**c**). These transporters are distributed along the length of the gut (*arrows*) (Adapted from American Gastroenterological Association Teaching Slide Collection 25 ©, Bethesda, Maryland; Used with permission)

In the **distal colon**, active Na^+ absorption must occur against very large electrochemical gradients. To accomplish this, the rectum absorbs Na^+ by an electrogenic, amiloride-sensitive *Na^+ channel* present in the luminal membrane (Fig. 5.11). Because this region of the gut has the least paracellular permeability, back diffusion of ions is minimal, and large potential differences can be maintained. The latter is necessary for passive absorption of the counter ion Cl^-/K^+ absorption in the colon, in contrast to its exclusively passive transport in the small intestine, is actively absorbed predominantly in the recto-sigmoid area. This is probably mediated by *K^+/H^+ exchanger* of a luminal membrane H^+/K^+-ATPase, which has recently been cloned (Fig. 5.12). This protein has 63 % amino acid homology

Fig. 5.9 Nonnutrient-dependent sodium absorption in proximal small intestine mediated by Na^+/H^+ exchange (Adapted from American Gastroenterological Association Teaching Slide Collection 25 ©, Bethesda, Maryland; Used with permission)

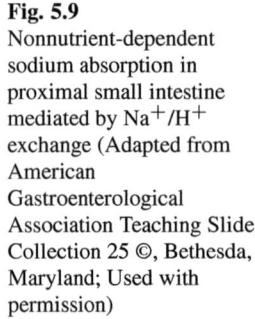
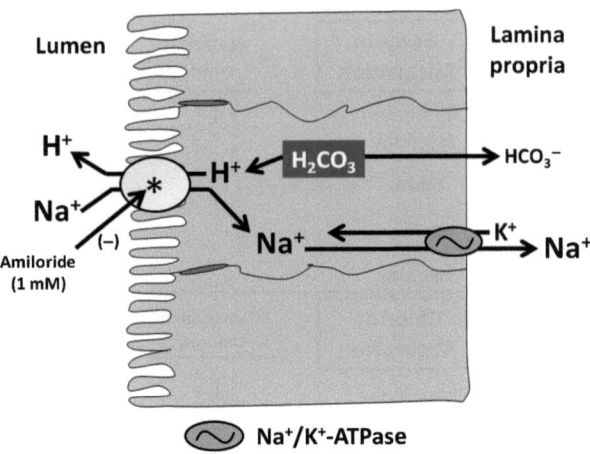

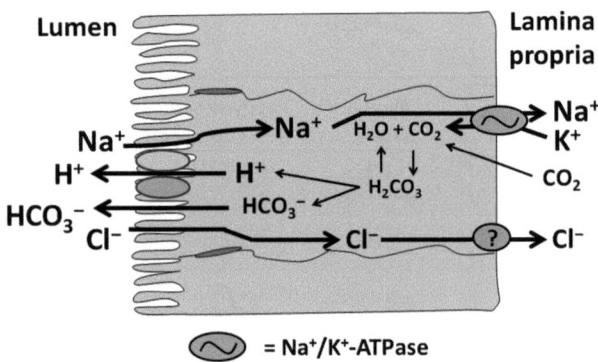

Fig. 5.10 Coupled Na^+-Cl^- transport in ileum and colon. The transport is achieved via the mediation of two distinct brush-border ion exchangers: Na^+/H^+ exchange and Cl^-/HCO_3^- exchange. This process is electroneutral (Adapted from American Gastroenterological Association Teaching Slide Collection 25 ©, Bethesda, Maryland; Used with permission)

with the gastric H^+/K^+-ATPase-α subunit and is most abundantly expressed in the distal colon.

Finally, the **colon** is also the major site for the generation and absorption of **short-chain fatty acids (SCFA)**. SCFA are products of bacterial metabolism of undigested complex carbohydrates derived from dietary sources such as *fruits* and *vegetables*. They are produced in large quantities by colonic bacterial flora and in fact represent the major luminal anions of the region. Although SCFA are believed to have many important trophic effects on colonic mucosa, they also appear to promote Na^+ absorption by the colonic mucosa. Unfortunately, the exact mechanism for Na^+-SCFA absorption remains controversial. Nevertheless, luminal SCFA may play an extremely important role in aiding colonic water and electrolyte absorption.

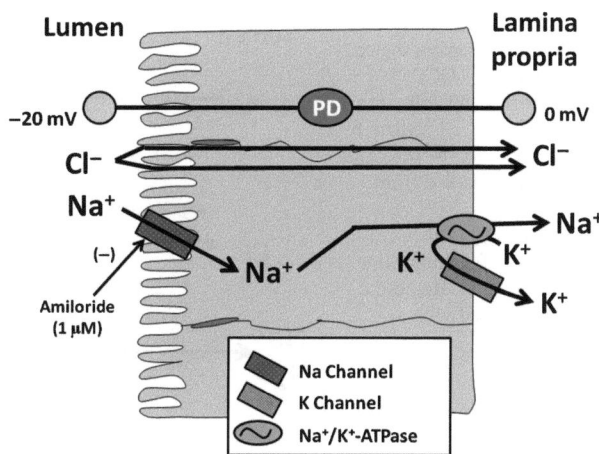

Fig. 5.11 Electrogenic Na^+ absorption in distal colon. The process involves an amiloride-sensitive Na^+ channel in the luminal membrane (Adapted from American Gastroenterological Association Teaching Slide Collection 25 ©, Bethesda, Maryland; Used with permission)

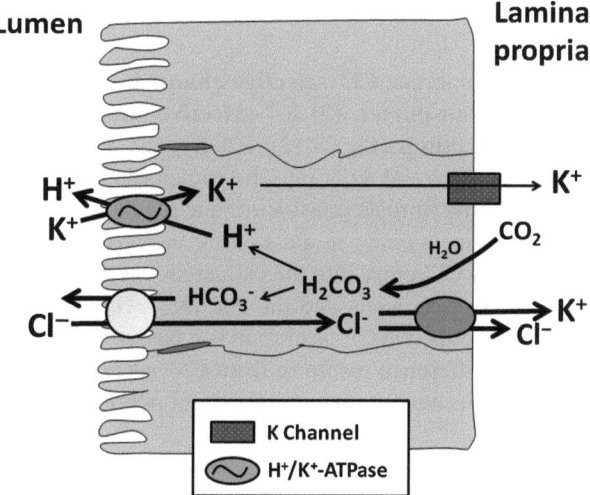

Fig. 5.12 K^+ absorption in the recto-sigmoid area. This electroneutral process is probably mediated by K^+/H^+ exchange by the K^+/H^+-ATPase in the luminal membrane (Adapted from American Gastroenterological Association Teaching Slide Collection 25 ©, Bethesda, Maryland; Used with permission)

3.2 Secretory Pathways for Water and Electrolyte

Active water and electrolyte secretion serve several important purposes. Its major physiological role is to provide the aqueous medium for proper digestion and absorption of luminal nutrients. Intestinal secretion is largely driven by active secretion of Cl^- or HCO_3^-.

Secretion of Cl^-

The secretion of Cl^-, in particular, has been well characterized and appears to involve the coordinated actions of **four membrane proteins: (1)** a

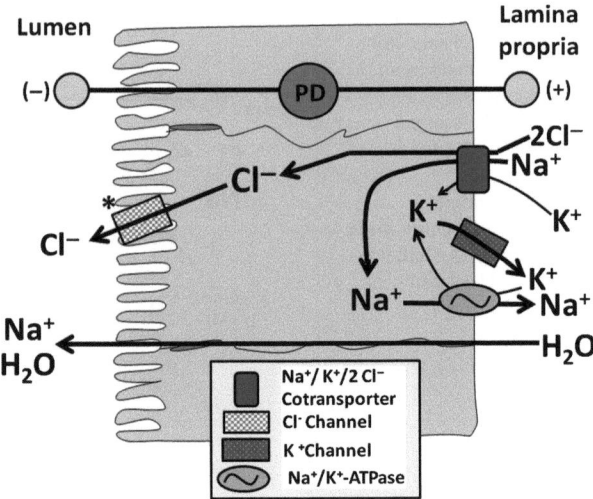

Fig. 5.13 Electrogenic Cl^- secretion. Cl^- enters the cells by the $Na^+/K^+/2Cl^-$ cotransporter and exits into the lumen via Cl^--sensitive channels (*). PD = potential difference (Adapted from American Gastroenterological Association Teaching Slide Collection 25 ©, Bethesda, Maryland; Used with permission)

luminal-membrane **Cl^- selective channel**, (2) the basolaterally located **$Na^+/K^+/2Cl^-$ co-transporter**, (3) **K^+-selective channels**, and (4) the **Na^+/K^+-ATPase**, or **sodium pump** (Fig. 5.13). Briefly, Cl^- enters the cells by the $Na^+/K^+/2Cl^-$ co-transporters and exits into the lumen via Cl^--selective channels (one of which may be cystic fibrosis transmembrane regulator channel, CFTR), with the openings of the channels being modulated by various protein kinases. K^+ and Cl^-, which accompany Cl^- entry into the cell, are recycled across the basolateral membrane by K^+-selective channels and the Na^+/K^+-ATPase, respectively. Na^+ secretion that accompanies active Cl^- secretion is a passive process, driven by the trans-epithelial potential difference resulting from Cl^- secretion. Cl^- secretions can be regulated by numerous neurotransmitters and gut peptides (discussed below).

Secretion of HCO_3^-

In contrast to Cl^- secretion, much less is known about HCO_3^- secretion. In rabbit ileum, HCO_3^- secretion appears to be electrogenic and vectorially transported across the epithelium, rather than generated by the production of HCO_3^- from the action of carbonic anhydrase. Furthermore, HCO_3^- secretion appears to be dependent on the presence of serosal Na^+ and is postulated to involve at least **three different transporters: (1)** an apical membrane **anion-selective** (HCO_3^- and possibly Cl^-) **channel**; (2) a coupled, electrogenic **Na^+/HCO_3^- cotransporter** in the basolateral membrane; and (3) a **Cl^-/HCO_3^- exchanger** in the apical membrane. In the **distal colon**, HCO_3^- secretion appears to be electroneutral and may involve a Na^+-independent Cl^-/HCO_3^- exchanger located in the apical membrane. Along with HCO_3^- secretion, it is important for providing the aqueous phase for luminal digestion and nutrient absorption.

Secretion of K^+

Active K^+ secretion is also found throughout the **colon**, apparently mediated by a barium-sensitive K^+ conductance in the apical membrane. K^+ enters the basolateral membrane of the colonocyte via the Na^+/K^+-ATPase pump, and possibly by the $Na^+/K^+/2Cl^-$ co-transporter. This is a process that can be stimulated by increases in cellular cAMP and Ca^{2+}.

4 Physiological Regulation of Gut Water and Electrolyte Transport

The intestines contribute 12–14 % of the secretions needed to maintain the chyme in a fluid state. In addition to allowing for efficient digestion and absorption of the complex components of mammalian meals, the secretions serve as a conduit for the luminal delivery of secretory **immunoglobulin A (IgA) glycoproteins** and for flushing out infectious agents and noxious stimuli. The regulation of intestinal secretion is multifactorial, involving *luminal* and *systemic* influences. It has long been recognized that simple **mechanical stimulation** of the gut lumen can increase secretion. Similarly, **chemical stimuli**, no matter in the form of the normal breakdown products of dietary intake, toxigenic elaborations of the microflora, or noxious chemicals and antigens, elicit secretions. **Systemic metabolic changes**, including *volume overload, dehydration,* and *acidosis* or *alkalosis*, also influence secretion. Finally, in addition to local regulation of neurohumoral factors, there is increasing evidence that *higher centers of the brain* have modulatory effect on intestinal secretion. All these stimuli act on an intricate complex of neural, hormonal and autocrine modulators in the mucosa and gut wall that govern both basal and stimulated states of secretion. As described above, these agents act on specific receptors on the target cells, which can be on either the enterocytes or underlying neural, vascular, or immune elements.

4.1 Neurohormonal Agents That Cause Net Intestinal Secretion or Increase Net Absorption

Factors that regulate intestinal epithelial ion-transport properties can be broadly classified into **two groups:** the **secretagogues** and the **pro-absorptive agents**. As their names imply, secretagogues cause net accumulation of fluid in the intestinal lumen by either stimulating active secretory processes and/or inhibiting absorption; pro-absorptive agents promote absorption by having the opposite effect. The major secretagogues and pro-absorptive agents known to act on enterocytes are listed in Fig. 5.5. The source of regulatory agents varies. Some, such as *neurotransmitters*

Fig. 5.14 Intracellular processes mediating agonist-stimulated net secretion

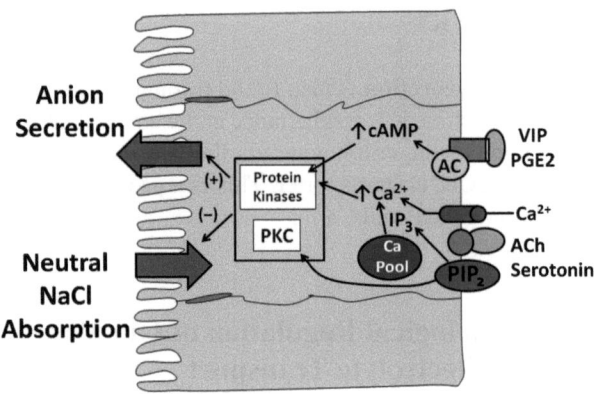

and *paracrine hormones*, are released locally near the basolateral membrane of the enterocyte. Other substances, such as *guanylin*, are released into the lumen, and substances such as *adrenalocorticoids* reach the enterocyte via the systemic circulation. Enterocytes have receptors on both apical and basolateral membranes and, in the case of the steroids, intracellular receptors as well. In addition to **species and regional differences** in receptor distribution, **multiple receptor isotypes** for each neurotransmitter/hormone appear to exist. Likewise, several *histamine-, muscarinic-,* and *adrenergic-receptor subtypes* have been identified in different regions of the gut. The type of ligand-receptor interaction is a major determining factor of the duration of the response. Thus, the effects of a number of **calcium-dependent secretagogues** and **prostaglandins** are *short-lived,* and some of these are due to **receptor-associated tachyphylaxis**. These short-lived responses are probably crucial for dealing with the minute-by-minute challenges in the gut milieu. In contrast, the **steroid-mediated responses** are much *longer lived*. For example, in the distal colon, they act through specific intracellular receptors in the enterocytes initially to increase the recruitment of transporters to the epithelium by promoting the synthesis of various proteins needed for transport; they include apical membrane Na^+ channels Na^+/H^+ exchange, as well as increasing the number of Na^+, K^+-ATPase pumps in the basolateral membrane to provide the necessary driving force for transport. Long-term changes are more beneficial in an adaptation-type of response, as seen in exposure to a low-sodium diet or dehydration.

Mechanisms of Regulation of Ion Transport

As with many types of cells, a number of signal transduction pathways are involved in the regulation of ion transport. **Secretagogues** act mainly by stimulating one of the classic signal transduction cascades; the pathways dependent on *cAMP, cyclic guanosine monophosphate (cGMP), Ca^{2+},* or *phosphatidylinositol (PI)* (Fig. 5.14). These second messengers activate their respective protein kinases that phosphorylate proteins directly involved in ion transport or critical in regulating

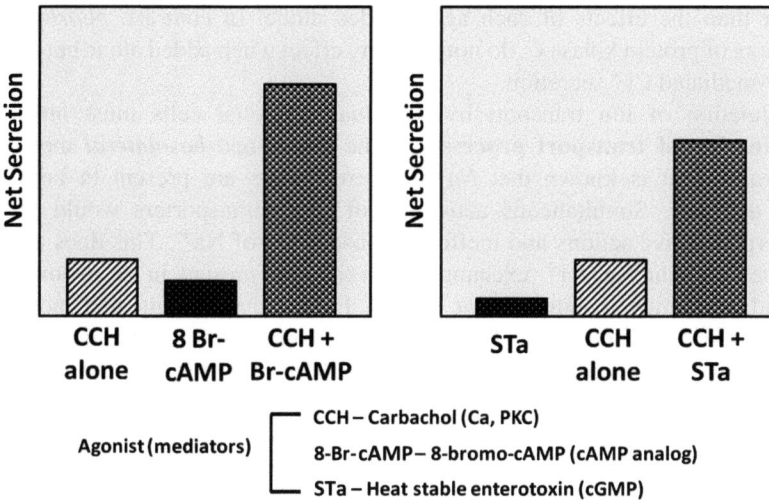

Fig. 5.15 Synergistic effect of intracellular mediators that regulate intestinal electrolyte transport

transport activities. The mechanisms of action of the **pro-absorptive agents** vary and include an *activation of the inhibitory G protein (G_i) cascade*, the *inhibitory arm of the adenylate cyclase cascade*, and the *PI cycle*. The mechanism of activation of these second messengers could also influence the duration and type of the biological response. This is best exemplified by comparing the effects of cholera toxin, vasoactive intestinal peptide (VIP), and prostaglandins, all three of which stimulate secretion via the cAMP cascade. The activation of secretion by *cholera toxin* involves an essentially irreversible covalent modification of the stimulatory G protein (G_s)-type of guanosine triphosphate (GTP)-binding proteins. Activated Cl^- secretion is only "shut off" when the enterocyte is sloughed off at the end of its normal life span. Secretagogues such as *VIP* can cause sustained stimulation of secretion in vitro; secretion returns to baseline as the hormone/neurotransmitter is degraded. The effects of prostaglandins are limited by receptor tachyphylaxis. The general dogma is that second messengers activate specific protein kinases to cause a biological effect and that endogenous phosphoprotein phosphatases reverse this effect to return the system to the basal state. There is considerable evidence to indicate a role for protein phosphorylation in regulating intestinal ion transport, but recent evidence suggests that there may be direct, non-kinase activation of Cl^- channels by cyclic nucleotides as well. Although a regulatory role for protein phosphatases in signal transduction gains acceptance, to our knowledge, a role for them as a primary trigger of intestinal ion transport has not yet been demonstrated.

Considerable cross-talk also exists among the different signaling mechanisms and among different steps of the same signaling pathway. Two examples are provided here and illustrated in Fig. 5.15. Synergistic actions between *cAMP and Ca^{2+}-mediated secretion*, as well as *cGMP (ST_a) and Ca^{2+}-mediated secretion*, have been demonstrated. In both cases, the effects of the combined additions are greater than the effects of each agent added alone. In contrast, *phorbol esters*, activators of protein kinase C, do not have any effect when added alone but attenuate cAMP-mediated Cl^- secretion.

Regulation of ion transport by intestinal epithelial cells must involve the **coordination of transport processes** at the *apical* and *basolateral* membranes. For example, it is known that *Na^+/H^+ exchangers* are present in both membrane domains. Simultaneous activation of these transporters would result in counterproductive actions and ineffective absorption of Na^+. This does not occur because different Na^+/H^+ exchanger isoforms are present in each domain that respond differently to intracellular signals. For example, stimulated increases in cytosolic Ca^{2+} and activation of protein kinase C appear to inhibit luminal *Na^+/H^+ exchangers*, whereas they may activate basolateral isoforms. Another example of coordinated regulation of cellular transport processes is the stimulation of active Cl^- secretion. Here, the opening of the *Cl^- channel* is accompanied by activation of the *$Na^+/K^+/2\ Cl^-$ co-transporter*, the latter event required for bringing in additional Cl^- ions for sustained secretion. Again, this coordination of cellular processes requires the actions of intracellular second messengers and protein kinases that simultaneously activate co-dependent transport pathways.

Finally, the regulation of intestinal transport processes may in part involve **modulation of paracellular permeability**. For instance, the activation of protein kinase C by various agonists or by the **zonula occludens toxin (ZOT)** of *Vibrio cholera* may cause cytoskeletal alterations that affect the peri-junctional actin ring, a major component of the tight-junction apparatus. Alterations in paracellular permeability markedly affect both absorptive and secretory processes.

5 Abnormality in Water and Electrolyte Transport

Alterations in intestinal transport function in response to a variety of pathophysiological processes can be mediated by many different mechanisms. In some instances, the responses are appropriate, i.e. as part of defensive or healing mechanisms or in response to increased metabolic and nutritive demands caused by disease. In other instances, however, the response is an aberration or manifestation of the disease and serves no clear physiological role. This section will briefly discuss a few illustrative examples of mechanisms by which disease processes affect intestinal water and electrolyte transport.

5.1 Adaptive Transport Mechanism in Diabetes

Diabetes has significant and complex effects on nutrient and electrolyte transport functions of the intestinal mucosa, some of which have now been well characterized. An uncommon but debilitating complication of chronic diabetes is the development

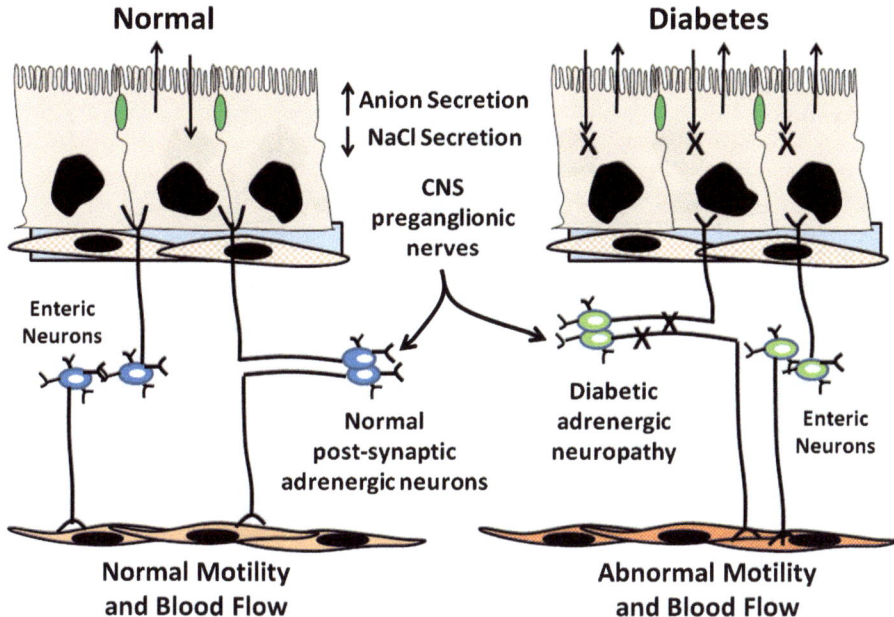

Fig. 5.16 Cell mechanisms of diabetic diarrhea. Autonomic and possibly enteric neuropathy in diabetes leads to impaired homeostatic regulation of intestinal fluid and electrolyte transport, abnormal motility, and alterations in capillary blood flow (*right side*), compared with normal condition (*left side*) (Adapted from American Gastroenterological Association Teaching Slide Collection 25 ©, Bethesda, Maryland; Used with permission)

of *diabetic diarrhea*. This condition is invariability associated with *autonomic neuropathy* and arises from an aberrant neural regulation of salt and water transport, as well as abnormal motility. In streptozotocin-treated diabetic rats, for instance, a selective impairment of adrenergic innervation of the mucosa results in an imbalance in factors that regulate absorptive and secretory processes, favoring the latter (Fig. 5.16). Thus, the development of autonomic and possibly enteric neuropathy in diabetes causes *impaired homeostatic regulation of intestinal fluid and electrolyte transport, abnormal motility*, and *alterations in capillary flow*.

5.2 Transport of Inflamed Mucosa

In mucosal inflammation, the interactions and effects of the factors and cellular components that normally regulate ion transport functions are dramatically altered or diminished as a result of the large number of infiltrating inflammatory cells. These cells secrete large amounts of immune and inflammatory mediators and cause net intestinal secretion; they can also adversely affect mucosal integrity and function

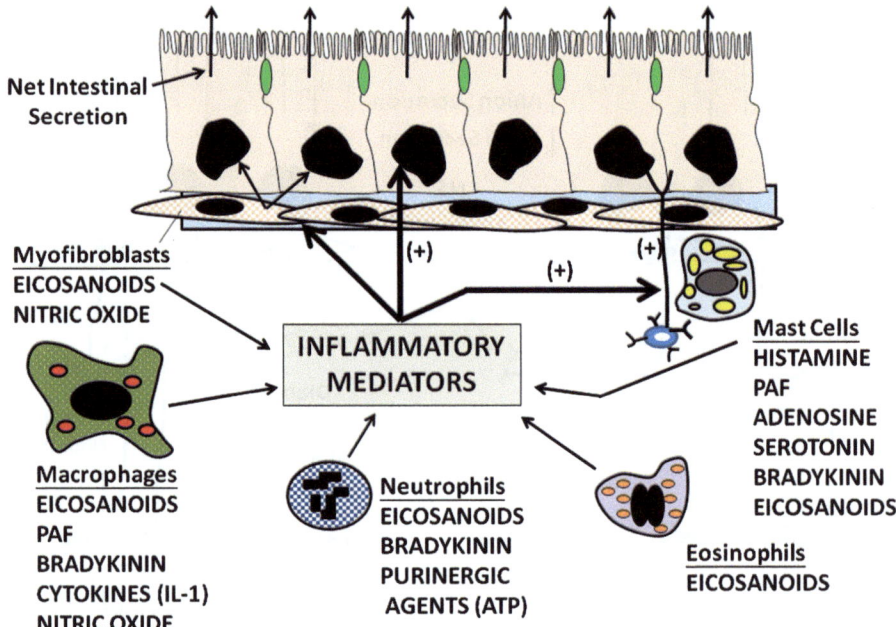

Fig. 5.17 Effects of inflammatory mediators on ion transport. These cells from several sources cause direct and indirect stimulation of intestinal secretion (Adapted from American Gastroenterological Association Teaching Slide Collection 25 ©, Bethesda, Maryland; Used with permission)

(Fig. 5.17). While net secretion serves to purge the gut of noxious agents and pathogens, it may also cause severe fluid losses and metabolic imbalances. **Arachidonic acid metabolites** from *neutrophils, fibroblasts* and *activated macrophages* play a major role in stimulating net secretion. While many prostaglandins such as PGE_2 appear to activate specific epithelial cell receptors to stimulate anion secretion and inhibit Na^+ and Cl^- absorption, agents such as PGD_2, PGI_2, and the *peptido-leukotriene LTC₄* activate secretomotor neurons. Similarly, other inflammatory products such as *radical oxygen metabolites, serotonin, histamine,* and *platelet-activating factor* that are made and secreted by activated phagocytic and inflammatory cells have multiple sites of action. Many of these agents cause immediate changes in motility and in blood capillary and lymphatic flow conducive to net secretion.

Some immune and inflammatory mediators appear to stimulate changes in electrolyte transport via stimulation of intestinal arachidonate metabolism. For example, *interleukin-1* has been shown to stimulate anion secretion by activating arachidonic acid metabolism of submucosal cells. The nonapeptide bradykinin is liberated from the precursor kininogen by the enzyme **kallikrein**, found either in plasma or tissue. **Bradykinin** is a potent secretagogue, stimulating large increases in prostaglandin production from the intestine; however, it may also have a direct effect on intestinal epithelial transport. **Nitric oxide**, a non-adrenergic and non-cholinergic

(NANC) neurotransmitter and product of endothelial cells, also has many unique actions that may play a role in inflammation-induced secretion. It is produced by activated phagocytic cells such as neutrophils and macrophages and is probably produced in large amounts in inflamed tissue. Nitric oxide activates the iron- and sulfur-containing soluble form of guanylate cyclase, which, in intestinal mucosa, is predominantly found in lamina propria cells. However, it is not clear whether this effect mediates the secretory actions of nitric oxide. On the other hand, nitric oxide does appear to stimulate arachidonate metabolism and enteric secretomotor neurons, which mediate the secretory action of this agent. Purinergic agonists such as *adenosine* and *ATP* are also major products of inflammatory cells and stimulate net intestinal secretion. They may be important mediators stimulating secretion after their release from neutrophils in crypt abscesses, as their secretory effects are most pronounced on the luminal side. Adenosine may act through stimulation of adenosine type 2 receptors, increasing cAMP but at low concentrations; its secretory effects may be mediated by other mechanisms.

A common denominator of the actions of many inflammatory mediators is their ability to activate the secretomotor neurons of the enteric nervous system. Enteric nerves may therefore be important in amplifying the actions of numerous immune and inflammatory mediators, either by potentiating their actions at target tissues through the release of neurotransmitters or by enlarging the domain of action and affecting numerous target tissues simultaneously. Stimulated secretomotor neurons would affect smooth muscle function and the flow and permeability of blood capillaries and lymphatic in a manner that would help sustain or enhance the secretory effects.

5.3 Effects of Other Conditions and Disease Processes on Transport

Infectious Pathogens

Diarrhea, a common sequela of infections by enteric pathogens, arises from numerous mechanisms (Fig. 5.18). Enterotoxigenic organisms cause net secretion by stimulating intestinal anion secretion and inhibiting Na^+ and Cl^- absorption. *Cholera toxin*, for example, specifically binds to the GM_1-ganglioside receptor of the enterocyte luminal membrane and activates epithelial adenylate cyclase following ADP-ribosylation of G_s the GTP-binding regulatory protein (Fig. 5.19). In contrast, *E. coli* heat-stable enterotoxin (ST_a) binds to the putative guanylin receptor and activates guanylate cyclase to increase cellular cGMP.

Many cytotoxic bacterial toxins such as those from *Salmonella* and *Shigella* and viral pathogens obviously cause **mucosal destruction** and loss of absorptive surface area, leading to luminal accumulation of fluid and loss of plasma proteins.

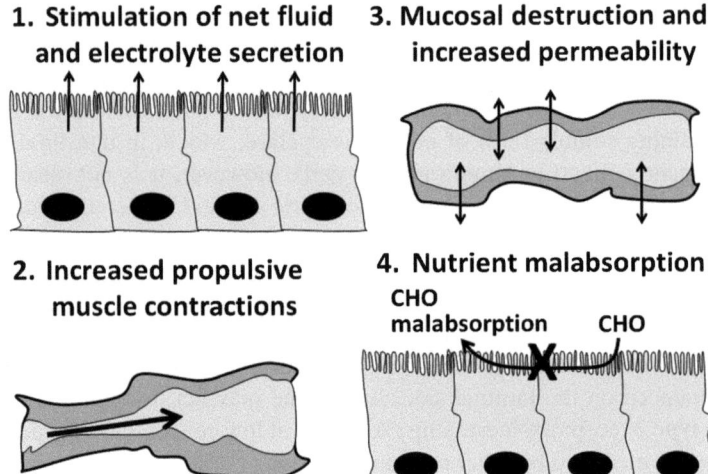

Fig. 5.18 Cellular mechanisms of action by which enteric pathogens cause diarrhea (Adapted from American Gastroenterological Association Teaching Slide Collection 25 ©, Bethesda, Maryland; Used with permission)

The induced inflammatory response plays a key role in stimulating net intestinal secretion, propulsive motor contractions, and alterations in blood flow. Others, such as **toxin A** of *Clostridium difficile*, cause net secretion by directly stimulating epithelial anion secretion and by increasing paracellular permeability. Toxin A also appears to have chemo-attractant properties and is capable of stimulating immune and inflammatory cells by a Ca^{2+}-mediated mechanism. Finally, several **non-invasive enteric pathogens**, such as entero-adherent *E. coli*, *Giardia lamblia*, and *Cryptosporidium*, may cause diarrhea by affecting luminal membrane function. These organisms adhere to the brush-border membrane and can cause subtle structural alterations, best appreciated at the ultrastructural level. Often there is an effacement of microvilli and disruption of the microvillus and terminal-web cytoskeleton. These abnormalities may cause or contribute to defective ion transport and maintenance of tight-junction integrity. In *Giardia* infections, other mucosal alterations may be present; they include a significant decrease in crypt depth, decreased villus height in the duodenum, increased villus height in the ileum, and a generalized shortening of microvillus height.

Adaptation Following Extensive Small Bowel Resection

Following extensive resection of the small bowel, several morphologic and functional alterations of the intestinal mucosa occur and thus increase the absorptive capacity of the remaining gut. There is a significant (about 30 %) increase in villus size caused by **mucosal hyperplasia**. The net effect of these structural changes

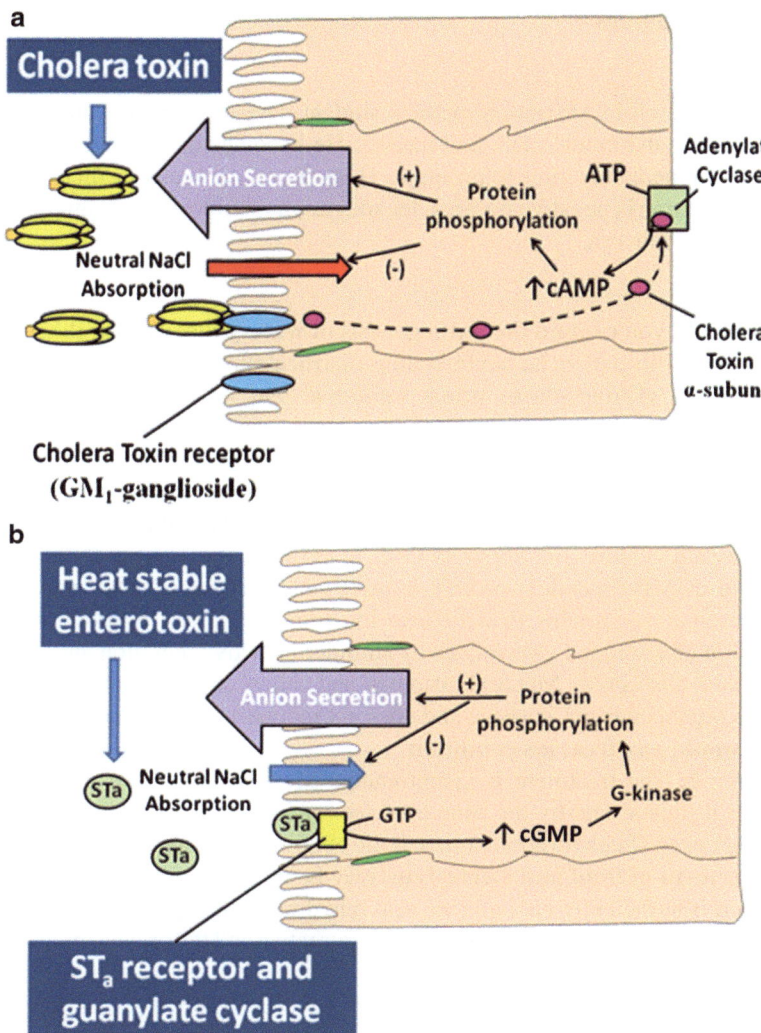

Fig. 5.19 Cellular mechanisms of action by which bacterial enterotoxins causes diarrhea. (**a**) Cholera toxin binds to a specific membrane receptor, enters the cell, and activates adenylate cyclase. (**b**) *E. coli* heat-stable enterotoxin causes diarrhea by stimulating guanylate cyclase and increasing cyclic GMP (Adapted from American Gastroenterological Association Teaching Slide Collection 25 ©, Bethesda, Maryland; Used with permission)

is to increase the surface area and the absorptive and digestive capacities of the remaining mucosa. The mechanisms underlying these adaptive changes are not fully elucidated, but they are believed to involve stimulation by luminal factors such as *dietary nutrients, pancreatic and biliary secretions,* and *intestinal growth and trophic factors.*

Clinical Correlations

Case Study 1

While on vacation in Mexico, a medical student develops severe watery diarrhea without blood, abdominal pain, and nausea. He becomes increasingly dehydrated and weak. Immediately after returning to Hong Kong, he seeks medical attention. Stool examination is negative for fecal leukocytes, fat, or blood, and sigmoidoscopy reveals normal mucosa.

Questions

1. **How would you explain the pathogenesis of watery diarrhea in this patient?**
   Answer 1: This patient has an infectious diarrheal illness frequently experienced by travelers visiting places where *sanitation and water supply may be suboptimal*. Because of the absence of blood and fecal leukocyte and the normal appearance of gut mucosa, the most likely pathogen in this instance is an **enterotoxigenic *E. coli*** that elaborates a **heat-stable enterotoxin (ST_a)**. STa mainly binds to luminal receptors for guanylin in the colon. This receptor is a guanylate cyclase and converts GTP to cGMP. Increases in cellular cGMP activate cGMP-dependent protein kinase, which phosphorylates a number of target substrate proteins that mediate ion transport. Thus, ST_a stimulates net secretion of water and electrolytes by stimulating active secretion and inhibition of sodium absorption. Mucosal histology and endoscopic appearances are intact, as this organism is not invasive and does not induce an inflammatory reaction.

2. **How would you treat his condition?**
   Answer 2: The treatment is symptomatic. Most patients have a self-limiting diarrheal illness, and by the time they seek medical attention, they are usually feeling better. If dehydration or metabolic abnormalities are severe, **intravenous replacement of fluid and electrolytes** may be indicated. Otherwise, **oral rehydration** is sufficiently enough (see oral rehydration therapy below). Antibiotics are *not* usually required, as the diarrhea is purgative and the organisms are cleared within a few days.

Case Study 2

While in rural Bangladesh, you witness an outbreak of cholera during flooding which is caused by the rainy season. One of the major clinical manifestations of the affected people they experience is profuse watery diarrhea that can cause dehydration and electrolyte imbalance, thus death ensues. Because of the limited health resources in your area, you must treat patients with whatever is available to you.

Questions

1. **What is the underlying mechanism of cholera that causes watery diarrhea?**
   Answer: Cholera is a major cause of morbidity and mortality in the world which is caused by ***Vibrio cholerae***, a bacterial organism that contaminates water

and food supplies and it is acquired by **fecal-oral transmission.** *V. cholerae* elaborates a heat-labile toxin called *cholera toxin*, which is one of the most potent secretagogue substances known. The toxin causes a functional derangement of sodium and water transport. After the toxin binds GM_1-ganglioside receptors of the luminal membrane of enterocytes, the alpha subunit is inserted into the cell. The subunit catalyzes the ADP-ribosylation of the alpha subunit of stimulatory G protein (G_s), irreversibly activating it and adenylate cyclase activity. Resultant increases in cAMP activate cAMP-dependent protein kinase and phosphorylation of proteins involved in mediating active anion secretion or neutral sodium chloride absorption. The effects of the toxin are only diminished after the enterocyte population turns over (several days). During this time, no mucosal lesions are seen, but patients experience profuse watery diarrhea of such severity that many die of *dehydration* and *metabolic disturbances*.

2. **What is the most effective treatment option that you recommend for this condition?**
   Answer: Oral replacement with water and sodium is *not* effective because cholera toxin inhibits absorption by the gut, probably by inhibiting luminal Na^+/H^+ exchangers. However, by taking advantage of other Na^+-transport pathways not affected by increases in cellular cAMP, oral replenishment of lost fluid and electrolytes is possible. Thus, by **adding sugar (e.g. *glucose*) to oral salt replacement fluids**, one can dramatically increase intestinal absorption of salt and water. Since the function of the luminal carrier protein, called **sodium-dependent glucose co-transporter 1 (SGLT1),** is intact and unaffected by cholera toxin, hexoses such as glucose will promote the absorption of sodium and water. With this type of oral formulation, millions of lives each year are saved.

   In this instance, where glucose may not be readily available, one could try the **traditional remedy of rice water supplemented with salt,** which contains starch or complex carbohydrates. Starch is made up of glucose polymers assembled by 1,4- and 1,6-glycosidic linkages. Starch is rapidly hydrolyzed by pancreatic and intestinal membrane-bound enzymes or disaccharidases (which are intact in cholera) to form glucose. Luminal glucose (as described above) is co-transported with Na^+.

Case Study 3

A newborn infant develops abdominal distention and vomiting within 48 hours of birth. Physical examination of the abdomen is consistent with intestinal obstruction. Barium enema shows a microcolon and meconium, normally excreted after birth, in the distal ileum, suggesting a diagnosis of meconium ileus.

Questions

1. **What might this child have and how does it predispose to meconium ileus?**
   Answer: This child has **cystic fibrosis (CF),** which was later confirmed by an *abnormal sweat test*. The defects in CF involve mutations of the cystic fibrosis transmembrane regulator (CFTR) gene, resulting in the failure of CF-protein

insertion into the cell membrane and activation by cAMP-dependent protein kinase. In the intestinal mucosa, CFTR is a cAMP-regulated Cl^- channel of the luminal membrane that mediates active water and electrolyte secretion. Its defect in CF results in inadequate luminal hydration of intestinal contents, especially in utero. As a consequence, meconium (consisting of ingested contents of amniotic fluid, secretions that collect in the intestinal lumen, and mucosal, pancreatic, and other enteric proteins) become inspissated and extremely viscous. This can eventually result in obstruction and failure of luminal contents to enter the colon (hence a **microcolon**). This patient may require **surgical intervention or medical treatment** aimed at *reducing meconium viscosity* (e.g. acetylcysteine) or *increasing its hydration and breakup* (Gastrografin).
2. **Would a patient with CF be susceptible to cholera?**
   Answer: Probably not. In fact, it has been hypothesized that the development of CF-genetic lesions arose from a selection process that favored resistance to cholera toxin. Since these patients have **defective regulation of Cl^- secretion by cAMP**, they are virtually immune to the actions of cholera toxin.

Further Reading

1. Althaus M (2012) Gasotransmitters: novel regulators of epithelial Na^+ transport. Front Physiol 3 (Article 83):1–10
2. Ghishan FK, Kiela PR (2012) Ion transport in the small intestine. Curr Opin Gastroenterol 28:130–134
3. Gill RK, Alrefai WA, Borthakur A, Dudeja PK (2012) Intestinal anion absorption. In: Johnson LR (ed) Physiology of the gastrointestinal tract, 5th edn. Academic, New York, pp 1819–1847
4. Heitzmann D, Warth R (2008) Physiology and pathophysiology of potassium channels in gastrointestinal epithelia. Physiol Rev 88:1119–1182
5. Kato A, Romero MF (2011) Regulation of electroneutral NaCl absorption by the small intestine. Annu Rev Physiol 73:261–281
6. Keita AV, Söderholm JD (2010) The intestinal barrier and its regulation by neuroimmune factors. Neurogastroenterol Motil 22:718–733
7. Kiela PR, Ghishan FK (2009) Ion transport in the intestine. Curr Opin Gastroenterol 25:87–91
8. Kiela PR, Ghishan FK (2012) Na^+/H^+ exchange in mammalian digestive tract. In: Johnson LR (ed) Physiology of the gastrointestinal tract, 5th edn. Academic, New York, pp 1781–1818
9. Laforenza U (2012) Water channel proteins in the gastrointestinal tract. Mol Aspects Med 33:642–650
10. Malakooti J, Saksena S, Gill RK, Dudeja PK (2011) Transcriptional regulation of the intestinal luminal Na^+ and Cl^- transporters. Biochem J 435:313–325
11. Murek M, Kopic S, Geibel J (2010) Evidence for intestinal chloride secretion. Exp Physiol 95:471–478
12. Thiagarajah JR, Verkman AS (2012) Water transport in the gastrointestinal tract. In: Johnson LR (ed) Physiology of the gastrointestinal tract, 5th edn. Academic, New York, pp 1757–1780

Part II
Nutritional Physiology

Chapter 6
Digestion and Absorption of Carbohydrates and Proteins

Michael D. Sitrin

1 Introduction

The pathways used for the digestion and absorption of carbohydrates and proteins share several important common features. Proteins and starch, one of the major dietary carbohydrates, are both polymers that are initially broken down into smaller compounds by enzymes secreted into the intestinal lumen, principally by the pancreas. Further digestion occurs by a variety of enzymes present in the small intestinal brush-border membrane, generating the small molecules (monosaccharides, amino acids, di- and tripeptides) that are capable of being absorbed by the enterocytes. Transport of these molecules through the intestine occurs mainly via various membrane carrier proteins. In some cases, there is an **active transport mechanism** by which the movement of the nutrient through the membrane is coupled directly or indirectly to a source of metabolic energy, permitting concentrative transport against an electrochemical gradient. Other carrier systems accelerate the rate of movement of a nutrient through the membrane, but they are not linked to an energy source and therefore do not result in concentration against an electrochemical gradient. This type of transport process is known as **facilitated diffusion**. **Simple passive diffusion** either through the plasma membrane or between the enterocytes (the *paracellular pathway*) is quantitatively important only when the luminal concentration of nutrients is very high.

M.D. Sitrin, M.D. (✉)
Department of Medicine, University at Buffalo, The State University of New York, Buffalo, NY, USA
e-mail: mdsitrin@buffalo.edu

2 Carbohydrates

Dietary carbohydrates consist of complex carbohydrates, starch or dietary fiber, and various sugars, including ***monosaccharides*** (glucose, galactose, fructose) and ***disaccharides*** (sucrose, lactose, maltose). Adult males consume approximately 300 g/day of carbohydrates, and adult females, approximately 200 g/day, half as starch and half as various sugars. Carbohydrates provide about 45–50 % of the energy contained in the average American diet.

2.1 Structure of Dietary Starch

Two principal types of dietary starch are present in the human diet. **Amylose** is composed almost entirely of **α-1,4-linked linear chains** of glucose, with an average chain length of about 600 glucose residues and a molecular weight of about 100 kd. **Amylopectin** is a branched starch with a molecular weight of approximately 1,000 kd and contains about 6,000 glucose residues. Amylopectin has both **α-1,4-linked straight chains** and **α-1,6-linked branch points** occurring approximately every 20 glucose residues (Fig. 6.1). The ratio of amylose to amylopectin varies in different foods, but it is usually about 1:4.

In food, starches are present in association with proteins, many of which are hydrophobic. The starch tends to be in the interior of the complex particle, limiting access to enzymes responsible for starch hydrolysis. Starch digestion is facilitated by physical processing, such as cracking grains and milling, and by cooking, which converts the starch from a crystalline to a gel structure. Certain foods contain protein inhibitors of starch digestion.

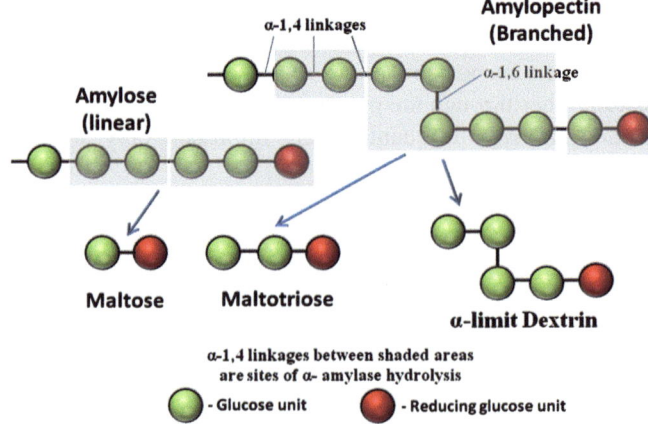

Fig. 6.1 Digestion of starch by pancreatic of alpha-amylase. The action of alpha-amylase on amylose and amylopectin, the two major forms of starch (Adapted from Alpers [1])

2.2 Starch Digestion by α-Amylases

The enzymes responsible for intraluminal starch digestion, α-**amylases**, are present in both *salivary* and *pancreatic* secretions. Human salivary amylase is 94 % identical with pancreatic amylase, although the two enzyme forms are the products of different genes. **Salivary amylase** tends to be degraded at acidic pH (less than pH 3-3.5), but the rate of degradation decreases when salivary amylase is accompanied by its starch substrate or by its oligosaccharide breakdown products. The substrate interacts with the active hydrolytic site and maintains the amylase in a more favorable conformation that allows some active enzyme to survive passage into the duodenum. Thus, while **pancreatic amylase** is responsible for most of the intraluminal starch hydrolysis in adults, some active salivary amylase is usually present in jejunal fluid. In neonates with immature pancreatic function and limited secretion of pancreatic amylase, salivary amylase supports a significant fraction of starch digestion.

The five glucose residues adjacent to the terminal-reducing glucose unit of amylose and amylopectin bind to specific catalytic subsites of α-amylase. There is then cleavage between the second and third α-1,4-linked glucose residues. The disaccharide **maltose** and the trisaccharide **maltotriose** are therefore the *main products of amylose digestion* (Fig. 6.1). α-**amylase** has no activity against the α-1,6-linkages in amylopectin and the capacity to break α-1,4 links adjacent to branch points is sterically hindered. In addition to maltose and maltotriose, therefore, about one third of the *degradation products of amylopectin* are α-**limit dextrins** and **branched oligosaccharides** with one or more α-1,6 bonds and an average molecular weight of 800–1,000 kd (5–10 glucose units) (Fig. 6.1).

2.3 Brush-Border Membrane Digestion of Oligosaccharides

The oligosaccharide products of α-amylase starch digestion are hydrolyzed by enzymes on the brush-border membrane surface, and only free glucose is transported into the enterocyte. These oligosaccharidases are large glycoproteins, with most of the protein, including the catalytic domain, residing at the lumen-cell interface. The enzymes are anchored into the membrane by a short hydrophobic segment. The activities of the oligosaccharidases are greatest in the proximal small intestine and decline distally; starch digestion is normally complete in the jejunum. Membrane hydrolysis of starch-derived oligosaccharides is normally very rapid and is not rate-limiting in the overall process of starch digestion and absorption.

Three enzymes are responsible for the brush-border membrane hydrolysis of oligosaccharides to glucose. (**1**) **Glucoamylase** (**maltase**) removes single glucose residues sequentially from the nonreducing end of the α-1,4 chain, preferring substrates of 5–9 saccharide units; however, like α-amylase, it is blocked when an α-1,6-linked glucose is at the terminal end of the saccharide. *Sucrase-isomaltase* is initially synthesized in the enterocyte as a single glycoprotein chain and, after

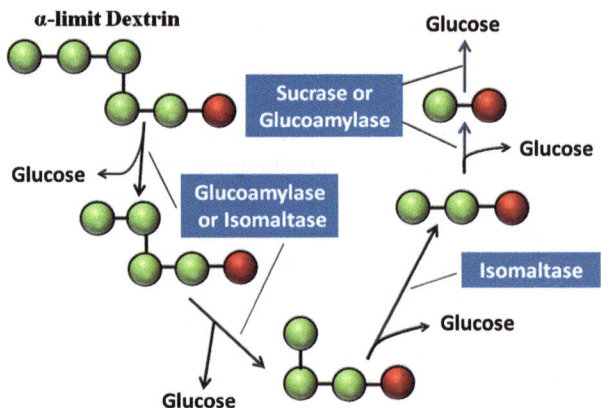

Fig. 6.2 Digestion of α-limit dextrins by brush-border membrane oligosaccharidases (Adapted from Gray [8])

insertion in the brush-border membrane, is cleaved by pancreatic proteases into sucrase and isomaltase units, which then reassociate noncovalently at the intestinal surface (see below). (**2**) **Sucrase** is highly-efficient at hydrolyzing short α-1,4-linked oligosaccharides, such as *maltose* and *maltotriose*. (**3**) **Isomaltase** is the only enzyme that can cleave the nonreducing, terminal α-1,6 bond once it becomes uncovered. As shown in Fig. 6.2, the digestion of a typical α-limit dextrin involves the initial removal of glucose residues from the nonreducing end mainly by glucoamylase, the cleavage of the α-1,6 branching link by isomaltase, and finally the digestion of maltotriose and maltose to glucose principally by sucrase. In general, glucoamylase activity determines the rate of starch digestion.

The **oligosaccharidases** are synthesized as **large glycosylated precursors** that are then processed to the mature enzyme form. For example, sucrase-isomaltase is first synthesized as a single polypeptide chain with a molecular weight of 240–260 kd. O-linked and N-linked sugar chains are added and the precursor is transferred to the brush-border membrane by a vesicular pathway moving along microtubules and inserted into the membrane through its amino terminus. Pancreatic enzymes, mainly *trypsin*, cleave the protein into **two subunits**, the *isomaltase* subunit that is membrane-anchored near its amino-terminus and *sucrase* that is noncovalently bound to isomaltase. Various mutations in the sucrase-isomaltase gene affecting the processing of the precursor protein, the intracellular transport to the apical membrane, and the active catalytic site have been reported. In contrast to sucrase-isomaltase, human glucoamylase is synthesized as a single polypeptide chain that is N- and O-glycosylated and inserted into the brush-border membrane without intracellular or extracellular proteolytic processing.

Brush-border membrane oligosaccharidase activities are affected by many factors and their regulation is complex, involving alterations in transcriptional and post-transcriptional events. In humans, sucrase and glucoamylase mRNA and protein appear in the fetal intestine at 10–14 weeks' gestation and are fully developed at birth. The regulation of enzyme activities along the crypt-villus axis also appears to generally reflect the mRNA abundances. For example, sucrase-isomaltase mRNA is

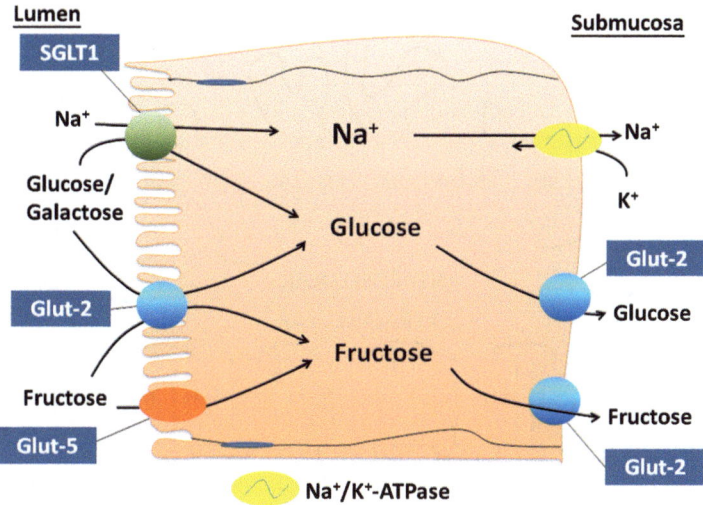

Fig. 6.3 Enterocyte transport of monosaccharides (Adapted from Wright et al. [19])

absent in **crypt cells**, but *detectable* at the **crypt-villus junction**, and enzyme activity is *greatest* in the **mid-upper villus**. Dietary modifications produce alterations in oligosaccharidase activities that may be caused by changes in enzyme synthesis, enzyme degradation, or the activity of the catalytic site. For example, *sucrose feeding* increases sucrase mRNA content, whereas feeding a *carbohydrate-free diet* has been shown to cause conversion of brush-border membrane sucrase to an inactive molecular species. Pancreatic enzymes participate in the degradation of oligosaccharidases and may account, at least in part, for the observed diurnal variation in enzyme activities. **Epidermal growth factor** (**EGF**), an important peptide regulator of intestinal function, affects the intracellular processing and degradation of sucrase.

2.4 Intestinal Glucose Transporters

The efficient intestinal absorption of glucose occurs through the coordinated action of transport proteins located in the **brush-border** and **basolateral membranes** (Fig. 6.3). One pathway of glucose uptake into the enterocyte occurs by a **Na^+-dependent active transport mechanism**. Two sodium ions enter the enterocyte for each molecule of glucose transported, and this absorptive process is therefore electrogenic (net charge transfer across the membrane). Na^+ increases the affinity of the transporter for glucose, and the energy for glucose accumulation is derived from the transmembrane electrochemical Na^+ gradient. The Na^+ gradient is maintained by the action of **Na^+,K^+-ATPase** (**sodium pump**) located on the basolateral

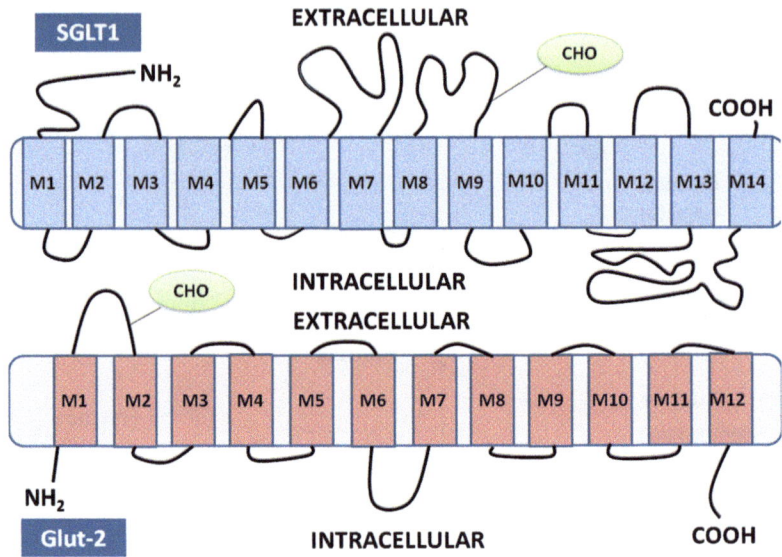

Fig. 6.4 Proposed membrane structure for SGLT1 and Glut2. CHO, glycosylation site; M1-14, membrane-spanning domains (Adapted from Wright et al. [19])

membrane surface, and glucose uptake is therefore a secondary active transport system in which the ATPase indirectly provides the energy for uptake.

The transport protein responsible for **Na$^+$-dependent glucose transport** (**SGLT1**) has been cloned and characterized (Fig. 6.4). The human protein contains 664 amino acids and has a molecular weight of about 73 kd. It has **14 membrane spanning domains** with both extracellular amino- and carboxyl-termini. SGLT1 contains an **N-linked glycosylation site** in a hydrophilic, extracellular domain, but glycosylation does not appear to alter the properties of the carrier protein significantly. SGLT1 has several **phosphorylation sites** that appear to play regulatory roles. Activation of protein kinase A or protein kinase C alters the glucose transport rate. SGLT1 is a high affinity, but rather low capacity glucose transporter. In addition to glucose, galactose and other sugars, but not fructose, are transported by SGLT1. SGLT1 expression is high in neonatal intestine, and increases as enterocytes migrate from the crypt to the villus tip (Fig. 6.5). SGLT1 mRNA is increased by feeding a high sugar diet and in *diabetes mellitus*.

Glucose-galactose malabsorption is a rare genetic defect in monosaccharide absorption that is manifested as *severe watery diarrhea* in newborn children when milk or sugar water is ingested. A variety of mutations ranging from missense mutations to mutations that impair trafficking of SGLT1 to the brush-border membrane causes glucose-galactose malabsorption.

Oral rehydration therapy is used to treat the dehydration associated with cholera and other diarrheal diseases, and this simple approach has markedly reduced mortality due to childhood diarrhea. Patients are given solutions containing *glucose*,

6 Digestion and Absorption of Carbohydrates and Proteins

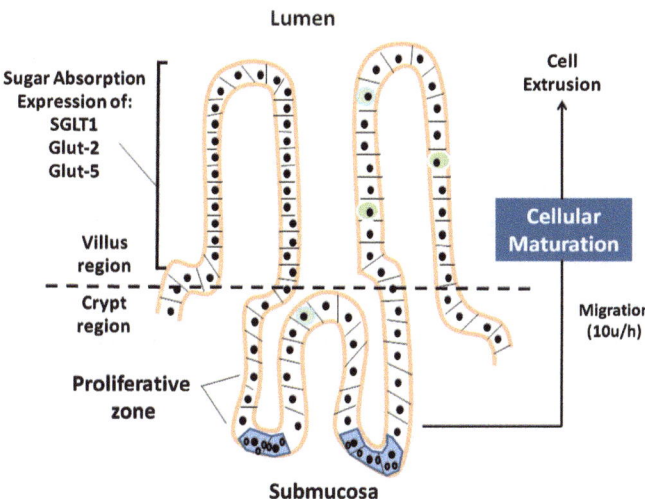

Fig. 6.5 Sites of sugar absorption along the intestinal crypt-villus axis (Adapted from Wright et al. [19])

sodium, *chloride*, *and potassium*. Oral administration of glucose upregulates SGLT1 and absorption of glucose and Na$^+$ results in increased water absorption by a mechanism that is still uncertain.

Glucose and other sugars are transported out of the enterocyte into the portal circulation by a **Na$^+$ independent facilitative diffusion mechanism**. Kinetic studies using basolateral membrane vesicles have shown that this process has a relatively low affinity and high capacity for glucose transport. This membrane transporter, **Glut-2**, is a 524-amino-acid protein that has no homology to SGLT1 (Fig. 6.4). Glut-2 has **12 membrane-spanning regions** with intracellular amino and carboxyl terminals and **sites for glycosylation and phosphorylation**. In addition to glucose, Glut-2 can also transport galactose, fructose, and mannose. Glut-2 expression increases as cells migrate from the intestinal crypt to the villous tip, and increases with sugar feeding (Fig. 6.5). Evidence exists for a second pathway for glucose exit from the enterocyte involving glucose phosphorylation, transport to the endoplasmic reticulum, dephosphorylation, and exit via microsomal transport and trafficking.

Glut-2 can also be detected in the enterocyte brush-border membrane. Current evidence suggests that transport of glucose into the enterocyte via SGLT1 promotes the rapid insertion of Glut-2 into the brush-border membrane via a signal transduction pathway involving **protein kinase C βII** (**PKCβII**). The hypothesis is that brush-border membrane Glut-2 is low when the luminal glucose level is low. When luminal glucose increases there is transport of glucose into the enterocyte by SGLT1 which induces recruitment of Glut-2 to the brush-border membrane, providing a high capacity system for glucose uptake. This model explains the **severe sugar malabsorption** observed with mutation of SGLT1, as recruitment of Glut-2 to the brush-border membrane would be impaired in this condition. Apical Glut-2 insertion is regulated by many factors, including sugars, long-term dietary carbohydrate intake, glucocorticoids, the gastrointestinal hormone GLP-2, cellular metabolic needs, and certain medications used for treating diabetes.

2.5 Digestion and Absorption of Dietary Sugars

Dietary sugars represent approximately half of the digestible carbohydrate in the human diet. Important dietary sugars include the **monosaccharides** glucose and fructose and the **disaccharides** lactose, sucrose, and maltose. The disaccharides must be digested by brush-border membrane **disaccharidases** to their component monosaccharides that are then transported through the enterocytes.

The milk sugar **lactose** (glucose-β-1,4-galactose) is the major carbohydrate consumed in the neonatal period. Human milk contains 7 % lactose, more than most other mammalian species. The brush-border membrane disaccharidase **lactase** is responsible for the digestion of lactose to glucose and galactose, which is then absorbed using the brush-border and basolateral-membrane glucose transporters, SGLT1 and Glut-2. In contrast to other disaccharides, hydrolysis by lactase is the rate-limiting step in the overall process of intestinal lactose absorption. Lactase also is responsible for the hydrolysis of glycolipids present in the intestinal lumen, utilizing a different catalytic site on the protein than that responsible for lactose hydrolysis.

Expression of lactase begins in late gestation, and high levels of the enzyme are expressed in the neonatal period, particularly in the proximal and mid-jejunum. In most mammalian species, lactase activity markedly declines after weaning. In humans, however, two general patterns of lactase expression are observed post-weaning. In most humans, a pattern similar to other mammals is observed, with lactase activity declining after weaning, and by age 5-10 years enzyme specific activity drops to the adult level, approximately 10 % of the neonatal value. Individuals with nonpersistence of lactase (**adult-type hypolactasia**) frequently develop *abdominal pain*, *cramping*, *distention*, *flatulence*, *diarrhea*, *nausea*, and occasionally *vomiting* after consuming significant amounts of lactose-containing dairy products. These symptoms are caused by malabsorbed lactose drawing water into the intestinal lumen, producing an *osmotic diarrhea*, and by the gut bacterial flora metabolizing the unabsorbed lactose, forming gases such as *hydrogen*, *methane*, and *carbon dioxide* (Fig. 6.6). A minority of humans, however, demonstrate persistence of a high level of lactase activity in the adult small intestine. This is a genetically determined trait with an **autosomal recessive mode** of inheritance of lactase persistence. It has been suggested that lactase persistence offers a *selective advantage* to individuals living in population groups that practice dairy farming. The prevalence of lactose malabsorption therefore varies widely in different ethnic or racial groups (Table 6.1).

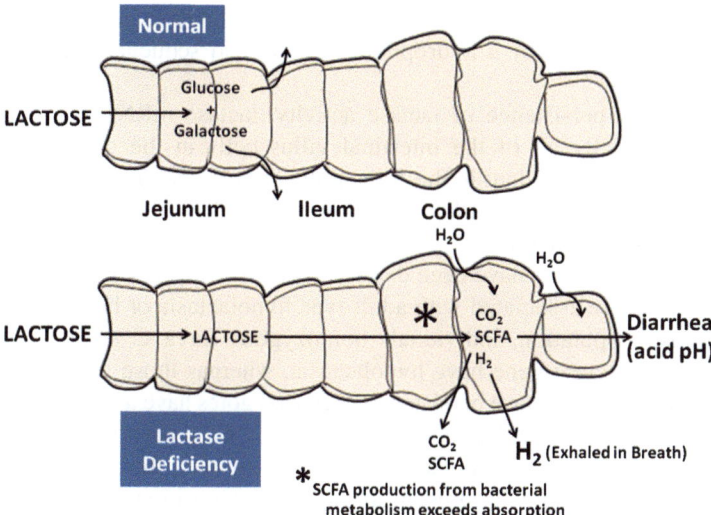

Fig. 6.6 Pathophysiology of lactose intolerance. SCFA, short chain fatty acids

Table 6.1 Prevalence of late-onset lactose malabsorption among various ethic and racial groups

Group	Prevalence of lactose malabsorption (%)
Orientals in the United States	100
American Indians (Oklahoma)	95
Ibo and Yoruba (Nigeria)	89
Black Americans	81
Italians	71
Aborigines (Australia)	67
Mexican Americans	56
Greeks	53
White Americans	24
Danes	3
Dutch	0

(From Buller and Grand [4])

The molecular regulation of lactase expression has been intensely investigated, particularly with respect to the factors that determine the postweaning decline in lactase levels or lactase persistence. The proenzyme is synthesized as a single polypeptide containing 1,927 amino acids, including a 19-amino-acid signal peptide. **Prolactase** undergoes a series of N- and O-glycosylations that affect intracellular transport and enzymatic activity. Prolactase is converted by intracellular proteolysis and extracellular cleavage by pancreatic proteases to the mature enzyme form (approximately 160 kd, 1,060 amino acids) present in the brush-border membrane. The C-terminus is intracellular and the N-terminus is found on the luminal side of the brush-border membrane. Lactase is anchored into the brush-border membrane by a hydrophobic amino acid sequence located near the carboxyl-terminus.

In adults with persistence of lactase activity, lactase mRNA and protein are present in essentially all of the intestinal villus cells in the proximal jejunum. The **mRNA** is most abundant in the cells at the *crypt-villus junction*, whereas the **protein** is highest in the *midvillus region*. Lactase mRNA is highest in the distal jejunum and proximal ileum.

Gene polymorphisms have been described in the five-prime region flanking the lactase gene that are associated with adult-type hypolactasia or lactase persistence. In the Finnish population, individuals homozygous for a C at position -13910 upstream of the lactase gene have hypolactasia, whereas those homozygous for a T at that position have lactase persistence. Heterozygotes have an intermediate level of lactase that is sufficient for lactose digestion. A G at position -22018 is more than 95 % associated with hypolactasia, whereas an A at that position correlates with lactase persistence. Chromosomes with a T at position -13910 all have an A at -22018. The C -13910 allele has been reported to result in reduced transcription of the lactase gene, but this is controversial. The T-13910 and A-22018 alleles, however, are not found in some African and Bedouin populations with lactase persistence; other mutations in the five prime region have been identified. A small number of individuals with non-persistence of lactose have been showed to have an abnormality in the intracellular processing of newly synthesized lactase.

Lactase is expressed at high levels in the human neonatal intestine. **Congenital lactase deficiency** is a rare **autosomal recessive** disorder characterized by a marked reduction in intestinal lactase activity resulting in *watery diarrhea*, *dehydration*, *acidosis*, and *weight loss* when the neonate is fed breast milk or lactose-containing formula. Mutations in the lactase gene causing premature truncation of lactase and missense mutations have been described in children with congenital lactase deficiency.

In humans, lactose ingestion has little effect on intestinal lactase activity. Hormones, particularly **thyroid hormone** and **glucocorticoids**, appear to play modulatory roles in the developmental regulation of lactase expression.

Fructose is present in the diet in several forms, as part of the disaccharide sucrose (see below), as a monosaccharide in foods such as honey and certain fruits, and as fructose sweeteners formed by the enzymatic conversion of glucose in cornstarch to fructose. Fructose is taken up into the enterocyte by a Na^+**-independent, facilitated diffusion mechanism**. **Glut-5** has been identified as the intestinal brush-border membrane fructose transporter (Fig. 6.3). This transporter is a 501-amino-acid protein that has 41 % amino acid identity and a similar structure to Glut-2. Glut-5 knockout mice have severely impaired fructose absorption. Glut-2 in the brush-border membrane may also be responsible for some fructose uptake, particular at high fructose loads. Glut-5 is expressed at low levels at birth and increases post-weaning, perhaps modulated in part by glucocorticoids. Glut-5 mRNA expression is increased by fructose feeding. Glut-5 expression increases as enterocytes migrate from the crypt to the villus (Fig. 6.5). Fructose is transported from the enterocyte into the portal circulation via the basolateral-membrane Glut-2 transporter.

Fructose is not as well absorbed as is glucose, and there is considerable interindividual variation in the efficiency of fructose absorption. Patients with *chronic gastrointestinal symptoms*, such as *diarrhea*, *gas*, and *bloating*, due to a very limited capacity for fructose absorption have been identified.

Sucrose (**table sugar**) is the disaccharide glucose-α-1,2-fructose. The brush-border membrane **disaccharidase sucrase** hydrolyzes sucrose into the monosaccharides glucose and fructose, which are then absorbed as previously described. Intestinal fructose absorption is considerably more efficient when the fructose is administered as sucrose rather than as an equivalent amount of the monosaccharide. Furthermore, co-administration of glucose with fructose enhances fructose absorption. These findings may be explained by recruitment of Glut-2 to the apical membrane during glucose absorption.

2.6 Dietary Fiber

The term *dietary fiber* operationally refers to plant material that resists digestion by the enzymes of the human GI tract. With the *exception* of *lignin*, the components of dietary fiber are carbohydrates that are mainly constituents of **plant cell walls**, such as cellulose, hemicellulose, and pectins, among others. These polysaccharides, with the exception of wood cellulose, are partially metabolized by enzymes present in gut bacteria, producing short-chain fatty acids (acetate, propionate, and butyrate), hydrogen, carbon dioxide, and methane. The short-chain fatty acids are absorbed to some extent by the distal small intestine and colon and can be utilized as a metabolic fuel. **Butyrate**, in particular, appears to be the principal energy source for the large intestine. Starches that resist efficient hydrolysis by α-amylases and some poorly absorbed dietary sugars and sugar alcohols will also be converted to short-chain fatty acids by gut bacteria, permitting salvage of some of the energy contained in those compounds.

The components of dietary fiber have many diverse effects on gastrointestinal function. For example, fiber can alter intestinal absorption of various nutrients by several mechanisms, including changing the physical characteristics of the luminal contents via the hydratability of fiber and by binding cations and organic compounds. Certain types of fiber, such as wheat bran, decrease gastrointestinal transit time, whereas other fiber components, such as pectin and guar gum, increase gastrointestinal transit time. Short-chain fatty acids produced from dietary fiber stimulate colonic salt and water absorption and regulate colonocyte proliferation and differentiation. Dietary fiber alters the gastrointestinal flora, influencing the diversity and numbers of microorganisms.

Epidemiologic studies have repeatedly demonstrated an association between low fiber diets and common diseases such as *cardiovascular disease, type 2 diabetes, colorectal* and other *cancers*, and *obesity*. The specific components of dietary fiber and the mechanisms of action resulting in these putative health effects remain controversial.

3 Protein

The normal adult human requires approximately 0.75 g/kg body weight per day of highly digestible, high-quality protein to ensure maintenance of nitrogen balance and an adequate supply of essential amino acids. Higher protein intakes are needed to support normal **growth, pregnancy,** and **lactation**. Nine of the 20 amino acids commonly found in food, histidine, isoleucine, leucine, lysine, methionine, phenylalanine, threonine, tryptophan, and valine are not synthesized by mammals. They are therefore termed **essential amino acids**, because they must be obtained from dietary sources. Other amino acids may become conditionally essential in disease because of impaired conversion from precursors, and protein requirements are generally increased with illnesses such as *sepsis* and *trauma*.

Individual dietary proteins differ in their amino acid compositions, but all food proteins, *except gelatin*, contain some of each amino acid. The digestibility of different proteins varies considerably, with plant proteins generally digested more poorly than animal proteins. Protein digestibility is influenced by methods of food processing. In addition to proteins contained in the diet, approximately 35 g/day of protein is secreted into the intestine in various digestive juices, and 30 g/day enter the lumen as part of desquamated intestinal cells. These endogenous proteins are digested and absorbed along with the dietary proteins. Overall, protein absorption is quite efficient, with the normal adult excreting fecal nitrogen equivalent to only 6–12 g/day of protein. The digestion of protein begins with intraluminal hydrolysis by proteases secreted by the stomach and pancreas and continues as peptidases in the brush-border membrane of the small intestine digest the protein to amino acids and di- and tripeptides that are taken up by the enterocytes (Fig. 6.7).

3.1 Acid-Peptic Digestion

Gastric juice contains a family of proteolytic enzymes known as **pepsins**. In the acid environment of the stomach, these enzymes are capable of hydrolyzing a broad range of peptide bonds. Proteolysis within the stomach not only contributes to overall protein digestion, but the peptides and amino acids that are generated are important stimulators of gastrin and cholecystokinin release and are regulators of gastric acid secretion and gastric emptying. In the neonatal stomach, **chymosin**, which very effectively digests milk protein, is the predominant enzyme. In adults, two families of enzymes, **pepsin A** and **pepsin C**, predominate. Pepsin A activity comprises five closely related isoenzymes coded for by genes on chromosome 11, whereas pepsin C represents two isoenzymes coded for on chromosome 6. These enzymes are irreversibly denatured in a mildly alkaline fluid, and therefore little peptic digestion occurs past the duodenum. Acid-peptic hydrolysis is not essential for efficient protein digestion, as patients who have undergone total gastrectomy have little impairment of protein utilization. In patients with pancreatic disease and limited secretion of pancreatic proteases, however, acid-peptic proteolysis is an important facilitator of protein digestion.

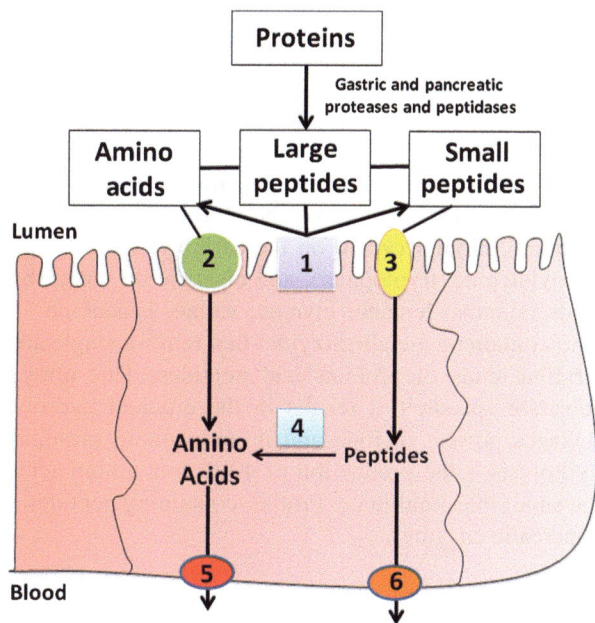

Fig. 6.7 Protein digestion and absorption in the small intestine. (*1*) Brush-border membrane peptidases. (*2*) Brush-border membrane amino acid transporters. (*3*) Brush-border membrane di- and tripeptide transporters. (*4*) Intracellular peptidases. (*5*) Basolateral-membrane amino acid carriers. (*6*) Basolateral membrane di- and tripeptide carriers (Adapted from Ganapathy et al. [7])

Pepsins are synthesized in the **gastric chief cells** as inactive **pre-proenzymes**. A signal sequence of about 15 amino acids directs the protein to the interior of the endoplasmic reticulum, where the signal sequence is cleaved. The resulting **proenzyme (pepsinogen)** is transferred to the Golgi apparatus and condensed into secretory granules; in some cases it is also glycosylated. Secretion of the granules occurs by **exocytosis**, stimulated by agents that *increase cyclic adenosine monophosphate (cAMP)* (secretin, vasoactive intestinal polypeptide, epinephrine, prostaglandins) or that *increase intracellular calcium concentration* (acetylcholine, cholecystokinin, gastrin). Secretion of pepsinogen is initially very rapid, but then declines to a slower rate even before the chief cells are depleted of granules, suggesting feedback regulation or receptor desensitization. Synthesis of new preproenzyme occurs following stimulation of secretion. In the acid environment of the stomach, there is cleavage of an N-terminal peptide of 44 amino acids from pepsinogen, forming the active pepsin with a molecular weight of approximately 35 kd. The acid milieu appears to cause a conformational change in pepsinogen that permits intramolecular cleavage of the N-terminal activation peptide.

3.2 Pancreatic Proteases

Proteases are secreted by the **pancreas** as proenzymes (Chap. 4 for a detailed discussion of the regulation of pancreatic secretion). Activation of pancreatic proteases is initiated by the intestinal brush-border enzyme enteropeptidase (**enterokinase**), which is stimulated by **trypsinogen** contained in pancreatic juice and released from the brush-border membrane by bile salts. Enteropeptidase has a single substrate, trypsinogen, and cleaves an N-terminal hexapeptide forming trypsin, which can then activate more trypsinogen and other pancreatic proenzymes.

Pancreatic secretions contain a variety of **endopeptidases** (trypsin, chymotrypsin, elastase) and **exopeptidases** (carboxypeptidase A and B). **Trypsin** cleaves peptide bonds on the carboxyl side of basic amino acids (lysine, arginine), **chymotrypsin** on the carboxyl side of aromatic amino acids (tyrosine, phenylalanine, tryptophan), and **elastase** on the carboxyl side of aliphatic amino acids (alanine, leucine, glycine, valine, isoleucine). The **carboxypeptidases** are zinc-containing metalloenzymes that remove single amino acids from the carboxyl-terminal ends of proteins and peptides. This array of enzymes with different substrate specificities results in the efficient hydrolysis of the different peptide linkages present in food and in endogenous proteins. The result of intraluminal hydrolysis is the production of 40 % free amino acids and 60 % oligopeptides in the small intestinal juice. Proline-containing peptides are resistant to hydrolysis by pancreatic enzymes.

3.3 Brush-Border Membrane Peptidases

Peptidases within the enterocyte brush-border membrane hydrolyze the oligopeptides generated by intraluminal digestion to free amino acids and the di- and tripeptides transported into the enterocyte (Fig. 6.7). About 20 different peptidases have been identified in the brush-border membrane, in contrast to the relatively few enzymes involved in brush-border carbohydrate digestion. These peptidases recognize specific amino acids at the amino- or carboxyl-terminus or internally within the oligopeptide and are necessary for the hydrolysis of the wide variety of peptide linkages found in food proteins. Many of these peptidases are **dimeric metalloenzymes** that are anchored into the brush-border membrane by either a hydrophobic membrane-spanning domain of the protein or through covalent attachment to membrane glycosyl-phosphatidylinositol. The active site of the enzyme is generally contained within a large hydrophilic portion of the protein that projects into the intestinal lumen. The brush-border membrane peptidases are generally classified into **four groups**: *endopeptidases*, *aminopeptidases*, *carboxypeptidases*, and *dipeptidases*. The most abundant peptidase is probably aminopeptidase N, which sequentially removes N-terminal amino acids from oligopeptides. The brush-border membrane contains at least four peptidases that are active against peptides with proline at the cleavage site. These enzymes are therefore complementary to the pancreatic proteases that have little ability to hydrolyze peptide bonds containing proline.

The enterocyte also contains at least **four intracellular peptidases** that are involved in the digestion of absorbed di- and tripeptides (Fig. 6.7). Ninety percent or more of the products of protein digestion appear in the portal blood as free amino acids.

In addition to participating in overall dietary protein digestion, several of the intestinal peptidases have other specific functions. **Folate conjugase** is a brush-border membrane carboxypeptidase that specifically degrades the polyglutamate side chain of dietary folates, producing the monoglutamate form that is transported into the enterocytes (Chap. 9). Other intestinal peptidases have been shown to be involved in the activation or catabolism of gastrointestinal peptide hormones or in the turnover of endogenous intestinal proteins. **Dipeptidyl aminopeptidase IV (DPP IV)** plays an important role in the metabolism of the **incretin GLP1** that enhances insulin secretion, and **DPP IV inhibitors** are used in the treatment of *type 2 diabetes.*

The mechanisms for biosynthesis and intracellular sorting of brush-border membrane peptidases are complex and differ significantly for the various enzymes. Complex glycosylation events, changes in protein folding, metabolism of proenzyme forms, and dimerization are important steps in the synthesis and incorporation into the brush-border membrane of peptidases. Most peptidases apparently have direct transport routes to the brush-border membrane, but aminopeptidase N has been shown to first appear in the basolateral membrane, followed by transit through the Golgi apparatus before insertion in the brushborder membrane.

The activities of brush-border membrane peptidases are highly regulated during development and cellular differentiation and by starvation or feeding, with specific patterns of enzyme expression observed for each peptidase. Aminopeptidase N is abundant in the *proximal jejunum*, whereas dipeptidyl aminopeptidase IV is higher in the *ileum*.

3.4 Brush-Border Membrane Amino Acid Transport

The intestinal brush-border membrane contains multiple transport systems for the uptake of amino acids. Systems capable of transporting classes of amino acids were defined based upon the physical-chemical properties, stereospecificity, size, and charge of the amino acids transported, as well as the ion and pH dependencies. **Transport systems** that have been identified include:

1. "**Neutral system**" or "**methionine preferring system**" that transports *all neutral amino acids*;
2. "**Basic system**" that transports *cationic amino acids* and *cysteine*;
3. "**Acidic system**" that transports *glutamate* and *aspartate*;
4. "**Iminoglycine system**" that transports *proline, hyroxyproline*, and *glycine*; and
5. "**β-amino acid system**" that transports *taurine*.

It should be recognized that there is significant overlap of substrate specifies among the different transport systems. The various carrier systems may require Na^+, H^+, Cl^-, or K^+ as co-factors for amino acid transport, and may be electrogenic or neutral. Some of the transport proteins corresponding to these systems have been identified, and inherited defects in these transporters correspond to genetic abnormalities in amino acid metabolism. The association of some of the amino acid transporters to the apical membrane requires partner proteins, including members of the renin-angiotensin system, **collectrin** (Tmem27) and **angiotensin-converting enzyme 2**. Amino acid transport proteins also can transport orally administered amino-acid based drugs and derivatives.

3.5 Brush-Border Membrane Peptide Transport

Extensive evidence now exists that small peptides are transported across the brush-border membrane by a mechanism that is independent of the systems used for amino acid uptake (Fig. 6.7). A patient with a genetic defect in amino acid transport, such as *Hartnup disease* or *cystinuria*, may absorb certain free amino acids poorly, but absorption of those amino acids as part of small peptides is normal. There is a lack of competition between amino acid and peptide transport, often a faster absorption rate for amino acids presented in peptides rather than in the free forms, and differences in the developmental and dietary regulation of amino acid versus peptide uptake. The absorptive capacity for small peptides is greater in the jejunum than the ileum, whereas for amino acids it is greater in the ileum than the jejunum. Mainly di- and tripeptides are absorbed by the enterocyte, with very limited uptake of tetrapeptides and essentially no absorption of larger peptides.

PEPT1 is the major, if not only, di-tripeptide carrier expressed in the small intestinal brush-border membrane. The gene for PEPT1 is on chromosome 13, and the open reading frame encodes 708 amino acids. There are 12 transmembrane spanning domains with both the amino- and carboxy termini facing the cytoplasmic side. PEPT1 is an electrogenic symport system in which the inwardly directed proton gradient allows peptides to enter the cell against a concentration gradient. In some experimental situations, a **partial Na^+ dependency of peptide transport** was identified. This is apparently an indirect effect, as brush-border Na^+/H^+ exchange via the Na^+-proton antiporter NHE-3 is needed to maintain the inwardly directed proton gradient. A remarkable feature of PEPT1 is the ability of this one molecule to transport a myriad of possible di- and tripeptides generated from protein digestion. Evidence suggests that water in the vicinity of the carrier's binding domain plays a central role in shielding the electric charges of amino acid side chains, allowing charged, polar, and large apolar molecules to bind. In addition to peptide digestion products, PEPT1 can transport certain drugs, such as some *β-lactam antibiotics* and *ACE inhibitors*.

PEPT1 is acutely regulated by intracellular signaling pathways involving protein kinase A, protein kinase C, and intracellular calcium concentration. PEPT1 abundance and activity is also regulated by hormones such as insulin, growth hormone, thyroid hormone, and leptin, and is altered in diabetes. Dietary protein, certain amino acids, and peptides also alter PEPT1 expression. Normally PEPT1 is expressed throughout the **small intestine**, but is present at very low levels in the **colon**. In patients who have undergone major resections of the small intestine and have the short bowel syndrome and malabsorption, PEPT1 mRNA and protein is expressed in the **colonic mucosa**, and may permit colonic peptide absorption. Multiple mutations affecting various aspects of PEPT1 carrier function have been described.

3.6 Basolateral-Membrane Transporters

Amino acids transported across the brush-border membrane and generated intracellularly from proteolysis of absorbed peptides are transferred into the portal blood via multiple carrier systems with different electrolyte and pH requirements. As much as 10 % of absorbed protein may be delivered into the portal blood as small peptides (Fig. 6.7). A transport system similar, but not identical to PEPT1 has been found in the basolateral membrane. There is not likely to be a substantial H^+ gradient from the cytoplasm to the portal blood to drive transport, and it is possible that the high postprandial concentration of di- and tripeptides within the enterocyte cytoplasm compared with the minimal concentration of these peptides in blood is sufficient to cause efficient transfer into the circulation.

Transporters in the basolateral membrane can also mediate uptake of amino acids from the circulation into enterocytes during fasting that supports enterocyte protein synthesis.

3.7 Regulation of Amino Acid and Peptide Absorption

The absorption of amino acids and peptides is developmentally regulated and influenced by diet, hormones, and growth factors. The transport systems are generally present at birth, and transport rates for peptides and most amino acids decline with age, although the magnitude of change differs among individual carriers. **High dietary levels of protein or amino acids** generally result in upregulation of peptide and most amino acid transport. **Short-term fasting** also tends to increase the absorption rates, but **long-term starvation** causes decreased amino acid transport with little change in peptide absorption.

3.8 Role of Absorbed Amino Acids in Enterocyte Energy and Protein Metabolism

Absorbed amino acids, particularly *glutamine*, *glutamate*, and *aspartate*, are the major fuels for the small intestine. About 10 % of the absorbed amino acids is used for protein synthesis within the enterocyte. In fact, there is preferential utilization of absorbed amino acids versus those entering the enterocyte from the blood for energy metabolism and protein synthesis.

Clinical Correlations

Case Study 1

A 1-month-old infant is admitted with dehydration and a history of severe diarrhea since birth. Her pediatrician places her on lactose-free and sucrose-free infant formulas with no significant improvement in her condition. Stool examination demonstrates the presence of a large quantity of reducing substances, which are confirmed to be mainly glucose; the stool pH is acidic. There is no family history of a similar illness, but her parents are first cousins.

Questions

1. **What is the likely diagnosis?**
   Answer: This child most likely has **genetic glucose-galactose malabsorption**, an *autosomal recessive* disorder. Malabsorption of these monosaccharides causes a severe *osmotic diarrhea*, resulting in dehydration. Bacterial metabolism of the unabsorbed sugar produces organic acids and results in an *acidic stool pH*. **Congenital deficiencies of the disaccharidases sucrase-isomaltase** or **lactase** have similar clinical presentations, but this child failed to improve when sucrose or lactose were eliminated from her diet. **Monosaccharide malabsorption** can also occur following certain intestinal infections, but the onset of this child's symptoms at birth makes a genetic defect more likely.
2. **What is the molecular basis of this disorder?**
   Answer: Children with glucose-galactose malabsorption have *impaired Na^+-dependent glucose and galactose uptake* into the small intestine. **Mutations in the glucose transporter SGLT1** have been identified. The lack of glucose and Na^+ uptake via SGLT1 impairs recruitment of GLUT2 to the brush-border membrane, further impeding glucose and galactose absorption.
3. **What would you recommend as dietary treatment for this condition?**
   Answer: These children need to be placed on **infant formulas** that have *markedly reduced contents of glucose, galactose, disaccharides* that are *hydrolyzed* to glucose and galactose (lactose, sucrose), and starches or glucose polymers. Since fructose is absorbed by a different glucose transporter (Glut5 rather than SGLT1), **fructose-containing formulas** are effectively absorbed and are life-saving.

Case Study 2

A 30-year-old Chinese-American man complains of excessive gas and mild chronic diarrhea for many years. He has noted that eating ice cream and drinking milk tend to provoke these symptoms. He is otherwise healthy, has no other GI complaints, and has not lost weight.

Questions

1. **What is the most likely cause of this patient's symptoms?**
   Answer: The patient most likely has **adult-type hypolactasia**, resulting in *malabsorption of lactose* and *intolerance of dairy products*. The malabsorbed lactose causes an *osmotic diarrhea* and is metabolized by gut bacteria to *gases* such as carbon dioxide, hydrogen, and methane. *Hypolactasia* is extremely common in Chinese adults. Various small-intestinal diseases, for example, *celiac sprue*, can result in a reduction in brush-border lactase (*secondary hypolactasia*), but this patient's lack of other signs or symptoms of generalized malabsorption makes this diagnosis less likely.
2. **What tests could you use to make a definitive diagnosis?**
   Answer: Several tests have been developed to diagnose lactose malabsorption. A standard **lactose tolerance test** measures the rise in blood glucose after oral administration of lactose; the increase in blood glucose is diminished in patients with hypolactasia. **Breath-hydrogen testing** is currently the most convenient and accurate method for evaluating lactose absorption. A dose of lactose is given orally, and breath samples are serially collected for measurement of breath hydrogen. Individuals with lactase-persistence will digest the lactose to the component monosaccharides, which are efficiently absorbed, and there will be little change in breath hydrogen. Patients with hypolactasia will malabsorb lactose, which is then metabolized by gut bacteria to hydrogen gas that is absorbed and appears in the breath. Other tests that are used less commonly to diagnose lactose malabsorption include **intestinal biopsy** with measurement of lactase activity and **assessment of the appearance of galactose in blood or urine** after oral lactose administration.
3. **What treatment would you prescribe?**
   Answer: Avoidance of lactose-containing dairy foods will relieve the patient's symptoms. Dairy foods, however, are important sources of calcium, vitamin D, protein, and other nutrients. Some dairy foods, such as aged cheeses, have relatively little lactose. **Microbial lactases** are used to prepare **low-lactose milk**, and they can also be taken by mouth prior to eating dairy foods to enhance lactose digestion and diminish GI symptoms.

Case Study 3

A 7-year-old boy presents with recurrent scaly rashes on his face and arms, mild diarrhea, and ataxia. Laboratory testing demonstrates elevated urinary and fecal excretion of neutral amino acids, including tryptophan, and indoles. His parents are first cousins, and testing of his family demonstrates that his brother also has increased urinary and fecal amino acids and indoles, but is asymptomatic.

Questions

1. **What are the diagnosis and the pathogenetic basis of this syndrome?**

   Answer: The child has **Hartnup disease**, an *autosomal recessive* disorder in which there is a reduction of small intestinal and renal transport of certain neutral amino acids, including *tryptophan,* causing elevated amino acid excretion in urine and feces. The disease is caused by a **mutation** in the **neutral amino acid transporter SLC6A19**. **Malabsorbed tryptophan** is metabolized by gut bacteria to various indoles that are excreted in the feces and urine. The patient's symptoms are believed to be due, at least in part, to niacin deficiency that produces a pellagra-like syndrome, since tryptophan is a niacin precursor. **Abnormal production of serotonin** in the nervous system because of reduced tryptophan and altered protein synthesis because of amino acid deficiencies may also contribute to the patient's symptoms.

2. **How would you explain the observation that his brother has a similar metabolic abnormality but is asymptomatic?**

   Answer: There are multiple amino acid transport systems in the intestinal brush-border membrane with overlapping specificities. In addition, there is a separate carrier for di- and tripeptides PEPT1. A defect in a single transporter therefore is unlikely to result in a gross deficiency of essential amino acids, and most individuals with the Hartnup abnormality are asymptomatic. It is believed that other genetic, nutritional, or environmental factors are required in addition to the mutation in the amino acid transporter to cause clinically apparent metabolic abnormalities.

3. **What treatment would you recommend?**

   Answer: The clinical features of Hartnup disease, particularly the skin rash, will sometimes respond to **niacin supplementation**. These children will also often improve on a **high-protein diet**, which may result in the production of a large quantity of di- and tripeptides that are taken up via PEPT1, providing a sufficient supply of essential amino acids. **Special peptide-containing formula diets** that would be efficiently absorbed in patients with Hartnup disease may also be useful supplements for affected children.

Case Study 4

A 45-year-old woman complains of many years of excessive gas, bloating, flatulence, and watery diarrhea. Extensive evaluation has not revealed any gastrointestinal disease and a lactose hydrogen breath test was normal. A review of dietary records shows that her diarrhea and other symptoms occur when she eats fruits and other high fructose foods.

Questions

1. **What is the basis for her symptoms?**

   Answer: The patient likely has poor intestinal absorption of fructose. Studies have shown wide variation in the efficiency of absorption of fructose in healthy individuals, and fructose malabsorption can produce symptoms of the irritable bowel syndrome. The malabsorbed fructose results in an osmotic diarrhea and is metabolized by intestinal bacteria forming hydrogen and other gases causing the bloating and flatulence.

2. **How could you make the diagnosis of fructose malabsorption?**
   Answer: A hydrogen breath test can be performed after fructose ingestion similar to the breath test that is done to detect lactose malabsorption. Excessive breath hydrogen after fructose ingestion is indicative of fructose malabsorption.
3. **Why is this patient able to tolerate eating large amounts of sucrose (a disaccharide containing glucose and fructose) without developing gastrointestinal symptoms?**
   Answer 3: Sucrose is hydrolyzed in the intestine to the monosaccharides glucose and fructose. Some fructose is absorbed across the brush-border membrane by the transporter GLUT-5. Glucose will be absorbed by the transporter SGLT-1. Glucose transport by SGLT-1 initiates a signal transduction cascade that recruits another transporter, GLUT-2, to the brush-border membrane that is a high capacity transport system for glucose and fructose. Fructose absorption as sucrose or when co-administered with glucose is, therefore, greater than when fructose alone is ingested.

Further Reading

1. Alpers DH (1994) Digestion and absorption of carbohydrates and protein. In: Johnson LR (ed) Physiology of the gastrointestinal tract, 3rd edn. Raven Press, New York, pp 1723–1749
2. Broer S (2008) Amino acid transport across mammalian intestinal and renal epithelia. Physiol Rev 88:249–286
3. Broer S, Palacin M (2011) The role of amino acid transporters in inherited and acquired disease. Biochem J 436:193–211
4. Buller HA, Grand RJ (1990) Lactose intolerance. Ann Rev Med 41:141–148
5. Douard V, Ferraris RP (2008) Regulation of the fructose transporter GLUT5 in health and disease. Am J Physiol Endocrinol Metab 295:E227–E237
6. Drozdowski LA, Thompson ABR (2006) Intestinal sugar transport. World J Gastroenterol 12:1657–1670
7. Ganapathy V, Brandsch M, Leibach FH (1994) Intestinal transport of amino acids and peptides. In: Johnson LR (ed) Physiology of the gastrointestinal tract, 3rd edn. Raven Press, New York, pp 1773–1794
8. Gray GM (1992) Starch digestion and absorption in nonruminants. J Nutr 122:172–177
9. Gudmand-Hoyer E, Skovbjerg H (1996) Disaccharide digestion and maldigestion. Scandanavian J Gastroenterol 31(Suppl 216):111–121
10. Hannelore D (2004) Molecular and integrative physiology of intestinal peptide transport. Annu Rev Physiol 66:361–384
11. Jarvela I, Torniainen S, Kolho K-L (2009) Molecular genetics of human lactase deficiencies. Ann Med 41:568–575
12. Jones HF, Butler RN, Brooks DA (2011) Intestinal fructose transport and malabsorption in humans. Am J Physiol Gastrointest Liver Physiol 300:G202–G206
13. Kellett GL, Brot-Laroche E (2005) Apical Glut-2: a major pathway of intestinal sugar absorption. Diabetes 54:3056–3062

14. Lattimer JM, Haub MD (2010) Effects of dietary fiber and its components on metabolic health. Nutrients 2:1266–1289
15. Lunn J, Buttriss JL (2007) Carbohydrates and dietary fibre. Nutr Bull 32:21–64
16. Montgomery RK, Krasinski SD, Hirschhorn JN, Grand RJ (2007) Lactose and lactase: who is lactose intolerant and why? J Pediatr Gastroenterol Nutr 45:S131–S137
17. Robayo-Torres CC, Nichols BL (2007) Molecular differentiation of congenital lactase deficiency from adult-type hypolactasia. Nutr Rev 65:95–98
18. Robayo-Torres CC, Quezada-Calvillo R, Nichols BL (2006) Disaccharide digestion: clinical and molecular aspects. Clin Gastroenterol Hepatol 4:276–287
19. Wright EM, Hirayama BA, Loo DDF, Turk E, Hager K (1994) Intestinal sugar transport. In: Johnson LR (ed) Physiology of the gastrointestinal tract, 3rd edn. Raven Press, New York, pp 1751–1772

Chapter 7
Digestion and Absorption of Dietary Triglycerides

Michael D. Sitrin

1 Introduction

The mechanisms for digestion and absorption of dietary triglycerides have been extensively studied, reflecting the importance of triglycerides in normal human nutrition and in disease states. The Western diet contains a large amount of triglycerides, typically 60–120 g/day. Normally, 95 % or more of consumed triglycerides is digested and absorbed, providing 30–40 % of the total energy requirement. In addition, essential polyunsaturated fatty acids contained in dietary triglycerides are the precursors for important lipid-derived mediators such as prostaglandins and leukotrienes involved in regulation of diverse cellular functions. The processes for triglyceride digestion and absorption also significantly influence the efficiency of absorption of other important dietary lipids such as cholesterol and the fat-soluble vitamins.

The **stomach**, **liver**, **biliary tract**, **pancreas**, and **small intestine** all participate in normal triglyceride digestion and absorption. Inadequate triglyceride digestion and absorption result in excessive loss of fat in the stool (steatorrhea). The presence of *steatorrhea* is therefore the hallmark of a diverse group of digestive disorders that cause malabsorption. Dietary triglycerides also play important pathogenetic roles in some of the most prevalent diseases in Western society, including *obesity*, *atherosclerotic cardiovascular disease*, and several common types of *cancer*.

M.D. Sitrin, M.D. (✉)
Department of Medicine, University at Buffalo, The State University of New York, Buffalo, NY, USA
e-mail: mdsitrin@buffalo.edu

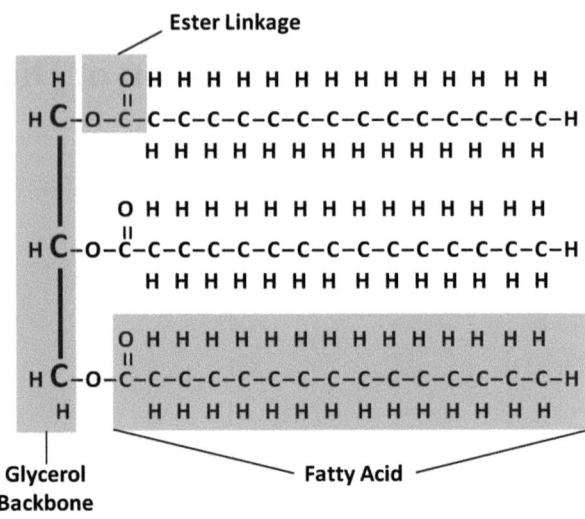

Fig. 7.1 Basic structure of triglycerides

2 Structure of Triglyceride

Dietary triglycerides consist of **three fatty acids** attached by ester linkage to a **glycerol backbone**, forming **triacylglycerol** (Fig. 7.1). Most dietary triglycerides are composed of various **long-chain** (C_{14}-C_{22}) saturated or unsaturated fatty acids, with only small amounts of triglycerides containing **medium-chain** (C_6-C_{12}) or **short chain** (less than C_6) fatty acids. Polyunsaturated fatty acids tend to be preferentially located in the 2 or β position of the glycerol backbone.

3 Triglyceride Digestion

3.1 Intragastric Events

Several important steps in triglyceride digestion occur within the stomach. **Lipase** activity is readily detected in the gastric fluid, and in normal adult humans appears to account for 10–30 % of triglyceride hydrolysis. Lipase is produced in **serous glands** at the *base of the tongue* or in **acinar cells** in *pharyngeal tissue*, and the lipase is then swallowed along with the food bolus. Lingual lipase generating free fatty acids in the mouth and pharynx may play a role in **gustatory fatty acid signaling (fatty acid taste)**. In humans, the lipase found in gastric fluid appears to originate mainly within **chief cells** of the *gastric mucosa*. Secretion of gastric lipase can be stimulated by cholinergic agents, gastrin, and possibly other gastrointestinal hormones such as *cholecystokinin*. Gastric lipase has a **broad pH optimum (pH 2–6)** and is **resistant to pepsin** and therefore active within the environment of the stomach. Gastric lipase emptied into the duodenum continues to be active within the environment of the small intestine, although pancreatic lipase is the enzyme responsible for most of the triglyceride hydrolysis (see below). Gastric lipase preferentially hydrolyzes the ester bond in the 3 position of triacylglycerol, generating mainly fatty acids and diglycerides.

Within the stomach, there is acid-peptic digestion of the protein component of food lipoproteins and liberation of oil droplets. The churning action of gastric contractions against the closed pylorus results in the formation of an emulsion containing small lipid particles (less than 2 nm) that can then be emptied from the stomach into the duodenum. These small oil droplets with a large surface-to-volume ratio are the preferred substrate for the **colipase-pancreatic lipase enzyme system** that is responsible for the majority of triglyceride digestion in adults (see below).

3.2 Hormonal Regulation of Biliary and Pancreatic Secretion

As the gastric fluid empties into the duodenum, there is initiation of a coordinated series of events designed to properly deliver biliary and pancreatic secretions into the small intestine to continue the process of triglyceride digestion. Exposure of the duodenal and upper jejunal mucosa to gastric acid stimulates secretion of the peptide hormone **secretin** from neuroendocrine cells in the mucosa into the portal circulation. Secretin, in turn, stimulates *production of a bicarbonate-rich fluid* by the pancreatic ductular cells. Delivery of this alkaline pancreatic fluid into the duodenum partially neutralizes the gastric acid secretion. The pH of the upper small intestine is therefore adjusted to approximately pH 6.5, which is optimal for the activity of the colipase-pancreatic lipase enzyme system. Fatty acids and certain amino acids in the intestinal fluid stimulate the release of a second peptide hormone, **cholecystokinin (CCK)** (also called **pancreozymin**) from the mucosa into the portal circulation. CCK induces *digestive enzyme secretion* by the pancreatic acinar cells, including the secretion of *pancreatic lipase* and *colipase*. In addition, CCK also stimulates *contraction of the gallbladder* and *relaxation of Oddi's sphincter*, resulting in the delivery of concentrated bile into the small intestine. Secretin, CCK, and other hormones produced in the intestinal mucosa also *regulate gastric motility*, preventing excessively rapid delivery of the food bolus into the intestine and overwhelming of the digestive processes. For a detailed discussion of the regulation of gastric emptying, pancreatic, and biliary secretion, see Chaps. 3, 4 and 12.

3.3 Pancreatic Lipase

The enzyme responsible for most of the triglyceride digestion in adult humans is pancreatic lipase. This 449-amino-acid glycoprotein is secreted from the pancreas in its active form. Pure pancreatic lipase has a pH optimum of pH 8–9, but in the

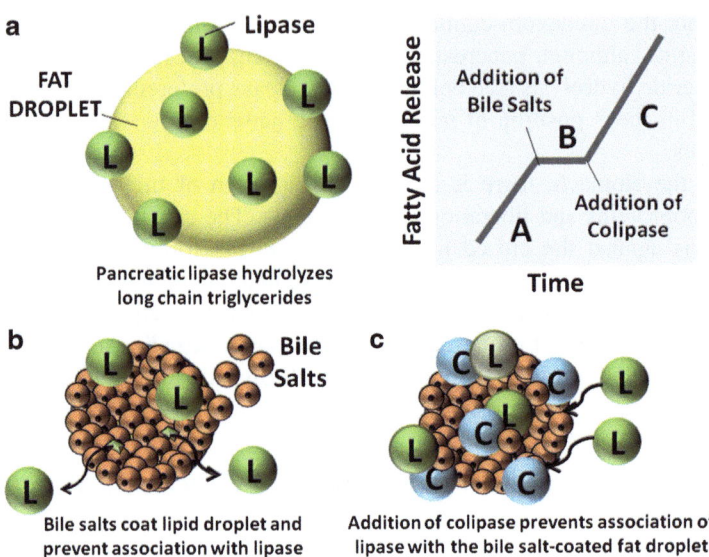

Fig. 7.2 Actions of pancreatic lipase, bile salts and colipase in fat digestion. (**a**) Pancreatic lipases break down triglycerides. (**b**) Bile salts coat lipid droplet and displace lipases. (**c**) Addition of colipases anchor lipases to bile salt-coated fat droplet (Adapted from American Gastroenterological Association Teaching Slide Collection 19 ©, Bethesda, Maryland, slide 29; Used with permission)

presence of bile salts the pH optimum is reduced to pH 6–7, the typical pH of the upper small intestine. At **acid pH** (less than pH 4.0), pancreatic lipase is *irreversibly inactivated*; therefore, neutralization of gastric acid by bicarbonate in pancreatic juice is essential for preserving enzyme activity. Pancreatic lipase is an **interfacial enzyme** that is most active at an oil-water interface and thus efficiently hydrolyzes triglyceride present in the small lipid droplets that are emptied from the stomach into the small intestine. In the small intestine, the oil droplets are coated on their surfaces with bile salts, phospholipids and other compounds that inhibit the binding of pure pancreatic lipase (Fig. 7.2). In order to achieve effective association of pancreatic lipase with the surface of these oil droplets, a protein cofactor known as **colipase** is required which forms a complex with pancreatic lipase and bile components that associate with lipid droplets. Colipase is secreted into pancreatic fluid as a proprotein, and an amino-terminal pentapeptide is cleaved by **trypsin** in the small intestinal fluid to produce the 96-amino-acid form that can associate with the lipid droplet. In addition, fatty acids also appear to facilitate association with pancreatic lipase and the oil-water interface; therefore **partial triglyceride hydrolysis** by gastric lipase may be an important mechanism to enhance the efficiency of the colipase-pancreatic lipase enzyme system. The amounts of both pancreatic lipase and pro-colipase in pancreatic fluid are increased by dietary fat and secretin, and fat feeding increases the mRNA abundances for both pancreatic lipase and pro-colipase in the pancreatic acinar cells.

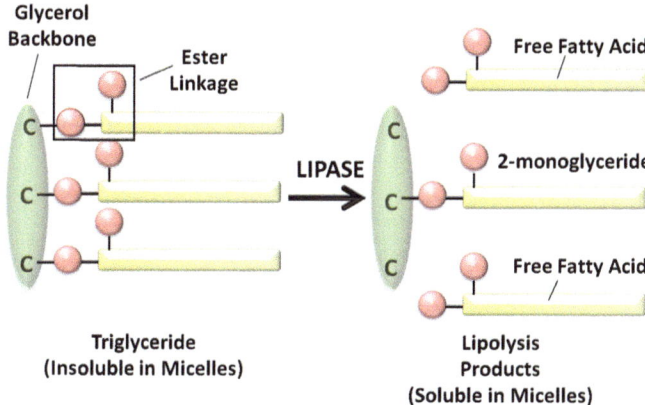

Fig. 7.3 Digestion of triglyceride by pancreatic lipase (Adapted from American Gastroenterological Association Teaching Slide Collection 5 ©, Bethesda, Maryland, slide 10; Used with permission)

Pancreatic lipase preferentially hydrolyzes the first and third ester bonds in triglyceride, forming **fatty acids** and **2- or β-monoglycerides** (Fig. 7.3). The colipase pancreatic lipase system is highly efficient, and triglyceride digestion is nearly complete within the first 100 cm of the proximal jejunum. There is a great excess of pancreatic lipase in pancreatic secretions, and lipase activity must be less than 10 % of normal for triglyceride absorption to be impaired and for steatorrhea to occur.

3.4 Other Lipases

In addition to pancreatic lipase, pancreatic secretions contain another lipase that requires bile salts for activity. In contrast to pancreatic lipase, which has a high degree of substrate specificity for triacylglycerol, **bile salt-activated lipase** will catalyze the hydrolysis of carboxyl ester bonds not only in acylglycerols, but also in other dietary fats such as cholesteryl esters, fat-soluble vitamin esters, and phospholipids. Human bile salt-activated lipase, which has a molecular weight of approximately 100 kd, represents about 4 % of total pancreatic juice protein. Bile salts likely induce conformational changes in the enzyme that enhance access of the active site to bulky lipid substrates and provide additional lipid-binding capability. **Trihydroxylated bile salts**, such as *cholate*, *taurocholate*, or *glycocholate*, are more potent activators than **dihydroxylated bile salts**, such as *taurochenodeoxycholate*; and the 7a-hydroxyl group of bile salts is extremely important for activation. Bile salt-activated lipase catalyzes the complete hydrolysis of triglyceride to fatty acids and glycerol. Patients with *congenital deficiency of pancreatic lipase or colipase*

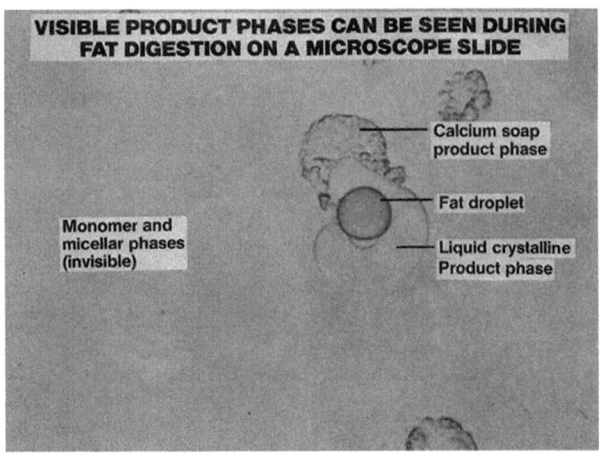

Fig. 7.4 Product phases of fat digestion seen by light microscopy (Adapted from American Gastroenterological Association Teaching Slide Collection 19 ©, Bethesda, Maryland, slide 31; Used with permission)

have been described, and although they have steatorrhea, some fat digestion and absorption continue due to the lipolytic activities of the gastric and bile salt-activated lipases.

Human milk contains a bile salt-activated lipase that is essentially identical to the pancreatic enzyme, differing principally in the glycosylation pattern. The concentration of bile salt-activate lipase in milk is approximately 100 mg/L. Because of the enzyme's bile-salt requirement, the enzyme will not catalyze hydrolysis of lipids in breast tissue or milk until it reaches the duodenum and is exposed to bile.

In adults, pancreatic lipase has 10–60 times the lipolytic activity of bile salt-activated lipase and is responsible for the majority of triglyceride digestion. In the neonate and especially in premature infants, however, pancreatic function is immature, and secretion of pancreatic lipase is inadequate to support efficient triglyceride hydrolysis. In addition, the colipase-pancreatic lipase system is relatively inefficient in digesting milk-fat globules. **Milk bile salt-activated lipase** and **gastric lipase** (which is present in neonates at about the same level as adults), therefore, play important roles in ensuring adequate triglyceride digestion during early development. Other **pancreatic lipase-related proteins** (**PLRP1 and 2**) in pancreatic juice may also play a role in neonatal triglyceride digestion. As noted above, the generation of some fatty acids by the actions of milk bile salt-activated lipase and gastric lipase on milk triglycerides will increase colipase binding to the lipid droplet and to pancreatic lipase, thereby enhancing pancreatic lipase activity.

3.5 Phase Contrast Studies

The process of triglyceride digestion has been studied *in vitro* using **phase contrast microscopy** (Fig. 7.4). Small oil droplets are suspended in buffer containing concentrations of bile salts and phospholipids similar to those in intestinal fluid, and hydrolysis is initiated by addition of pancreatic juice. Initially, one observes the formation of calcium fatty acid soaps on the surface of the lipid droplet, which, if unchecked, would limit access of colipase and lipases to the surface of the droplet and limit further lipolysis. Formation of calcium fatty acid soaps, however, is inhibited by β-monoglycerides. This has been offered as a potential teleologic explanation for the structural specificity of pancreatic lipase, as the partial breakdown of triglyceride to fatty acids and β-monoglyceride would limit production of fatty acid soaps and favor continued digestion of the lipid droplet.

The **second, or viscous isotropic, phase** involves the formation of liquid crystals containing protonated fatty acids and monoglycerides. Continued accumulation of fatty acids and monoglycerides on the surface of the lipid droplet, however, would result in product inhibition of the lipases and would limit lipolysis. The products of triglyceride digestion, however, are removed from the droplet surface by formation of vesicles and mixed micelles (see below).

4 Triglyceride Absorption

4.1 Bile Salts and Micelle Formation

Bile salts are a family of **amphiphilic** compounds that are important for the efficient absorption of the products of triglyceride hydrolysis. At concentrations of approximately 1–5 mmol/L, bile salts form **micelles**, macromolecular aggregates containing up to 20 or more bile salt molecules. These disc or spherical shaped structures have the polar aspect of the bile salts facing the outside of the structure and the hydrophobic portion pointing toward the interior of the micelle. Fatty acids and other lipids are incorporated between the bile salts and in the interior of these structures, forming **mixed micelles** (Fig. 7.5). This results in a remarkable increase

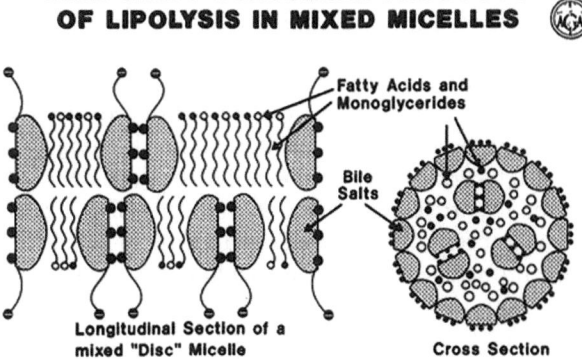

Fig. 7.5 Structure of a mixed disc micelle (Adapted from American Gastroenterological Association Teaching Slide Collection 19 ©, Bethesda, Maryland, slide 35; Used with permission)

Fig. 7.6 Solubility of fat digestion products in bile salt solution (Adapted from American Gastroenterological Association Teaching Slide Collection 19 ©, Bethesda, Maryland, slide 39; Used with permission)

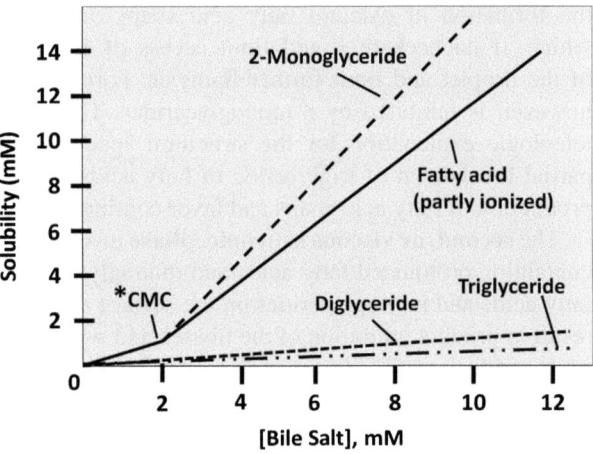

*CMC- Critical Micellar Concentration; Condition At pH 6.5, 37°C

in the solubilization of fatty acids, monoglycerides, and other lipids in the aqueous environment of the luminal contents (Fig. 7.6). It should be noted, however, than even in the complete absence of luminal bile salts, some fatty acid absorption (approximately 50–75 %) still occurs. Some data indicate that unilamellar vesicles formed in the intestinal contents at low bile salt concentrations may be a vehicle for some fat absorption. Adequate luminal bile salt concentration for micelle formation, however, is required for the normal highly efficient (95 % or greater) absorption of dietary triglycerides.

4.2 Uptake of Fatty Acids and β-Monoglycerides into the Enterocyte

Along the apical brush-border membrane surface of the enterocytes is a layer of structured water known as the **unstirred water layer**. Because of hydrophobic nature of fatty acids and β-monoglycerides, the unstirred water layer has been considered to represent the major permeability barrier limiting the rate of uptake of the products of triglyceride hydrolysis into the intestinal cells. The **rate of flux of a lipid** across the unstirred water layer will be determined by the product of a **diffusion rate constant** and the **concentration gradient** (Table 7.1). Because of their large size, the diffusion constant of mixed micelles is somewhat smaller than that of monomeric fatty acids in solution. This, however, is more than offset by the marked increase in the concentration of fatty acids achieved in mixed micelles compared with that possible with monomeric fatty acids in aqueous solution. The flux of fatty acids across the unstirred water layer in mixed micelles is therefore

Table 7.1 Effect of micellar solubilization of fatty acid on diffusive flux through unstirred water layer[a]

From	Fatty acid concentration (mmol/L)	$D \times 10^4$ (cm^2/s)	$J \times 104$ (mmol/L/cm-s)	Relative flux
Molecules	0.01	7	0.07	1
Micelles	10	1	10	142

[a]Fatty acid concentration x diffusion constant (D) = flux (J). All values are approximate

estimated to be more than 100 times greater than that achieved with fatty acid monomers. The fate of the mixed micelles at the apical brush-border membrane surface is still somewhat unclear. The most accepted theory is that the micellar fatty acids are in equilibrium with monomeric fatty acids in the unstirred water layer and that the monomeric fatty acids then cross the brush-border membrane.

The uptake of fatty acids across the enterocyte brush-border membrane occurs by both **passive diffusion** and by **protein-mediated mechanisms**. Protonate long chain fatty acids diffuse across the brush-border membrane by "flip flopping" through the membrane lipid bilayer down a steep concentration gradient. Several proteins have been identified that potentially play roles in fatty acid uptake. **CD36** is a 472 amino acid heavily glycosylated transmembrane protein that appears to be involved in fatty acid uptake in many tissues. In the small intestine, it is mostly expressed on the brush-border membranes of the *duodenum* and *jejunum*, with little expression in the ileum or colon. Enterocytes prepared from the jejunum of CD36 knockout mice demonstrate reduced fatty acid uptake compared with wild-type controls. CD36 knockout mice, however, do not have significant fat malabsorption, although fat absorption is shifted to the more distal small intestine. Current observations indicate that CD36 mainly plays a regulatory role in **intestinal triglyceride synthesis** and **chylomicron production** (see below). CD36 is also expressed in *lingual taste buds* where is mediates perception of fatty acids, and may be involved in satiety signaling. **FATP4** is high expressed in the *small intestine*, but is localized to the endoplasmic reticulum and subapical membranes. It, therefore, does not function as a true membrane transporter, but facilitates fatty acid uptake by virtue of its **acyl CoA synthase** activity. This results in metabolic trapping of fatty acids via conversion to fatty acyl CoA products that undergo rapid conversion to complex lipids, maintaining a concentration gradient for fatty acid uptake. Other candidate proteins involved in enterocyte fatty acid uptake (FABPpm, etc.) have been identified, but their role has not been completely elucidated. Further research is needed to define the relative contributions of diffusion and protein-mediated transport to enterocyte fatty acid uptake.

Studies of the intestinal uptake of β-**monoglyceride**, the other major lipolytic product, indicate that uptake is saturable and reduced by trypsin digestion of the cell membrane, implying a role for a membrane protein in β-monoglyceride transport. β-monoglyceride has been found to inhibit fatty acid uptake, suggesting coordinated transport of these lipids.

4.3 Intracellular Triglyceride Transport

Cytosolic Fatty Acid-Binding Proteins

Within the enterocyte, the lipolytic products, **fatty acids** and **β-monoglycerides**, are directed to the smooth endoplasmic reticulum for resynthesis of triglyceride and other complex lipids, such as phospholipids and cholesteryl esters. These processes are thought to involve the association of the absorbed fatty acids with low molecular weight cytosolic **fatty acid binding proteins** (**FABPs**). The small intestine contains at least **two cytosolic FABPs**, *I-FABP* and *L-FABP*, named after the organs (intestine and liver, respectively) in which they were first identified. These proteins are expressed at high levels (1–2 % of cytosolic protein) and their respective mRNAs represent about 3 % of total small-intestinal mRNA. The amino acid structures of these proteins have been deduced from cloned DNA sequences, and they belong to a large gene family of cytosolic lipid-binding proteins. The mRNA abundances for both I-FABP and L-FABP increase around birth in the rat and achieve maximal expression in the adult. Both I-FABP and L-FABP are expressed only in villus cells, and expression is highest in the jejunum and declines distally.

The roles that these cytosolic fatty acid binding proteins play in intracellular lipid transport are uncertain, but proposed **functions** include:

1. Removal of fatty acids from the brush-border membrane and maintenance of the concentration gradient for fatty acid influx;
2. Protection of cellular organelles from toxic effects of free fatty acids;
3. Direction of fatty acids to the smooth endoplasmic reticulum for triglyceride synthesis and
4. Targeting fatty acids toward oxidation.

Mouse models with knockout of either fatty acid binding protein have normal fat absorption, but L-FABP knockouts have reduced chylomicron secretion and I-FABP knockout mice show weight loss compared to wild-type animals.

Triglyceride Synthesis in Enterocytes

Fatty acids in the enterocyte are converted to their acyl-CoA derivatives that are used for complex lipid synthesis by enzymes that are specific for long-chain or very long-chain fatty acids. There is evidence that the synthetase ACLS5 delivers acyl-fatty acids for triglyceride synthesis whereas the enzyme ACLS3 channels acyl-fatty acids for phospholipid production. Within the smooth endoplasmic reticulum there is the re-synthesis of triglyceride on the smooth endoplasmic reticulum via **two metabolic pathways** (Fig. 7.7). The *major pathway* that accounts for approximately 75 % of triglyceride secreted from the intestine involves the generation of diglycerides from fatty acyl-CoA and β-monoglyceride by **monoglyceride acyltransferase** (**MGAT**) **enzymes**. A *minor pathway* for diglyceride production

Fig. 7.7 Pathways of triglyceride synthesis in enterocytes

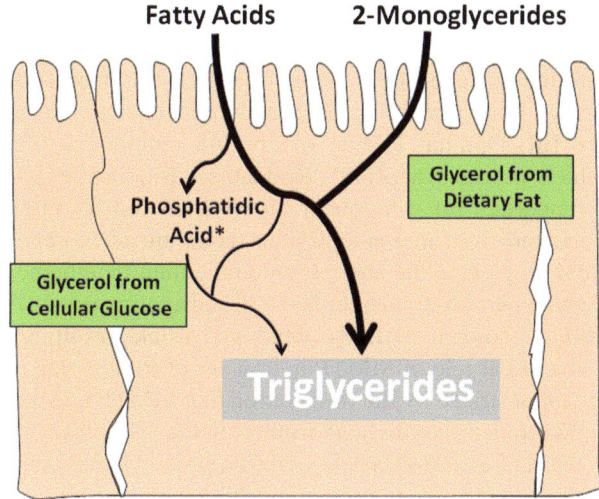

utilizes phosphatidic acid that is produced from α-glycerol phosphate derived from glycosis or absorbed glycerol. The phosphatidic acid is then dephosphorylated to produce diacylglycerol.

Fatty acyl-CoA is then transferred to diglyceride to form triglycerides by **diacylglycerol acyltransferases** (**DGATs**). The intestine expresses **two DGATs**, *DGAT1* and *DGAT2*. DGAT1 is thought to be responsible for most of the triglyceride secreted from the intestine, and DGAT1 knockout mice have reduced chylomicron triglyceride secretion. The two DGATs have different subcellular localization, with the active site of DGAT2 facing the cytosolic side of the smooth endoplasmic reticulum, whereas the active site of DGAT1 faces the lumen of the endoplasmic reticulum. Some MGATs also have DGAT activity.

4.4 Assembly of Chylomicrons

The small intestine is the site of synthesis of several classes of **lipoproteins**, which are macromolecular aggregates of lipids (triglyceride, cholesterol, cholesteryl ester, phospholipid, etc.) and proteins known as **apolipoproteins**. These lipoproteins are characterized on the basis of their flotation density, charge, and size. The *core* of these lipoproteins is composed of triglyceride, cholesteryl esters, and other lipids, whereas the apolipoproteins, phospholipid, and free cholesterol are arrayed on the *surface*. **Chylomicrons**, the largest lipoprotein class, are responsible for delivery of most of the absorbed triglyceride into the circulation. Two proteins are keys to the assembly of chylomicrons in the enterocyte, **apoB48** and the **microsomal triglyceride transfer protein** (**MTTP**). ApoB48 is the structural protein on the surface of chylomicrons. MTTP facilitates chylomicron production by shuttling triglycerides from within the endoplasmic reticulum to associate with apoB48.

ApoB48 is synthesized in the endoplasmic reticulum and is translocated to the endoplasmic reticulum lumen while still attached to the polysome. The amino terminus of apoB48 contains of region for interaction with MMTP, resulting in partial lipidation of apoB48 and organization into a primordial particle within the ER. Optimal folding of apoB48 requires physical interaction with MMTP which prevents apoB48 degradation. Binding of phospholipid to apoB48 also is important for subsequent for chylomicron secretion. Additional triglyceride is extracted from the endoplasmic reticulum membrane by MMTP, and brought to lipid droplets in the endoplasmic reticulum membrane lumen. These lipid droplets acquire phospholipids, cholesterol, and apolipoprotein AIV, and subsequently merge with the nascent apoB48-containing particle, resulting in expansion of the particle. The critical role of MMTP in transfer of triglyceride to the endoplasmic reticulum is demonstrated by the observation that MMTP knockout mice have large cytosolic lipid droplets, but no lipid droplets in the endoplasmic reticulum.

MMTP is a 97 kDa heterodimeric protein that is complexed with the endoplasmic reticulum chaperone protein disulfide isomerase. In addition to its primary role of transferring triglycerides to apoB48, MTTP also shuttles other lipid classes such as cholesteryl esters, free cholesterol, and phospholipids for lipoprotein formation. Loss of function mutations of MMTP results in the disease **abetalipoproteinemia**, an *autosomal recessive* disorder in which the heterozygous parents have normal plasma apoB levels. Affected children have very low plasma apo B, mild fat malabsorption, accumulation of triglyceride within enterocytes and hepatocytes, and neurologic and hematologic abnormalities secondary to vitamin E malabsorption and deficiency.

4.5 Intestinal Apolipoproteins

The term **apoB** refers to two large, hydrophobic proteins that are components of triglyceride-rich lipoproteins and their metabolic products. These two proteins, products of a single gene located on chromosome 2, are denoted on a centile scale as **apo B100** and **apo B48**. Apo B100 is produced in the liver and is the form present on **very low-density lipoproteins (VLDL)** and their metabolic products **intermediate-density lipoproteins (IDL)** and low-density lipoproteins (LDL). Apo B100 plays an important role in lipoprotein metabolism, as it is a ligand for the LDL receptor. Apo B48, which corresponds to the amino-terminal 48 % of apo B100, is produced in the intestine and is the major lipoprotein of chylomicrons.

The regulation of the tissue-specific production of **apo B48** in the *intestine* and **apo B100** in the *liver* has been extensively studied. In brief, the genomically-encoded mRNA contains as CAA codon encoding **glutamine**, which is modified in the intestinal transcript to UAA, an in-frame **stop codon**. This results in the termination of protein synthesis and the production of the smaller molecular form apo B48. This overall process is now referred to as **apoB-mRNA editing**. Apo B-mRNA editing in the intestine is mediated by a complex that contains

apobec-1, the catalytic deaminase, and an obligate RNA binding subunit, **apobec-1 complementation factor**. Aobec-1 knockout mice that have apoB100 in their chylomicrons have less efficient triglyceride absorption with fewer, larger chylomicron particles. In addition, there is evidence that apoB48 may protect the intestine from lipotoxicity when fat intake is high or MTTP content is reduced. ApoB-mRNA editing has several other important functional consequences. The region of apoB100 responsible for interaction with the LDL receptor is in the carboxyl-terminal portion that is not present in apoB48. Apo B48–containing chylomicron remnants are, therefore, not taken up via the LDL receptor, but instead are believed to undergo **receptor-mediated internalization** predominantly into hepatocytes via another receptor recognizing apolipoprotein, **apo E**. This results in significantly different plasma half-lives of LDL (2–3 days) versus chylomicrons (30 min). The carboxyl-terminal portion of apo B100 also contains the attachment site for another apolipoprotein known as **apo (a)**. This protein is a member of the **plasminogen multigene family**, and attachment of apo (a) to apo B100 produces a lipoprotein referred to as **Lp(a)**, which is an important, genetically determined risk factor for atherosclerotic disease.

The intestine is the site of other apolipoproteins that play crucial roles in the assembly and secretion of lipoproteins by enterocytes and are the major regulators of the metabolism and uptake of circulating lipoproteins by peripheral tissues. **Apo AI** is one of the most abundantly expressed apolipoproteins, as its mRNA represents 1–2 % of total intestinal mRNA. Apo AI is a 243-amino-acid protein with a molecular weight of approximately 28 kd. In addition to being an important protein component of **chylomicrons**, apo AI is the major protein of plasma **high density lipoproteins** (HDL). Apo AI is a cofactor for the enzyme **lecithin-cholesterol acyltransferase** (**LCAT**), which catalyzes the esterification of plasma cholesterol. It also has an important function in the removal of free cholesterol from cells for incorporation and esterification in HDL and transport to the liver for catabolism or secretion (**reverse cholesterol transport**). The abundance of intestinal apo AI mRNA increases at the time of birth, presumably in connection with suckling, and in adult animals there is a gradient of apo AI expression from the proximal to the distal intestine. The amount of protein, however, does not appear to be modulated by dietary fat intake. With triglyceride feeding, there is a shift of apo AI in the enterocyte to the lipoprotein-associated fraction and an increase in apo AI in intestinal lymph.

Apo AIV is a protein of 377 amino acids with a molecular weight of about 46 kd. It is abundantly expressed in the intestine, constituting about 3 % of protein synthesis. Apo AIV synthesis is highly-inducible by triglyceride feeding, and there is a gradient of expression within rat intestine, with the highest level in *proximal villus cells*. Apo AIV mRNA increases at birth in response to the initiation of suckling and fat intake, followed by a decline during the suckling period. Apo AIV increases MMTP expression, enhances intestinal triglyceride output by increasing the size of chylomicrons, and increases apo CIII secretion. In plasma, about 25 % of apo AIV leaves chylomicrons and is found in HDL or in the lipoprotein-free fraction. Apo AIV can activate LCAT, but is less active than apo AI, and may also participate in the transfer of **apo CII**, an **activator of lipoprotein lipase**, from HDL to triglyceride-rich lipoproteins. Apo AIV also appears to function as an **acute satiety factor**.

Apo CIII functions as an **inhibitor of lipolysis** of circulating triglyceride-rich lipoproteins, chylomicrons, and VLDL. Apo C-III displaces apo CII, an activator of lipoprotein lipase, from the surface of these particles. Apo C-II is expressed in the intestine, and synthesis of this protein is stimulated by triglyceride feeding. Evidence concerning the production of other apolipoproteins, apo AII, apo CI, and apo CII, in the intestine is somewhat conflicting, but if they are expressed, it is at very low levels. **Apo D** is found mainly in **HDL** and functions in the exchange of cholesteryl esters between lipoprotein species. The mRNA for apo D has been detected in *small intestine*.

4.6 Secretion of Chylomicrons

The nascent chylomicron particles leave the endoplasmic reticulum and are transported to the Golgi for secretion into the intestinal lymphatics. Several proteins appear to be involved in the movement of chylomicrons from endoplasmic reticulum to the Golgi, including several *COPII proteins*, *L-FABP*, *CD36*, and *VAMP7*. **CD36 signaling** via *Src kinases* and *extracellular regulated kinases* may be important for phosphorylating proteins required for endoplasmic reticulum processing of prechylomicron vesicles. CD36 signaling via a rise in intracellular calcium may also influence multiple events in lipid procession and secretion. **Mutations of the SARA2 gene** that codes SAR1B, one of the COPII proteins, is associated with *Anderson/chylomicron-retention disease* in which enterocytes assemble chylomicrons in the endoplasmic reticulum, but fail to transport them through the secretory pathway, causing intestinal lipid droplet accumulation. In the Golgi, apolipoprotein AI and other lipids are added to the particle and there is glycosylation of apolipoproteins. Golgi organelles containing chylomicrons fuse into secretory vesicles, which then move toward the basolateral membrane surface of the enterocytes by a process involving microtubules. These vesicles fuse with the basolateral membrane and release their contents to the extracellular space. The chylomicrons enter the intestinal lacteals and subsequently are transported through the intestinal lymphatics and into the circulation for further metabolism.

As they circulate through the mesenteric lymph and subsequently in the plasma, chylomicrons are modified by exchanging both surface and core components with other lipoprotein classes. Apolipoproteins such as *apo CII*, an activator of lipoprotein lipase, and *apoE*, a regulator of cellular lipoprotein uptake, are acquired by the chylomicrons. The enzyme lipoprotein lipase, found on capillary endothelial cells in muscle, adipose tissue, and other tissues, hydrolyzes triglyceride within the core of the chylomicron, forming smaller particles known as **chylomicron remnants**. These remnants undergo receptor-mediated uptake, thought to be mediated via apo E, mainly into **hepatocytes**, completing chylomicron metabolism.

4.7 Intestinal Production of Other Lipoproteins

The small intestine synthesizes and secretes lipoproteins corresponding in size and density to VLDL. The protein and lipid content of these particles is similar to that of chylomicrons, and it is possible that only the amounts of triglycerides and other complex lipids being produced and exported determine particle size. Some investigators, however, have presented data suggesting that intestinal chylomicrons and VLDL have distinct pathways of intracellular assembly.

The small intestine contains and secretes small amounts of LDL-sized lipoproteins. Since these particles have a protein composition similar to intestinal chylomicrons and VLDL and less triglyceride than phospholipid, they are thought to represent **hypolipidated VLDL**. Because the small intestine does not produce apo B100, the characteristic apolipoprotein of LDL, the gut is not felt to significantly contribute to the circulating plasma LDL pool.

HDL particles have been visualized as 6–13-nm spherical lipoproteins within Golgi vesicles of the rat small intestine. **Two populations** were noted: one containing *apo AI and apo AIV* as surface components, and the other containing *apo B48*. These particles were composed mainly of protein and phospholipid, with little triglyceride, and it has been speculated that they are the most under-lipidated precursors to intestinal chylomicrons. **Mesenteric lymph** has been demonstrated to contain both **spherical** and **discoidal HDL** that appears to arise by *de novo* synthesis in the intestine. In addition, HDL particles can be produced following lipolysis of chylomicrons or VLDL in the bloodstream. Figure 7.8 is a summary of assembly and secretion of chylomicrons.

5 Digestion and Absorption of Medium-Chain Triglycerides

Medium-chain triglycerides (**MCTs**), containing fatty acids of 6–12 carbon chain lengths, are present in only small amounts in the normal diet, but they are important components of nutritional supplements used in patients with GI disorders. They are *useful therapeutic agents* because the digestion and absorption of MCTs differs significantly from that of typical long-chain dietary triglycerides.

MCTs are hydrolyzed by various lipases, including gastric lipase, bile salt-dependent lipases, and pancreatic lipase, more rapidly than are long-chain triglycerides. The medium-chain fatty acids are much more water-soluble than are long-chain fatty acids and are efficiently taken up into the enterocytes even in the absence of intraluminal bile salts. Some MCTs can even be absorbed intact by the small intestine without requiring prior digestion. Within the enterocytes, medium-chain fatty acids are not utilized for the resynthesis of triglycerides and are therefore not packaged into chylomicrons. Instead, the medium-chain fatty acids are directly released from the enterocytes into the portal circulation, where they are rapidly taken

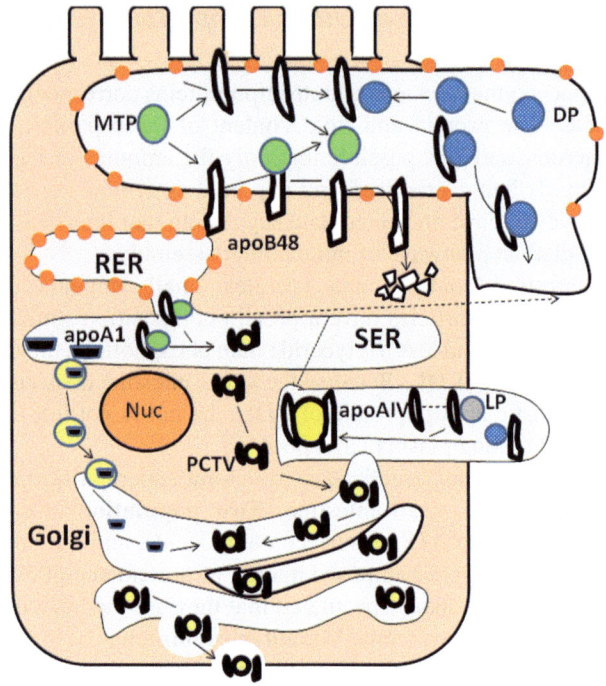

Fig. 7.8 Assembly and secretion of chylomicrons. Apo B48 is synthesized in the rough endoplasmic reticulum (RER). The microsomal triglyceride transfer protein (MTP) stabilizes apo B48, and promotes association with some lipid (DP). Triglyceride synthesized on the smooth endoplasmic reticulum (SER) membrane is brought by MTP to the SER lumen. The triglyceride droplet and apo AIV merge to form a light particle (LP), which fuses with the apo B48-containing particle to form a lipoprotein with a core of neutral lipid surrounded by a surface of apoB48, apoAIV, and phospholipid. This particle buds from the SER, and the prechylomicron transport vesicle (PCTV) fuses with the Golgi. Apo A1 associates to form a mature chylomicron that exits the Golgi and fuses with the basolateral membrane for secretion (Adapted from Mansbach and Gorelick [5])

up by the liver and other tissues and used as an energy source. The differences in digestive and absorptive mechanisms between MCTs and long-chain triglycerides are summarized in Fig. 7.9.

MCTs can therefore be employed as a ***well-absorbed source of calories*** in patients with a wide variety of GI diseases resulting in malabsorption. These disorders include **pancreatic insufficiency; intraluminal bile salt deficiency** due to cholestatic liver disease, biliary obstruction, or ileal disease or resection; **mucosal diseases** with impaired intracellular lipid metabolism and chylomicron assembly; and disorders causing **obstruction of intestinal lymphatics**. It must be remembered, however, that MCT preparations do not contain essential polyunsaturated fatty acids, and therefore some long-chain dietary triglycerides are required in patients with these malabsorptive disorders.

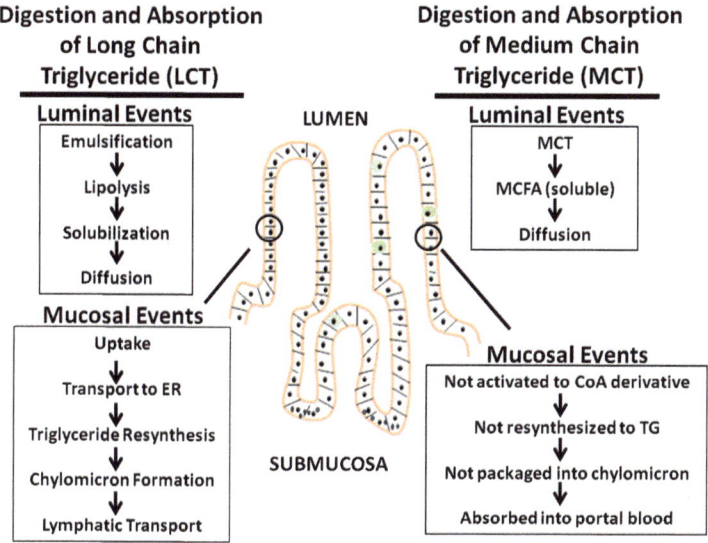

Fig. 7.9 Comparison of digestion and absorption of long-chain and medium-chain triglycerides

Clinical Correlations

Case Study 1

A 10-year-old girl with a history of recurrent pneumonias, chronic diarrhea, and poor growth is found to have cystic fibrosis. On a 100-g-fat diet, she is noted to excrete 40 g of fat in the stool. Analysis of duodenal fluid obtained after stimulation of the pancreas by injection of secretin and cholecystokinin shows undetectable pancreatic lipase. She is started on therapy with bovine pancreatic enzyme extracts that have been pressed into tablets given with meals and snacks; however, she continues to have marked steatorrhea.

Questions

1. **Why was this form of pancreatic enzyme supplementation of little therapeutic benefit?**
   Answer: At an **acid pH** (less than pH 4.0), **pancreatic lipase** is *irreversibly inactivated*. Therefore, when enzyme supplements are taken by mouth, almost all of the lipase is inactivated by gastric acid. Because of the lack of pancreatic bicarbonate secretion in this patient, the pH of the upper intestinal contents will be low, and pancreatic lipase present in the intestinal lumen will not be functioning at its pH optimum (pH 6–7).
2. **How would you alter the enzyme therapy to achieve better results?**
   Answer: The pancreatic enzyme supplements can be combined with measures to **decrease gastric acid**, such as administration of *sodium bicarbonate* with

meals, or treatment with **antisecretory drugs** such as *H_2 blockers* or *proton-pump inhibitors*. Pancreatic enzyme preparations have been prepared in which the enzymes are contained in **pH-sensitive microspheres** that do not dissolve at acid pH but release the enzymes at neutral pH. These measures will permit adequate delivery of active lipase into the small bowel and greatly improve triglyceride absorption. **Preparations with acid-resident microbial lipases** are being developed for clinical use.

3. **Why is the patient capable of absorbing some dietary triglyceride if her pancreas is secreting minimal pancreatic lipase?**

   Answer: **Lingual and gastric lipase activity** will be preserved in the patient with cystic fibrosis and accounts for much of the dietary triglyceride digestion. Pancreatic bile salt-activated lipase secretion will likely also be greatly diminished in this patient.

Case Study 2

A 65-year-old man presents with right upper quadrant abdominal pain, diarrhea, weight loss, and jaundice. He is found to have a common bile duct cancer and complete biliary obstruction. On a 100-g-fat diet, he is noted to have 32 g of steatorrhea.

Questions

1. **What is responsible for this patient's fat malabsorption?**

   Answer: **Biliary obstruction** in this patient will result in *lack of bile salts* and *phospholipids* in the intestinal lumen. Bile salts are required for the solubilization of large amounts of the lipolytic products, fatty acids and β-monoglycerides, in mixed micelles. Micellar solubilization results in a large concentration gradient for fatty acid and β-monoglyceride flux across the unstirred water layer. Bile salts are also needed for the **activation of pancreatic bile salt-activated lipase** that participates in digestion of triglyceride and other esterified lipids. Biliary phospholipid absorption is required to supply the phospholipid component of chylomicrons.

2. **A nutrition consultant recommends that the patient be put on a supplement containing medium-chain triglycerides to encourage weight gain. What is the rationale behind this therapy?**

   Answer: Medium-chain fatty acids derived from the lipolysis of medium-chain triglycerides are quite **water-soluble**, even in the absence of bile salts. Medium-chain triglyceride supplements would therefore be **well-absorbed** in this patient with biliary obstruction and intraluminal bile salt deficiency.

3. **Would it be advisable for this patient to be on a dietary program containing only medium-chain triglycerides and no long-chain triglycerides?**

   Answer: This patient continues to require some long-chain dietary triglyceride, at least 2–5 % of caloric intake. **Medium-chain triglyceride preparations** will *not* provide essential polyunsaturated fatty acids that are the precursors for prostaglandins, leukotrienes, and other important lipid-derived regulators of cellular function. Essential fatty acid deficiency can develop within a few weeks of dietary deprivation.

Case Study 3

A 2-year-old boy presents with diarrhea, poor growth, and multiple neurologic deficits. He is found to have steatorrhea, very low plasma levels of apo B (both apo B100 and apo B48), cholesterol, triglycerides, and vitamin E. His parents have reduced plasma apoB levels, and a diagnosis of homozygous hypobetalipoproteinemia is made.

Questions

1. **Why does this child have steatorrhea?**
   Answer: Children with homozygous hypobetalipoproteinemia have a **mutation in the apo B gene**, resulting in **defective apo B48 synthesis** in the *small intestine* and **apo B100** in the *liver*. Apo B48 is required for the proper assembly and secretion of chylomicrons. Intestinal biopsies of these children show enterocytes that are packed with triglyceride. Dietary triglyceride digestion, fatty acid absorption, and resynthesis of triglyceride in the intestine proceed normally; however, the triglyceride cannot be properly packaged into chylomicrons for secretion. The triglyceride is lost into the stool as the enterocytes are sloughed as part of normal mucosal turnover.

2. **Why does this child have low plasma levels of both apo B-100 and apo B-48?**
   Answer: Apo B100 and apo B-48 are both **products of a single gene**. Apo B48 is produced as a result of the process of apo B-mRNA editing, where a CAA codon is modified to UAA, an in-frame stop codon that results in the termination of protein synthesis. Children with homozygous hypobetalipoproteinemia have a **mutation in the apo B gene** that causes *reduced synthesis of both forms of apo B*.

3. **Would medium-chain triglyceride oil be a useful nutritional supplement for this patient? Why?**
   Answer: Yes. Absorbed medium-chain fatty acids are not utilized for the resynthesis of triglyceride in the enterocytes and are therefore not packaged into chylomicron particles. Medium-chain fatty acids are **released from the intestine into the portal circulation**. This child with a defect in chylomicron assembly and secretion would digest and absorb medium-chain triglycerides well, which could **provide an important source of energy**.

Case Study 4

A 40-year-old woman presents to the doctor with chronic diarrhea, a 10 lb weight loss, and a rash on her trunk and arms. Biopsy of the skin rash is diagnostic of dermatitis herpetiformis, and biopsy of the small intestine shows partial loss of the villous architecture and inflammation in the mucosa. Analysis of her stool shows increased fecal excretion of both fatty acids and triglyceride.

Questions

1. **What is the diagnosis of her intestinal disease?**
   Answer: The patient has gluten-sensitive enteropathy (**celiac disease**), a condition in which ingestion of certain grains, such as wheat, rye, and barley,

incites inflammation of the intestine in a patient with genetic susceptibility. The inflammation damages the small intestinal mucosa, and often results in malabsorption of fat and other nutrients. All patients with the skin disease ***dermatitis herpetiformis*** have gluten-sensitive enteropathy, although it is often milder than the damage seen in patients with gluten-sensitive enteropathy who do not have associated dermatitis herpetiformis.

2. **Why does this patient have increased fecal excretion of both fatty acids and triglyceride?**

   Answer: The damaged proximal small intestine will have reduced release of the hormones secretin and CCK in response to a meal, resulting in **diminished secretion of pancreatic fluid and enzymes**. This causes "functional" pancreatic insufficiency, even though the pancreas itself is normal. **Intestinal fatty acid absorption is impaired** because of the damage to the enterocytes, resulting in a reduced small intestinal surface area and lack of the transporters, binding proteins, and enzymes involved in fat absorption. **Bacterial metabolism of malabsorbed triglyceride** in the distal bowel also contributes to the increased fatty acid excretion.

3. **What is the treatment for this patient? Does she need a fat-restricted diet?**

   Answer: This patient should be placed on a **gluten-free diet**, eliminating foods containing wheat, rye, and barley. For most individuals, this results in resolution of the intestinal inflammation and return of the intestinal structure and function to normal. A minority of patients with gluten-sensitive enteropathy will require treatment with **immunosuppressive medication** to heal the mucosal injury. **Dietary fat restriction** is only needed in a small minority of patients to decrease steatorrhea and diarrhea while the intestine heals. Dietary fat restriction is not required in patients when the intestine has healed in response to a gluten-free diet.

Further Reading

1. Abumrad NA, Davidson NO (2012) Role of the gut in lipid homeostasis. Physiol Rev 92:1061–1085
2. Iqbal J, Hussain MM (2009) Intestinal lipid absorption. Am J Physiol Endocrinol Metab 296:E1183–E1194
3. Kindel T, Lee DM, Tso P (2010) The mechanism of the formation and secretion of chylomicrons. Atheroscler Suppl 11:11–16
4. Lowe ME (2002) The triglyceride lipases of the pancreas. J Lipid Res 43:2007–2016
5. Mansbach CM, Gorelick FC (2007) Development and physiological regulation of intestinal lipid absorption. II. Dietary lipid absorption, complex lipid synthesis, and intracellular packaging and secretion of chylomicrons. Am J Physiol Gastrointest Liver Physiol 293:G645–G650
6. Mu H, Hoy C-E (2004) The digestion of dietary triglycerides. Prog Lipid Res 43:105–133
7. Pan X, Hussain MM (2012) Gut triglyceride production. Biochim Biophys Acta 1821:727–735

Chapter 8
Digestion and Absorption of Other Dietary Lipids

Michael D. Sitrin

1 Introduction

The small intestine absorbs a variety of important lipids present in the diet or secreted in the bile using pathways similar to those for dietary triglycerides (Chap. 7). In this chapter, the mechanisms for absorption of cholesterol, phospholipid, bile acids, and the four fat-soluble vitamins will be reviewed. Absorption of these lipids will be considered with respect to intraluminal events, mechanisms for uptake across the brush-border membrane, metabolism within the enterocyte, and intracellular transport and secretion; the emphasis will be on the unique features of each lipid compound. In addition, the hepatic uptake and metabolism of the fat-soluble vitamins will be discussed.

2 Cholesterol

2.1 Sources of Intraluminal Cholesterol

The cholesterol present in the intestinal lumen comes from both the **diet** and from **endogenous sources**. The typical North American diet contains 300–500 mg/day of cholesterol. A variable portion (typically about 15 %) of dietary cholesterol is esterified with various fatty acids. Bile provides approximately 800–1,200 mg/day of unesterified cholesterol, and the turnover of the intestinal mucosa adds 250–400 mg/day of cholesterol to the luminal contents.

M.D. Sitrin, M.D. (✉)
Department of Medicine, University at Buffalo, The State University of New York, Buffalo, NY, USA
e-mail: mdsitrin@buffalo.edu

P.S. Leung (ed.), *The Gastrointestinal System: Gastrointestinal, Nutritional and Hepatobiliary Physiology*, DOI 10.1007/978-94-017-8771-0_8,

The dietary and biliary cholesterol do not appear to form a single pool in the intestinal lumen. Current data indicate that biliary cholesterol is more rapidly and perhaps more completely absorbed than is dietary cholesterol. Furthermore, while the jejunum absorbs more total cholesterol than the ileum, a greater fraction of dietary cholesterol may be absorbed in the distal small bowel than occurs with biliary cholesterol. Biliary cholesterol is secreted in micelles with bile acids and phospholipid. In contrast, cholesteryl esters require hydrolysis, and dietary cholesterol needs solubilization prior to intestinal absorption. Overall, approximately 30–60 % of the luminal cholesterol is absorbed.

2.2 Digestion of Cholesteryl Esters

Cholesteryl esters must be hydrolyzed to **free cholesterol** and **fatty acids** prior to intestinal absorption. **Pancreatic cholesterol esterase** (also known as **bile salt-activated lipase**) is the enzyme responsible for most of the cholesteryl ester hydrolysis. As previously discussed (Chap. 7), this enzyme requires bile salts (particularly glycocholate or taurocholate) for its activity; the bile salt appears to serve as a cofactor allowing for polymerization of the enzyme monomer into proteolysis-resistant polymers or for an activating conformational change.

2.3 Solubilization of Cholesterol

Cholesterol is nearly insoluble in an aqueous system, and bile salts are therefore absolutely required for cholesterol absorption. Cholesterol is only slightly soluble in a pure bile salt solution, but the **addition of other lipids**, such as phospholipid, monoglycerides, and fatty acids, **markedly increases cholesterol solubility in mixed micelles**. In the intestinal lumen, cholesterol partitions between the micellar phase and the lipid droplets.

Cholesterol in mixed micelles is in rapid equilibrium with cholesterol in monomolecular solution. The high cholesterol concentration in the micelles therefore ensures a maximal concentration of monomolecular cholesterol in the unstirred water layer lining the luminal enterocyte surface.

2.4 Brush-Border Membrane Cholesterol Transport

Traditionally cholesterol transport across the brush-border membrane was thought to occur by passive diffusion. It is now clear that several brush-border membrane proteins play key roles in regulating cholesterol movement across the apical cell membrane (Fig. 8.1). The protein **Niemann-Pick C1-Like 1 (NPC1L1)** is

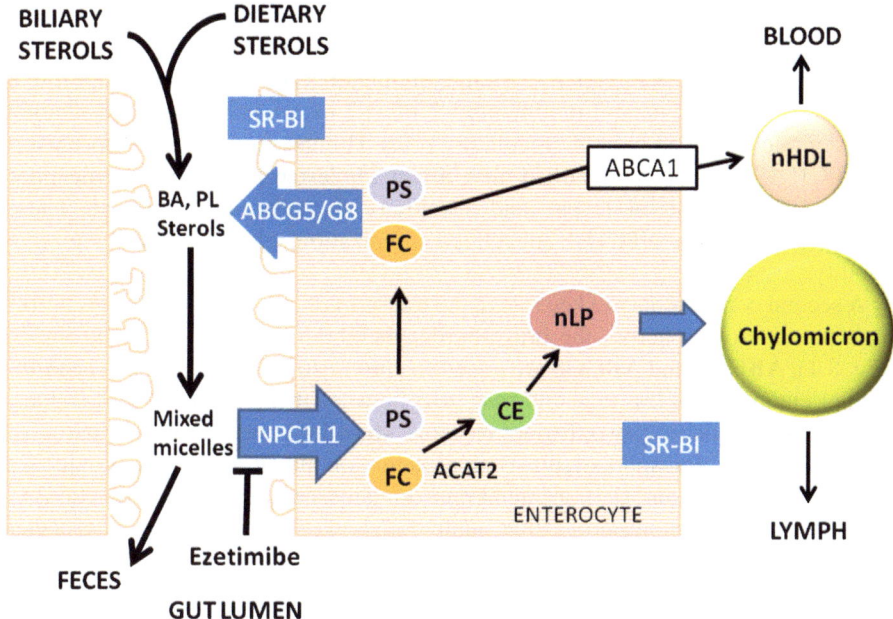

Fig. 8.1 Intestinal sterol absorption and secretion. Sterols including free cholesterol (*FC*) and free plant sterols (*PS*) from diet and bile are mixed with phospholipids (*PL*) and bile acids (*BA*) to form micelles. FC and PS solubilized in mixed micelles are transported into absorptive enterocytes via an NPC1L1-dependent and ezetimibe-inhibitable mechanism. FC is delivered to the ER for esterification by acyl-CoA: cholesterol acyltransferase-2 (*ACAT2*) to form cholesterol esters (*CE*) that is then packaged into nascent lipoprotein particles (*nLP*) and secreted as a constituent of chylomicrons. PS and FC that escapes ACAT2 esterification may be directly transported to nascent HDL (*nHDL*) through basolateral ABCA1, or back to the gut lumen via ABCG5/G8 (Adapted from Brown and Yu [2])

essential for intestinal cholesterol absorption. The gene for NPC1L1 is located on chromosome 7 and produces two alternatively spliced transcripts coding for proteins of 1332 and 724 amino acids; the significance of these two splice variants has not yet been defined. The protein has 13 transmembrane domains and sites for extensive N-glycosylation. NPC1L1 also contains a domain of about 180 amino acids that is a consensus sterol-sensing domain. Mice in which NPC1L1 has been knocked out demonstrate markedly reduced (>70 %) cholesterol absorption. NPC1L1 is the target of the **cholesterol absorption inhibitor ezetimibe** which is widely used in the treatment of patients with **hypercholesterolemia**. The mechanism by which NPC1L1 mediates brush-border cholesterol uptake is incompletely understood, but current evidence suggests a clathrin-mediated endocytic pathway. NPC1L1 is found in both the **apical plasma membrane** and in **intracellular compartments**, and cycling between these compartments is likely regulated by cholesterol. Cellular cholesterol depletion causes translocation of NPC1L1 to the brush-border membrane, and cholesterol repletion results in movement of NPC1L1 and cholesterol to the cell interior. Ezetimibe appears to inhibit sterol-induced internalization of

NPC1L1 via this endocytic pathway. NPC1L1 can facilitate the transport of various plant steroids, but with lower efficiency than cholesterol transport.

The regulation of NPC1L1 is incompletely understood. Some data suggest that cellular cholesterol availability influences NPC1L1 expression, and that activation of many nuclear receptors play regulatory roles. Human sequence polymorphisms in NPC1L1 have been identified that affect cholesterol absorption efficiency, plasma LDL-cholesterol levels, and sensitivity to ezetimibe treatment. These polymorphisms alter NPC1L1 subcellular localization, glycosylation, or protein stability.

A second complex plays a key role in the transport of cholesterol and plant sterols from the enterocyte across the brush-border membrane into the intestinal lumen (Fig. 8.1). **ABCG5** and **ABCG8** are genes on chromosome 2 arranged in a head-to-head orientation with less than 400 base pairs between their respective start codons. The two genes code for two distinct proteins, **sterolin-1** and **sterolin-2**, that must heterodimerize to transport sterols. Both proteins contain an ATP-binding sequence near the N-terminus and six transmembrane domains, and the complex functions as an ATP-dependent cholesterol efflux pump delivering sterols from the enterocytes into the intestinal lumen for disposal in feces. The two proteins heterodimerize in the endoplasmic reticulum, traffic through the Golgi, and subsequently move to the apical plasma membrane. ABCG5 and ABCG8 undergo N-glycosylation, and glycosylation at Asn-619 in ABCG8 is critical for proper trafficking of the complex.

The **ABCG5/ABCG8 complex** is largely responsible for the greater intestinal absorption of cholesterol compared with a variety of dietary plant sterols. β-**sitosterolemia** is a genetic disease characterized by *increased intestinal absorption* and *diminished biliary secretion of cholesterol and plant sterols* such as β-sitosterol. Affected patients have accumulation of cholesterol and plant sterols in the plasma and various tissues and often suffer from **premature atherosclerotic heart disease**. β-sitosterolemia is caused by **a mutation in either ABCG5 or ABCG8**, with the majority of mutants causing impaired transport of the heterodimer from the endoplasmic reticulum to the plasma membrane.

ABCG5 and ABCG8 both appear to be regulated primarily at the level of gene transcription. The sterol-sensing transcription factors **LXRα** and **LXRβ** up-regulate expression of both ABCG5 and ABCG8, and effects of other transcription factors have been identified.

Scavenger receptor class B type 1 (SR-B1) is thought to play a role in cholesterol homeostasis mainly through its mediation of the selective uptake of HDL cholesteryl esters into the **liver**. SR-B1, however, is also abundantly expressed in the **small intestine** where it is found in both the brush-border and basolateral membranes (Fig. 8.1). SR-B1 is expressed more in the proximal than distal small bowel, paralleling the site of most cholesterol absorption. SR-B1 may be mediate some cholesterol uptake across the brush-border membrane, but may have more important roles in the intestinal uptake of circulating HDL-associated cholesterol esters and subsequent transport across the apical membrane for disposal in the feces, and/or the secretion of cholesterol-containing lipoproteins across the basolateral enterocyte membrane.

2.5 Cholesterol Esterification, Incorporation into Lipoproteins, and Secretion

Within the enterocyte, 70–90 % of the absorbed cholesterol is esterified with fatty acids. Various lipid binding proteins may be involved in directing the absorbed cholesterol to the endoplasmic reticulum. The principal enzyme responsible for cholesterol esterification is **acyl CoA:cholesterol acyl transferase 2 (ACAT2)** that is present in the endoplasmic reticulum. Sitosterol and other plant sterols are less effective substrates for ACAT2 than cholesterol. The **microsomal triglyceride transfer protein (MTTP)** transfers cholesteryl esters to the nascent chylomicron particle, and chylomicrons are the principal lipoprotein delivering cholesterol into the circulation. Within chylomicrons and other intestinal lipoproteins, cholesteryl esters are incorporated into the oily lipoprotein core, whereas free cholesterol is present on the surface of the particle.

Another route of cholesterol exit from the enterocyte may be via incorporation into HDL. A **basolateral cholesterol efflux pump ABCA1** facilitates the transfer of cholesterol to an apolipoprotein A-I acceptor molecule forming discoidal HDL particles (Fig. 8.1). The amount of cholesterol absorbed via this pathway appears to be small, but ABCA1 may be a key role in HDL production. Patients with *Tangier disease* with **mutations in ABCA1** have virtual **absence of plasma HDL** and **increased enterocyte cholesterol content**. ABCA1 is regulated by the **transcriptional factor LXRα**.

3 Phospholipid

Most of the phospholipid that the intestine absorbs originates from biliary secretion (10–20 g/day) rather than the diet (1–2 g/day). In bile, phospholipid is present in mixed micelles along with bile acids and cholesterol. **Phosphatidylcholine** is the major phospholipid in the diet and in bile along with small amounts of **phosphatidylethanolamine, phosphatidylserine,** and **phosphatidylinositol**. In the intestinal lumen, the phosphatidylcholine distributes between the lipid droplets and the micellar phase, but favors the micelles.

3.1 Digestion of Phosphatidylcholine

The major enzyme responsible for the digestion of phosphatidylcholine is **pancreatic phospholipase A_2** (Fig. 8.2). This enzyme, which catalyzes the breakdown of phosphatidylcholine to *lysophosphatidylcholine* and *fatty acid*, is secreted as a **zymogen** and activated by tryptic cleavage of an N-terminal heptapeptide. Phospholipase A_2 has a molecular weight of about 14 kd and is activated by

Fig. 8.2 Digestion of phosphatidylcholine by pancreatic phospholipase A_2. Phospholipids are substrates of pancreatic phospholipase A_2

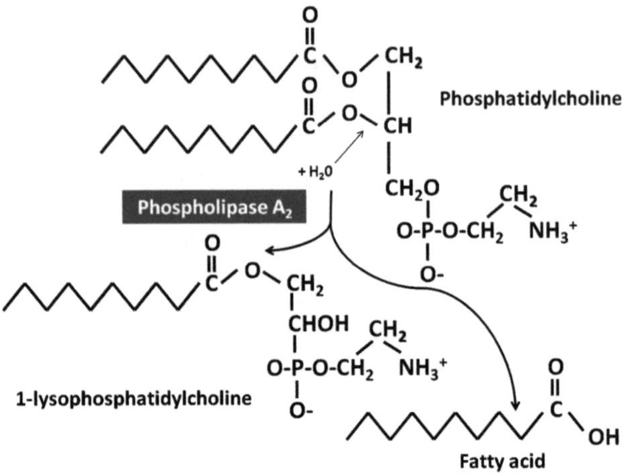

calcium and by bile salts. Phospholipase A_2 in the intestinal brush-border membrane and intracellular phospholipases within the enterocyte probably also participate in phospholipid digestion.

3.2 Absorption of Lysophosphatidylcholine

Lysophosphatidylcholine can be incorporated into both **mixed micelles** and **liquid crystalline vesicles**. Studies have also demonstrated that bile salts and lysophospholipids form submicellar aggregates than can transfer lysophospholipids between membranes. The relative contributions of these macromolecular structures to intestinal lysophospholipid absorption remain to be determined. It has been traditionally thought that lysophospholipids cross the enterocyte brush-border membrane by passive diffusion, but some data suggest a protein-mediated uptake mechanism.

3.3 Phospholipid Metabolism in the Enterocyte

The absorbed lysophosphatidylcholine can be metabolized by several pathways. Most of the lysophosphatidylcholine is **reacylated** to form *phosphatidylcholine*. Alternatively, the lysophosphatidylcholine can be **hydrolyzed** to *fatty acid* and *3-phosphorylcholine*, or two molecules of lysophosphatidylcholine can react to form one molecule of *phosphatidylcholine* and one molecule of *3-phosphorylcholine*. The fatty acids formed by these reactions can be used for triglyceride synthesis. In addition, the intestine is capable of synthesizing phosphatidylcholine from diglyceride (derived mainly from α-glycerophosphate) via the **Kennedy pathway**.

In the absence of absorbed lysophosphatidylcholine, there is a marked drop in the secretion of chylomicrons into lymph. The fatty acid composition of the phosphatidylcholine contained on the surface of chylomicrons is not greatly influenced by the dietary fatty acid content, but mainly reflects the fatty acid composition of biliary phosphatidylcholine. These data indicate that ***de novo* phosphatidylcholine synthesis by the enterocyte is inadequate to provide the phospholipid needed for optimal packaging and secretion of chylomicrons**, and that absorbed biliary phospholipid is preferentially utilized for chylomicron assembly.

4 Bile Acids

Bile acids are secreted by the liver almost exclusively as **conjugated forms**, mainly as conjugates with glycine (75 %) or taurine (25 %). A small fraction of these conjugated bile acids is absorbed in the *proximal small intestine*, mainly the glycine-conjugated dihydroxy bile acids, chenodeoxycholic acid-glycine and deoxycholic acid-glycine. These bile acids are likely protonated in the unstirred water layer of the jejunum, which is estimated to have a pH of approximately 5, and the protonated bile acids are then passively absorbed transcellularly. **Glycine-conjugated trihydroxy bile acids** and **taurine-conjugated bile acids** would be fully ionized at the small intestinal pH and therefore would not be absorbed by this mechanism. There is also the possibility of limited paracellular absorption of ionized and nonionized conjugated bile acids in the small bowel.

Most of the conjugated bile acids, however, remain in the lumen as the intestinal contents transit through the upper intestine, where they activate important enzymes and solubilize lipids to ensure efficient absorption. In the *ileum*, there is a high-capacity active transport mechanism for bile acids that results in the absorption of greater than 95 % of the secreted bile acids. The conserved bile acids are delivered via the portal blood to the liver, where they are taken up for secretion again into bile. This **enterohepatic circulation of bile acids**, which occurs several times during a meal, ensures an adequate delivery of bile acids into the intestine for highly efficient lipid absorption independent of the rate of hepatic bile acid synthesis from cholesterol.

4.1 Ileal Bile Acid Uptake

The active transport of bile acids across the ileal brush-border membrane is mediated by the transport protein **ASBT** (gene symbol SLC10A2) (Fig. 8.3). ASBT is a 348 amino acid membrane glycoprotein with a glycosylated extracellular amino terminus, a cytosolic carboxyl terminus, and 7 membrane spanning domains. ASBT is an **electrogenic Na^+-bile acid co-transporter**, with two sodium ions transported per molecule of bile acid. The driving force for transport is the inwardly

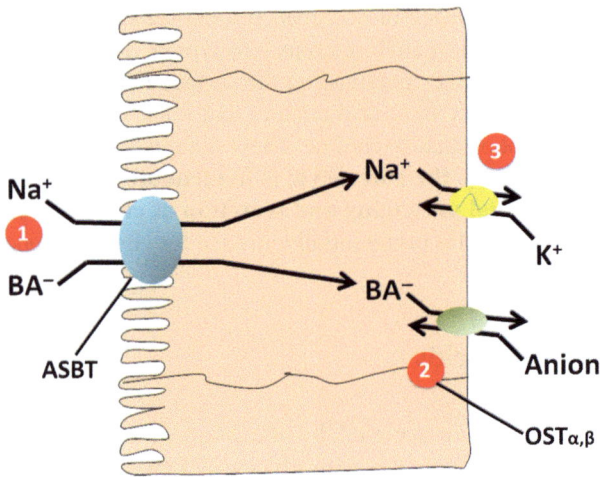

Fig. 8.3 Active transport of bile acids by the ileal enterocyte. (*1*) Apical Na^+-conjugated bile acid carrier ASBT. (*2*) Basolateral bile-acid-anion exchanger. (*3*) Basolateral Na^+/K^+-ATPase (Adapted from Hofmann [7])

directed sodium gradient. Conjugated bile acids are transported more efficiency than unconjugated forms, and the affinity for dihydroxy bile acids is higher than for trihydroxy bile acids. ASBT is expressed mainly in the *villus cells* of ileal enterocytes, with small amounts of ASBT in the proximal intestine and colon. The GATA4 transcription factor is essential for silencing ASBT expression in the proximal intestine.

ASBT expression is reduced by the bile acid **farnesoid X receptor (FXR)** that has chenodeoxycholate and other bile acids as ligands. The uptake of bile acids, therefore, can be altered to meet physiological requirements. ASBT is also regulated by the **PPARα receptor** and by **corticosteroids**. ASBT and Na^+-bile acid co-transport is developmentally regulated in mammals as they are absent at birth and develop in the first postnatal weeks at a variable, species-dependent rate.

Inactivating mutations in ASBT cause *bile acid malabsorption*. Beginning in early infancy these children have *interruption of the enterohepatic circulation of bile acids, chronic diarrhea,* and *malabsorption of triglycerides* and *fat-soluble vitamins*.

4.2 Intracellular Transport and Export from the Enterocyte

With the enterocyte bile salts associate with a 14 kDa protein called the **ileal bile acid binding protein (IBABP)** in a 2:1 stoichiometric complex. IBABP is postulated to facilitate transport of bile acids through the cell to the basolateral membrane. IBABP may also play a role in protecting the enterocyte from the toxic effects of high intracellular bile salt concentrations. IBABP expression is increased by FXR and is also developmentally regulated.

The movement of bile acids across the basolateral membrane occurs mainly via the **OSTα/β heterodimeric transporter** (Fig. 8.3). **OSTα** is a 340 amino acid protein with an extracellular amino terminus, 7 transmembrane domains, and an intracellular carboxyl terminus. **OSTβ** is a 128 amino acid protein with an extracellular amino terminus and a cytoplasmic carboxyl terminus. Both subunits are required for trafficking to the plasma membrane and for bile acid transport. This transport complex appears to function by **sodium-independent facilitated diffusion**. OSTα/β can transport a number of compounds in addition to bile acids, including estrone-3-sulfate, digoxin, prostaglandin E2, and dehydroepiandstosterone-3-sulfate.

OSTα/β is also increased by FXR. **Activation of FXR by bile salts** *decreases ASBT expression* and *increases IBABP and OSTα/β expression*, thereby preventing intracellular bile acid accumulation.

The **multidrug resistance protein 3 (MRP3)** plays a minor role in basolateral bile acid export, but may be more important for the small amounts of modified (glucuronidated or sulfated) bile acids present in the intestinal lumen. **Multidrug resistance protein 2 (MRP2)** may mediate transport of modified bile acids across the brush-border membrane into the intestinal lumen.

4.3 Absorption of Unconjugated Bile Acids

In the intestine, particularly the *colon*, bacteria modify bile acids in various ways. They **deconjugate bile acids** by enzymatic hydrolysis of the bond linking bile acids to their amino acid (glycine or taurine) conjugates. Further bacterial modifications include **7α-dehydroxylation** and **oxidation/epimerization of hydroxyl groups** at various sites. These modifications increase the hydrophobicity and pKA of the bile acids, permitting some passive absorption across the intestinal epithelium. In addition, limited amounts of unconjugated bile acids may be absorbed via the small amounts of transport proteins expressed in the colon.

Bile acids in the *distal intestine* and *colon* cause fluid and electrolyte secretion by directly altering enterocyte and colonocyte ion transport and indirectly by stimulating neuroendocrine mechanisms. **Excessive colonic bile salts** from ileal resection or disease or other processes result in *chronic diarrhea*.

4.4 Ileal FGF19 Secretion and Regulation of Hepatic Bile Salt Synthesis

Ileal enterocytes synthesize and secrete **fibroblast growth factor 19 (FGF19)**. FGF19 is responsive to FXR. FGF19 is released into the portal circulation, and in an endocrine manner actives **fibroblast growth factor receptor 4 (FGF4)** in hepatocytes, resulting in inhibition of bile acid synthesis. This pathway works in conjunction with the local intrahepatic regulation of bile acid synthesis by bile salts.

A group of patients have been described who have low circulating FGF19 levels and chronic watery diarrhea. It appears that these patients have disordered regulation of hepatic bile acid synthesis and an expanded bile acid pool size that overwhelms a normally functioning ileal absorption mechanism. Excessive colonic bile acids cause fluid and electrolyte secretion and chronic watery diarrhea.

5 Vitamin A

5.1 *Definitions*

The term *vitamin A* denotes a family of compounds that are **structurally related to all-trans-retinol** and are required for vision, growth, cellular differentiation and proliferation, reproduction, and the integrity of the immune system. Naturally occurring compounds such as retinol, retinaldehyde (retinal), and retinoic acid and a large number of synthetic analogues with or without vitamin A biological activity are termed **retinoids**. Retinoids vary both qualitatively and quantitatively with respect to specific biological actions. For example, retinoic acid in contrast to retinol cannot effectively maintain normal vision or reproductive function.

5.2 *Dietary Sources of Vitamin A*

Dietary sources of vitamin A include **preformed vitamin A**, present in animal tissues largely as long-chain fatty acyl retinol esters, and certain **carotenoid pigments** present in fruits, vegetables, and some **animal fats** that are precursors of vitamin A. Of more than 500 carotenoids found naturally, only about 50 are precursors of retinol. *All-trans-β-carotene* is the most active on a weight basis and the most important vitamin A precursor for humans. Because the bioavailability of food carotenoids is less than that of retinol, due to poorer intestinal absorption and limited conversion of carotenoids to vitamin A, 12 μg of dietary β-carotene and 24 μg of three other dietary carotenoids are assumed to be nutritionally equivalent to 1 μg of retinol. The major sources of vitamin A or provitamin A in the American diet are liver, carrots, eggs, vegetable-based soups, whole-milk products, and fortified milk and other foods. It is estimated that on average less than one third of total vitamin A activity comes from carotenoids, although there is certainly considerable interindividual variation based on dietary habits and the efficiency of carotenoid absorption and conversion to vitamin A. The roles of β-carotene and other carotenoids in the **prevention of cancers, heart disease, and other chronic degenerative diseases** is a topic of intense current interest. Many of the physiological effects of carotenoids are not due to their function as vitamin A precursors, but reflect antioxidant and other properties of the carotenoids themselves.

8 Digestion and Absorption of Other Dietary Lipids

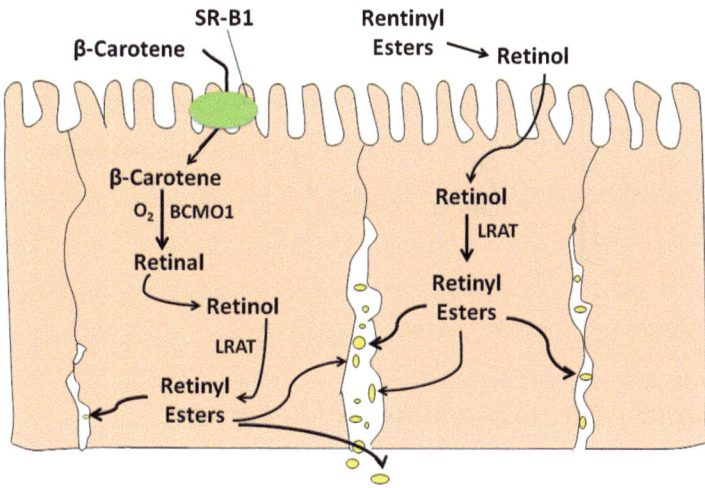

Fig. 8.4 Absorption and metabolism of β-carotene and retinyl esters (Adapted from Ong [10])

5.3 Digestion of Dietary Retinyl Esters

As with other esterified lipids, such as cholesteryl esters, retinyl esters must be hydrolyzed before intestinal absorption can occur (Fig. 8.4). Several pancreatic enzymes, including pancreatic lipase, pancreatic lipase-related protein 2, and pancreatic bile salt-activated lipase can hydrolyze retinyl esters, with pancreatic lipase appearing to be most important. Brush-border membrane phospholipase B (PLB) also participates in retinyl ester hydrolysis. The relative contributions of pancreatic and small intestinal enzymes to retinyl ester hydrolysis are currently uncertain.

5.4 Uptake of Retinol and β-Carotene by the Enterocyte

Both **retinol** (formed by hydrolysis of retinyl esters) and **β-carotene** are solubilized in mixed micelles within the luminal contents. At *low concentrations*, retinol is taken up by a saturable, energy-independent process that is consistent with **carrier-mediated facilitated diffusion** (Fig. 8.4). A specific carrier has not yet been identified. At *higher pharmacological retinol concentrations*, uptake is not saturable and likely occurs via **simple, passive diffusion**. Retinol uptake is greater in the jejunum than ileum and is greater in neonatal animals than adults, suggesting developmental regulation of a carrier. Uptake of β-carotene and other carotenoids across the brush-border membrane is mediated by the **scavenger receptor class B, type 1 (SR-B1)** (Fig. 8.4).

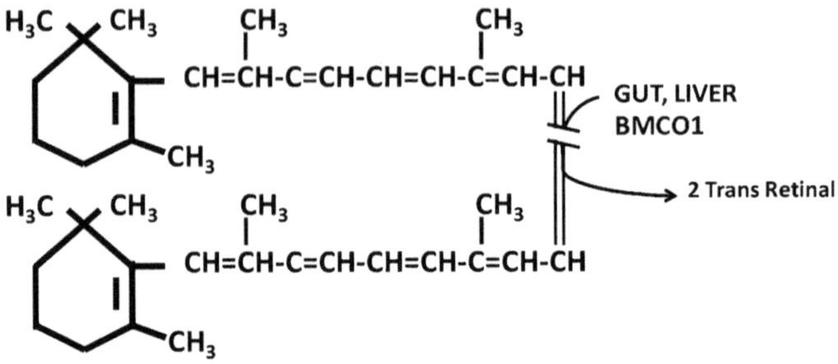

Fig. 8.5 Conversion of β-carotene to vitamin A

5.5 Carotene Cleavage and Retinal Reduction

Within the enterocyte, carotenes are cleaved, forming retinal. Two enzymes, **β-carotene-15,15'-monooxygenase** (**BCM01**) and **β-carotene-9',10'-monooxygenase** (**BCM02**) can cleave carotenoids (Fig. 8.5). BCM01 cleaves carotene centrally forming two molecules of retinal, whereas BCM02 cleaves asymmetrically forming **apocarotenals** that are subsequently shortened to produce retinal. Some intact β-carotene is absorbed and circulates in plasma. Liver and other tissues have the capacity to form vitamin A from carotene via BCM01. The retinal produced from carotenes is efficiently reduced to retinol in the enterocyte (Fig. 8.4). Both soluble and microsomal reductases have been identified. Within the enterocyte, most of the retinal is probably bound to a specific binding protein, **cellular retinol-binding protein, type 2** or **CRBP(II)** (see below). The **microsomal reductase** effectively utilizes retinal bound to CRBP(II) as a substrate, whereas the **soluble reductase** appears to prefer free retinal.

Intestinal BCM01 mRNA expression is increased when animals are fed a retinoid-deficient diet. Polymorphisms have been identified in the human BCM01 gene that decrease the conversion of β-carotene to vitamin A, and may explain some of the inter-individual variability in the ability to absorb and convert proretinoid carotenoids to retinoids.

5.6 Cellular Retinol-Binding Protein, Type 2

Several retinoid-binding proteins play crucial roles in the transport, metabolism, and cellular actions of vitamin A. In the enterocyte, one of these proteins, **CRBP(II)**, is key in the regulation of vitamin A absorption and metabolism. CRBP(II) is a 133-amino-acid protein with one retinoid-binding site that constitutes approximately 1 % of the soluble protein in the jejunum. CRBP(II) belongs to a large superfamily of lipid-binding proteins that includes the fatty acid-binding proteins and another retinoid-binding protein, CRBP(I). **CRBP(I)** is found at low levels in many tissues including the small intestine, but it is not thought to play a role in vitamin A absorption. In adults CRBP(II) is expressed only in the *small intestinal villus epithelial cells*. CRBP(II) mRNA and protein are greater in the jejunum than ileum and increase during pregnancy and lactation. In addition to its role in the reduction of retinal derived from carotenes, CRBP(II) also appears to regulate the esterification of retinol within the enterocyte.

5.7 Retinol Esterification

Both absorbed retinol and retinol generated from carotenes are esterified in the enterocyte with long-chain fatty acids. **Palmitate, stearate, oleate,** and **linoleate** account for most of the esterified fatty acids in an approximate ratio of 8:4:2:1, and this pattern is not significantly altered by changes in dietary fatty acid composition. **Two microsomal enzymes** capable of esterifying retinol have been identified. *Lecithin-retinol acyltransferase (LRAT)* effectively esterifies CRBP(II)-bound retinol (Fig. 8.4). This enzyme uses phosphatidylcholine as the fatty acid donor, and the fatty acid composition of the product retinyl esters is similar to that of position 1 of intestinal lymph phosphatidylcholine. LRAT has a low Km for CRBP(II)bound retinol that would be appropriate for activity at a low dietary vitamin A intake, and LRAT accounts for about 90 % of the esterification of physiological amounts of retinol. LRAT is greater in the proximal than the distal small bowel and is developmentally regulated in a pattern similar to CRBP(II). Levels of these proteins increase prior to birth, remain high during suckling, and then decline with weaning to the adult level. *Diacylglycerol acyltransferase 1 (DGAT1)*, a key enzyme in triglyceride absorption, also can esterify retinol, and plays an important role at high levels of retinol intake.

5.8 Export of Vitamin A and Carotenoids from the Enterocyte

Retinyl esters synthesized in the enterocyte and carotenoids that escape cleavage are incorporated into the lipid core of chylomicrons and released into the intestinal lymph. Retinyl esters and carotenoids appear to be inserted into chylomicrons by the MMTP at a late state of chylomicron production. Some retinol is released from the enterocyte into portal blood, perhaps mediated by the ABCA1 transporter in the basolateral membrane.

5.9 Hepatic Metabolism and Storage of Vitamin A

Most (65–75 %) of absorbed vitamin A stays associated with the lipid core of chylomicrons as the triglyceride in these lipoproteins is hydrolyzed, forming **chylomicron remnants**. Extrahepatic uptake of retinyl esters from chylomicrons

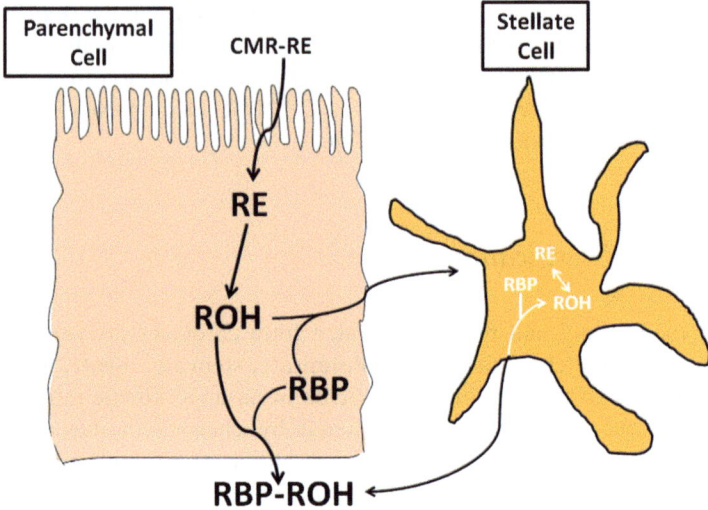

Fig. 8.6 Pathways of hepatic vitamin A metabolism and storage. *CMR RE* retinyl esters in chylomicron remnants, *RE* retinyl esters, *ROH* retinol, *RBP* serum retinal-binding protein

and chylomicron remnants in certain tissues is important in supplying vitamin A to these cells. Chylomicron remnants are taken up by hepatocytes using several receptor-mediated mechanisms. Once remnant-associated retinyl esters have been taken up by hepatocytes, rapid hydrolysis occurs; multiple hepatic enzymes can hydrolyze retinyl esters (Fig. 8.6). Retinol is transferred from hepatocytes to perisinusoidal hepatic stellate cells. **Cellular retinol binding protein 1 [CRBP1]** is important in the **transfer of retinol to hepatic stellate cells** and its subsequent esterification. Some, but not all data, suggests the involvement of the 21 kDa serum retinol binding protein (RBP) in this process. CRBP(I)-bound retinol is esterified by LRAT in the hepatic stellate cells. Normally, stellate cells contain about 90 % of liver vitamin A, and 98 % is in the form of retinyl esters. Retinyl esters are stored within the hepatic stellate cells along with other lipids in the form of lipid droplets.

In vitamin A-deficient animals, secretion of retinol and RBP from the liver is reduced and plasma concentrations are low. Repletion with small doses to vitamin A to a deficient animal does not result in accumulation of retinyl esters in stellate cells; instead, the retinol-RBP complex is rapidly secreted into the circulation. Stellate cells from vitamin A-deficient rats have been found to have reduced CRBP(I) and LRAT levels, a situation that may limit vitamin A storage and favor retinol-RBP delivery into the circulation.

Retinyl esters in hepatic cell lipid droplets are mobilized during dietary vitamin A deficiency to supply peripheral tissues. Several enzymes in stellate cells can hydrolyze retinol esters. After hydrolysis of retinyl esters, retinol is thought to be transferred back to hepatocytes where it binds to RBP. The retinol-RBP complex is secreted from the hepatocytes into the bloodstream. This process is highly regulated by vitamin A status, as retinol-RBP secretion is reduced in vitamin A deficiency and restored with vitamin A repletion. The ability of stellate cells to control retinol storage and mobilization ensures that plasma retinol is stable at about 2 μmol/L in spite of variations in daily intake of vitamin A.

5.10 Uptake and Metabolism of Retinol-RBP

A detailed consideration of the uptake and metabolism of retinol by tissues other than the intestine and liver is beyond the scope of this chapter; however, it will be briefly considered here. In plasma, most of the retinol-RBP is reversibly complexed with another protein, **transthyretin** (molecular weight, 55 kd), and this large complex is therefore less susceptible to glomerular filtration. A transmembrane-spanning protein **STRA6** avidly binds retinol and facilitates retinol uptake into many peripheral tissues. Cells that also express LRAT take up more retinol, indicating that conversion of retinyl to retinyl ester within the cells maintains the driving force for retinol uptake. STRA6 is apparently a **bidirectional retinol transporter** and can participate in retinol efflux from cells under certain circumstances. **Mutations of STRA6** result in *malformations* of the *eye, heart, lungs,* and *diaphragm*; similar abnormalities are noted in the offspring of mothers with vitamin A deficiency.

As mentioned above, some retinyl esters contained within chylomicrons can be taken up by peripheral tissues. The enzyme **lipoprotein lipase** that catalyzes chylomicron triglyceride hydrolysis can also catalyze the hydrolysis of retinyl esters and facilitate retinol uptake into certain tissues Many retinol binding proteins have been identified in interstitial fluids and in the cytosol of various cell types, and they appear to play important roles in regulating cellular uptake, storage, and metabolism of retinoids. Proteins with binding specificities for retinol, retinoic acid, and retinal have been identified. **Retinoic acid** is the form of vitamin A involved in the regulation of gene transcription. Some retinoic acid is produced in the intestine and is present at low concentration in plasma, bound to albumin. Most of the retinoic acid in target tissues, however, appears to derive from the **oxidation of retinol** in these cells. Some retinoic acid may also be synthesized from β-**carotene** in certain cells. **Two families of nuclear retinoic acid receptors** have now been identified, the *retinoic acid receptor (RAR)* and the *retinoid X receptor subfamilies* (Fig. 8.7). The RAR subfamily binds all-trans-retinoic acid and 13-cis-retinoic acid; in contrast, the RXR subfamily binds 9-cis-retinoic acid. These ligand-dependent transcription factors associate with response elements in the promoters of specific genes, regulating transcription. Retinoid X receptors frequently form heterodimers with other receptors, including the RAR receptors, vitamin D receptor, and others.

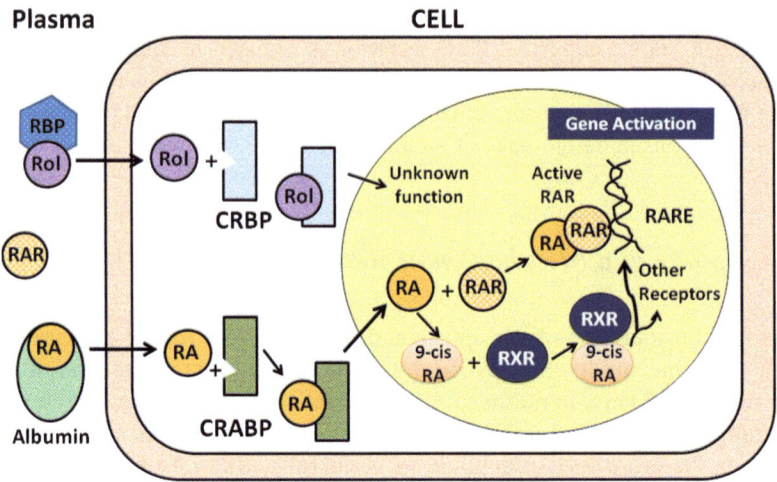

Fig. 8.7 Cellular uptake, metabolism, and mechanism of action of retinol and retinoic acid. *Rol* all-trans-retinol, *RBP* serum retinol-binding protein, *CRBP* cellular retinol-binding protein, type I, *CRABP* cellular retinoic acid-binding protein, *RAR* retinoic acid receptors, *RXR* retinoid X receptors, *9-cis-RA* 9-cis-retinoic acid, *RARE* retinoic acid response elements (Adapted from Wolf [17])

6 Vitamin D

The term vitamin D refers to a family of compounds involved primarily in the regulation of mineral and bone metabolism. **Vitamin D deficiency** results in **metabolic bone disease,** *rickets* in children and *osteomalacia* in adults, characterized by inadequate bone mineralization.

6.1 Sources of Vitamin D

Vitamin D_3 (cholecalciferol) is synthesized in **skin** from the precursor compound **7-dehydrocholesterol**. Ultraviolet light of wavelengths 290–320 nm opens the β ring of 7-dehydrocholesterol, forming **previtamin D_3**, which subsequently undergoes a slow, temperature-dependent, nonenzymatic isomerization to vitamin D_3. With enough sunlight exposure, sufficient vitamin D_3 can be synthesized in the skin to meet the vitamin D requirement. The amount of sunlight needed to produce adequate vitamin D_3, however, is influenced by skin pigmentation, age, season, distance from the equator, and conditions that filter out ultraviolet light such as industrial pollution. Because of these factors and customs of dress that limit sun exposure, vitamin D is considered an essential dietary nutrient.

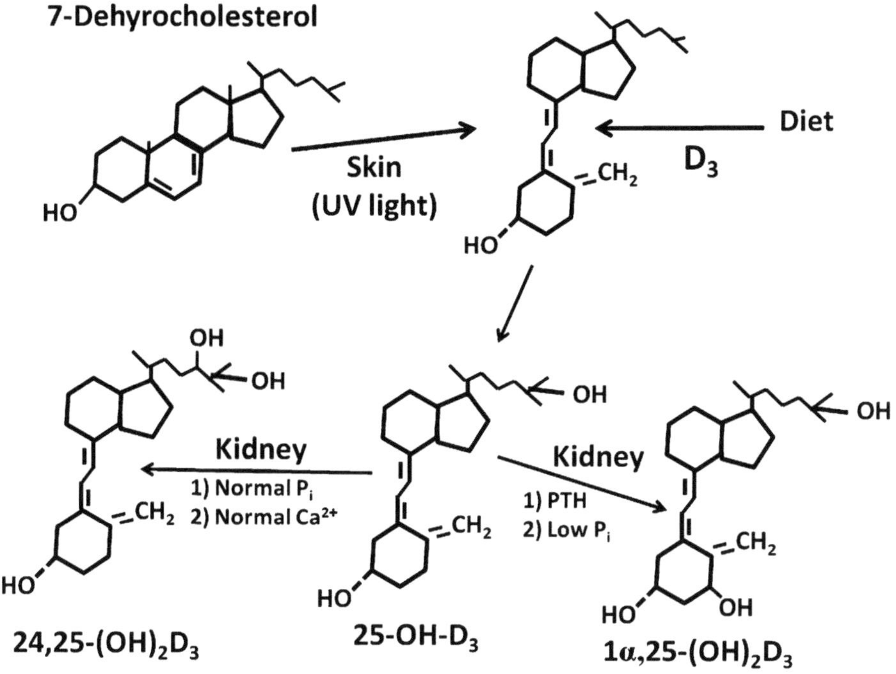

Fig. 8.8 Metabolism of vitamin D

The **major dietary forms of vitamin D** are *vitamin D_3* and *vitamin D_2 (ergocalciferol)*, which is formed by ultraviolet irradiation of the plant sterol **ergosterol**. In mammals, these two forms of the vitamin have similar intestinal absorption, metabolism, and cellular effects. In the United States, foods fortified with vitamin D_2 or vitamin D_3 are the major dietary sources of the vitamin. Processed cow's milk contains 10 μg/quart of vitamin D and is the major food source of the vitamin for children. Small amounts of vitamin D are also present in liver, eggs, poultry, fortified margarines, and other foods. Human milk contains relatively little vitamin D, 0.63–1.25 μg/L.

6.2 Overview of Vitamin D Metabolism and Biological Activity

At normal serum concentrations, vitamin D is not physiologically active. Two hydroxylation reactions are required to produce the **active** form of the vitamin, **1,25-dihydroxyvitamin D_3 (1,25(OH)$_2D_3$)** (Fig. 8.8). The first step occurs in the *liver* with the formation of **25-hydroxyvitamin D_3 (25(OH)D_3)**, which is the major form of the vitamin circulating in plasma. The principal hepatic **25-hydroxylase** is the **cytochrome P-450 CYP2R1**. The regulation of CYP2R1 is incompletely understood. 25 hydroxylase activity increases in vitamin D deficiency, but whether the enzyme is regulated by the substrate, product, other vitamin D metabolites, or serum calcium is unclear. The regulation of CYP2R1 is quite imprecise, as very high plasma levels of 25(OH)D$_3$ are seen with excessive vitamin D intake and toxicity. 25(OH)D$_3$ and all other vitamin D metabolites circulate in the plasma largely bound to a specific binding protein, the serum vitamin **D-binding protein (DBP)**. This glycoprotein is synthesized mainly in the liver and has a single high affinity binding site for vitamin D metabolites. 25(OH)D$_3$ bound to DBP is taken up by the kidney and certain other tissues by receptor-mediated endocytosis.

In the *kidney*, 25(OH)D$_3$ is converted to **1,25(OH)D$_3$**, the form of the vitamin that is biologically active in the regulation of calcium and bone metabolism. The renal **25(OH)D$_3$ 1-hydroxylase** is a mitochondrial enzyme that is highly regulated. Enzyme activity is increased by parathyroid hormones secreted in response to hypocalcemia. The elevated 1,25(OH)D$_3$ level then stimulates intestinal calcium absorption and together with parathyroid hormone enhances calcium mobilization from bone and renal tubular calcium resorption, thereby correcting the hypocalcemia and completing a feedback loop. 1,25(OH)$_2$D$_3$ production is also regulated by two other feedback loops. **FGF23** is a hormone that is secreted by bone in response to an elevation in the serum phosphate level. FGF23 acting through its **receptor (FGFR)** and a **coreceptor (klotho)** inhibits CYP27B1 in the kidney, suppresses parathyroid hormone, and like parathyroid hormone causes *phosphaturia*. 1,25(OH)$_2$D$_3$ inhibits intestinal phosphate absorption and bone FGF23 production. 1,25(OH)$_2$D$_3$ also directly represses CYP27B1 transcription and inhibits parathyroid hormone secretion. A variety of hormones, including insulin, estrogen, progesterone, prolactin, and growth hormone, have also been shown to raise plasma 1,25(OH)$_2$D$_3$ levels, and these responses are likely to be important for meeting the increased mineral requirements during growth, pregnancy and lactation. Thus, vitamin D can be considered as a component of a complex endocrine system geared to the defense of serum calcium and phosphorus levels and to the provision of adequate minerals for bone formation.

A **vitamin D$_3$ receptor (VDR)** has been identified in many tissues, including the intestine. This receptor belongs to a large superfamily of receptor proteins involved in **transcriptional regulation**, including steroid hormone, thyroid, and retinoic acid receptors. 1,25(OH)$_2$D$_3$ binds to the VDR, and the complex interacts with specific response elements in certain genes to either induce or repress transcription. Over 100 genes have been identified that are regulated by 1,25(OH)$_2$D$_3$, including calcium channels, calcium-binding proteins and calcium-transporting ATPases involved in enterocyte calcium transport. The VDR is expressed in many tissues other than the classic target organs of intestine, bone, and kidney involved in mineral physiology. In addition to its effects on mineral and bone metabolism, 1,25(OH)$_2$D$_3$ has also been demonstrated to influence cell proliferation and differentiation in many tissues, to modulate immune function, to alter insulin secretion, and to have other cellular effects. In addition to its regulation of gene transcription, $1,25(OH)_2D_3$ can also induce rapid cellular responses that do not require new mRNA synthesis.

6.3 Intestinal Absorption of Vitamin D_3

Vitamin D is present in the diet mainly as free cholecalciferol and ergocalciferol, and thus no intraluminal digestion is required. Vitamin D_3 has limited water solubility, but does aggregate to a micellar-like form at a concentration of about 10^{-8} mM/L. At low concentrations, therefore, vitamin D_3 can be absorbed in the absence of bile salts, if the intraluminal fluid contains only small amounts of other lipids. Under conditions reflecting the typical postprandial situation, where fatty acids, monoglycerides, and other lipids are present at high concentrations, absorption of vitamin D_3 is highly bile salt-dependent. Bile salts are needed to solubilize vitamin D_3 in the mixed micellar phase and to prevent it from portioning into the lipid droplets. Patients with **intraluminal bile salt deficiency** from *cholestatic liver disease* or from *ileal disease* or *resection* causing bile salt malabsorption have **very poor absorption of dietary vitamin D_3**.

Recent studies have shown that some of the transport proteins involved in the absorption of cholesterol and other lipids, such as SR-B1, CD-36, and NPC1L1, facilitate intestinal vitamin D_3 uptake. At high concentrations, vitamin D is likely absorbed by **passive diffusion**. Most of the absorbed vitamin D_3 is incorporated into chylomicron particles and secreted into the intestinal lymph. Within the lymph, there is transfer of the vitamin from chylomicrons to unoccupied DBP, a process that continues in the bloodstream. Some of the absorbed vitamin D_3 is released from the enterocyte by a **nonchylomicrondependent pathway** directly into the portal venous circulation, where it associates with DBP. In the absence of luminal bile salts, most of the absorbed vitamin D_3 appears in portal blood, whereas bile salts favor incorporation into chylomicrons.

6.4 Absorption of Vitamin D Metabolites

The hydroxyvitamin D metabolites **$25(OH)_2D_3$** and **$1,25(OH)_2D_3$** are not significant dietary sources, but are important as pharmacologic agents for the treatment of metabolic bone diseases and disorders of mineral metabolism. Studies in the rat have shown that these forms of the vitamin are more efficiently absorbed than vitamin D_3 itself and are less dependent on intraluminal bile salts and chylomicron production and secretion. These differences likely reflect the somewhat **greater water solubility** of these compounds. Clinical studies have also demonstrated that $25(OH)D_3$ is better absorbed than vitamin D_3 in normal humans and particularly in patients with digestive diseases causing intraluminal bile salt deficiency and steatorrhea.

6.5 Catabolism and Excretion of Vitamin D Metabolites

Renal mitochondria also contain a **24-hydroxylase enzyme (CYP24)** that produces **24,25-dihydroxyvitarnin D_3 [24,25(OH)$_2$D$_3$]**. Factors that stimulate the 1-hydroxylase (parathyroid hormone and low 1,25(OH)$_2$D$_3$) depress the 24-hydroxylase, whereas agents that inhibit the 1-hydroxylase (1,25(OH)$_2$D$_3$ and low parathyroid hormone) activate the 24-hydroxylase. This reciprocal relationship has led to the conclusion that the **formation of 24,25(OH)D$_3$ represents the initial step** in a pathway for degradation of 25(OH)D$_3$ and for regulation of the circulating 1,25(OH)$_2$D$_3$ level. 24,25(OH)$_2$D$_3$ is further metabolized in the kidney by a side-chain oxidation pathway that proceeds through several oxo- and keto-intermediates and culminates in oxidation and cleavage of the side chain. When radioactive vitamin D_3, 25(OH)D_3, 24,25(OH)$_2$D$_3$, or 1,25(OH)$_2$D$_3$ are injected intravenously, most of the radioactivity is recovered in bile, with lesser amounts in urine. In bile, the radioactivity is present in an array of lipid- and water-soluble products, probably generated mainly by the side-chain oxidation pathway described above, including conjugates with glucuronic acid, sulfate, and amino acids.

Many extra-renal tissues, including the intestine, also express CYP27B1 and CYP24. These enzymes in extra-renal tissues are not influenced by factors such as parathyroid hormone that act in the kidney to control the plasma 1,25(OH)$_2$D$_3$ concentration, but instead are involved in the regulation of 1,25(OH)$_2$D$_3$ at the local tissue level.

7 Vitamin E

7.1 Structure and Biological Activity

Vitamin E is a major lipid-soluble antioxidant present in plasma and cellular membranes. Vitamin E does not prevent the formation of carbon-centered radicals, but because it reacts more rapidly with peroxyl radicals than do polyunsaturated fatty acids, it traps peroxyl radicals and breaks the chain reaction of lipid peroxidation. **α-tocopherol**, the most active form of vitamin E, reacts with peroxyl radical to form an α-tocopheroxyl radical that is resonance stabilized and breaks the chain reaction. Subsequently, α-tocopherol is regenerated by ascorbic acid and potentially by other agents such as glutathione or uric acid. Alternatively, two α-tocopheroxyl radicals can react together, forming a dimer, or the radical can be completely oxidized to tocopherol quinone. Vitamin E also regulates gene expression and alters cell signaling and proliferation.

Vitamin E occurs in **eight forms: α-, β-, γ-,** and **δ-tocopherols** (which have a chromanol ring and a phytyl tail and differ in the number and position of methyl groups on the ring), and **α-, β-, γ-,** and **δ-tocotrienols** (which have unsaturated tails) (Fig. 8.9). The eight forms of vitamin E differ considerably in their biological

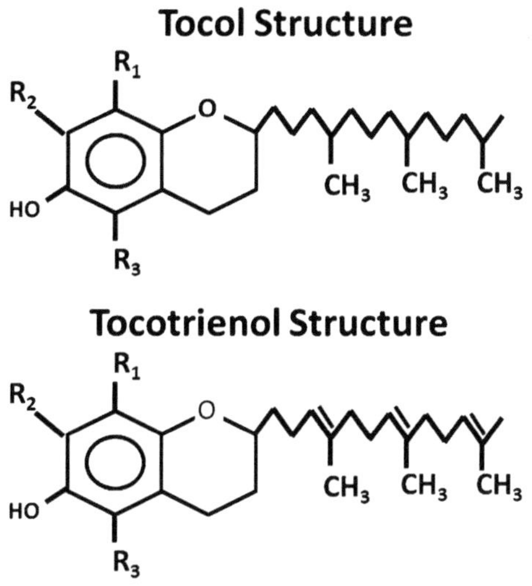

Fig. 8.9 Structures of tocol and tocotrienol forms of vitamin E

Tocol Structure

Tocotrienol Structure

activity. This is due both to differences in their intrinsic antioxidant activities and to important differences in bioavailability. α-tocopherol is the most important form of the vitamin for human nutrition, accounting for about 90 % of the vitamin E in tissues.

In experimental animals, vitamin E deficiency causes many different types of physiological impairment and tissue damage, depending on the species studied and on dietary and other factors producing oxidant stresses. **Severe vitamin E deficiency** in humans occurs mainly in premature infants and in patients with diseases leading to impaired fat absorption. In patients with malabsorption syndromes, vitamin E deficiency causes a *reduced red blood cell life span, neurologic dysfunction,* and *myopathies*. In premature infants, deficiency has been linked to *anemia, intraventricular hemorrhage* in the brain, *lung disease (bronchopulmonary dysplasia),* and *blindness (retrolental fibroplasia)*. Vitamin E nutritional status has also been associated with development of certain *cancers, heart disease,* and *impaired immune responses*.

7.2 Dietary Sources of Vitamin E

The richest sources of vitamin E in the American diet are **vegetable oils** (soybean, corn, cottonseed, and safflower) and products made from these oils, such as margarine and shortening. Wheat germ, nuts, and green leafy vegetables also contain appreciable amounts of this nutrient. Data from studies in experimental animals and to some extent in humans indicate that the requirement for vitamin E increases as dietary polyunsaturated fatty acid intake increases.

7.3 Intestinal Absorption of Vitamin E

The fractional intestinal absorption of vitamin E has varied from 20 % to 80 % in different studies. The absorptive efficiency declines with increasing dose and is influenced by ingestion of other dietary lipids. There do not appear to be major differences in the extent or rate of absorption of the different forms of vitamin E. Bile salts are required for solubilization of the vitamin in mixed micelles. Pancreatic secretions also facilitate vitamin E absorption, but this is likely due to effects on the formation of mixed micelles rather than a specific action on dietary vitamin E digestion or uptake. For pharmaceutical purposes, vitamin E is often given as **α-tocopheryl acetate** because of its greater stability. α-tocopheryl acetate is hydrolyzed prior to intestinal absorption, and only α-tocopherol appears in the circulation. Pancreatic and enterocyte enzymes are responsible for the hydrolysis of α-tocopheryl acetate.

Some of the same transport proteins that mediate cholesterol absorption are involved in vitamin E absorption. SR-B1 has been shown to mediate both the uptake and efflux of vitamin E across the brush-border membrane. NPC1L1 is involved in the apical membrane uptake of both α- and γ-tocopherol. Several intracellular vitamin E binding proteins have been identified in different tissues, but their role in intracellular vitamin E transport in the enterocyte has not been established.

Most of the absorbed vitamin E is secreted from the intestine in chylomicrons, but some is transported via the basolateral ABCA1 transporter and incorporated into HDL particles. Some absorption of vitamin E directly into the portal venous circulation has been demonstrated in laboratory animals.

7.4 Vitamin E Transport in Plasma, Tissue Uptake, and Catabolism

As chylomicrons are metabolized in muscle, adipose tissue, and other organs, some vitamin E is taken up by these tissues. **Lipoprotein lipase** plays a role in the transfer of vitamin E. There is also movement of vitamin E from chylomicrons to circulating **high-density lipoproteins (HDL)**. In turn, HDL particles can transfer the newly acquired vitamin E to all of the other plasma lipoproteins. The vitamin E remaining in chylomicron remnants is taken up into the liver and is re-secreted in **very low-density lipoproteins (VLDL)**. There is preferential incorporation of the stereoisomer RRR-α-tocopherol into nascent VLDL for secretion into the plasma, whereas other forms of vitamin E are largely metabolized in the liver and targeted for excretion in bile or urine. For example, γ-tocopherol is abundant in the human diet and is absorbed as efficiently as α-tocopherol, but relatively little is retained in tissues. In the hepatocyte, **α-tocopherol-transfer protein (α-TTP)** has binding specificity for α-tocopherol, and selectively directs that form of vitamin E for incorporation into VLDL and secretion from the liver. **Mutations in the gene encoding α-TTP** cause *vitamin E deficiency* and *neurological abnormalities* such as *ataxia*. Vitamin E absorption is normal in these patients, but incorporation of vitamin E into VLDL and hepatic secretion is impaired.

The α-tocopherol secreted in VLDL can have several metabolic fates. Some will be taken up by peripheral tissues or transferred to HDL during lipolysis of the VLDL triglyceride, as was described previously for chylomicrons. Some α-tocopherol will remain associated with VLDL during its metabolism and return to the liver with VLDL remnants or become part of low-density lipoproteins (LDL). Cells containing the LDL receptor will efficiently take up LDL α-tocopherol by a receptor-mediated process. Vitamin E is also incorporated into HDL secreted by the liver into the circulation using the ABCA1 transporter, and HDL likely can deliver some α-tocopherol to peripheral tissues. It should be appreciated therefore that multiple pathways exist for supplying peripheral tissues with α-tocopherol, using chylomicrons, VLDL, LDL, and HDL as transport vehicles. It is not surprising therefore that **in situations where one of these mechanisms is defective, normal tissue α-tocopherol levels are** *still* **attainable**. For example, patients with homozygous familial hypercholeserolemia, who have defective LDL-receptor activity, do not have clinical or biochemical vitamin E deficiency.

The **liver** is the primary site of α-tocopherol catabolism and excretion. The primary hepatic oxidation product is **α-tocopherol quinone,** which is further reduced to the **hydroquinone**. This compound is then conjugated with glucuronic acid and excreted in bile or degraded in the kidney to **α-tocopheronic acid**, followed by conjugation and elimination in the urine. Several transporters including SR-B1, NPC1L1, and the multidrug resistance 3 protein appear to mediate vitamin E transport across the canalicular membrane.

8 Vitamin K

8.1 Structure and Biological Activity

Vitamin K refers to a group of compounds that *all contain* a **2-methyl-l, 4-naphthoquinone-ring structure**, but *differ* in the structure of the **side chain** at the **carbon 3 position** (Fig. 8.10). The form found in plants, **phylloquinone** (also called **phytonadione, vitamin K_1**) contains a 20-carbon phytyl group at carbon-3. Bacteria synthesize a family of compounds known as **menaquinones (vitamin K_2)** that contain polyisoprenyl side chains 4–13 isoprenyl units long. Animal tissues contain small amounts of both phylloquinone and menaquinones. **Menadione (vitamin K_3)** is a synthetic compound that has no side chain, but its water-soluble derivatives are alkylated in the liver to biologically active menaquinones. Menadione, however, can combine with sulfhydryl groups in membranes and cause *hemolytic anemia, hyperbilirubinemia,* and *kernicterus* in infants and, therefore, should not be used as a therapeutic form of vitamin K.

The biological function of vitamin K is to serve as an essential **cofactor for a specific post-translational modification** of certain proteins in which selected

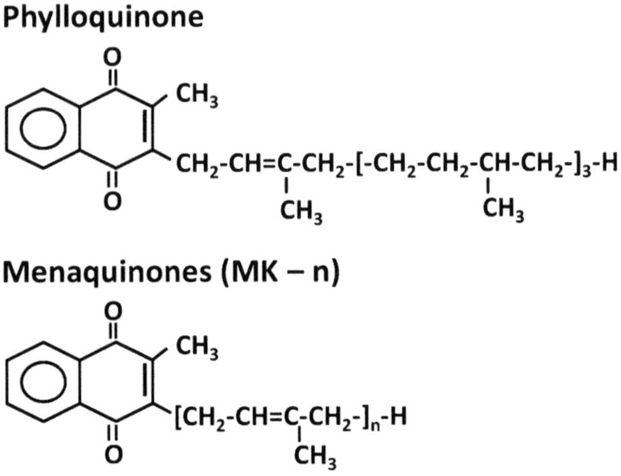

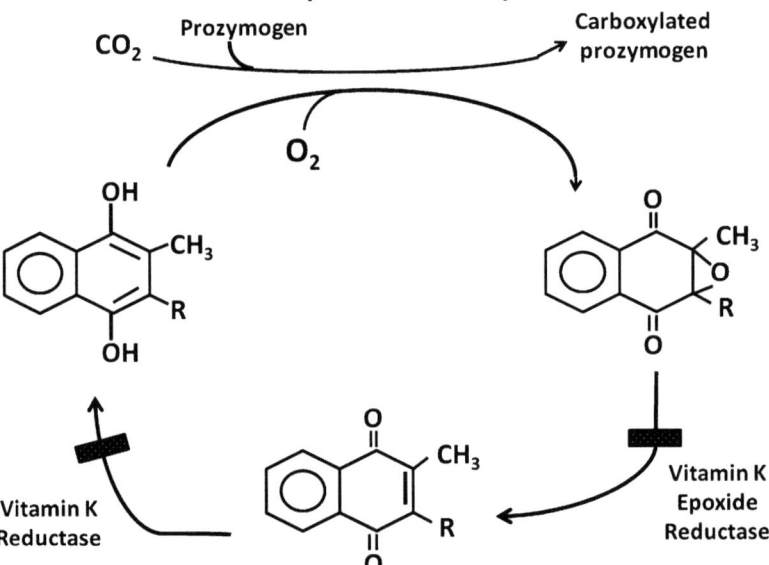

Fig. 8.10 Structures of phylloquinone and menaquinone forms of vitamin K

Fig. 8.11 Vitamin K cycle. Post-translational modification of prothrombin by vitamin K. *Shaded bars*, sites of inhibition by warfarin anticoagulants (Adapted from American Gastroenterological Association Teaching Slide Collection 13B ©, Bethesda, Maryland, slide 68; Used with permission)

glutamic acid residues are transformed to γ-carboxyglutamic acid (Gla) residues (Fig. 8.11). **Four clotting factors** synthesized in liver, *factors II (prothrombin), VII, IX,* and *X,* undergo this post-translational modification, which is required for these factors to bind calcium and interact with phospholipid in cellular membranes. **Vitamin K deficiency** is manifested clinically by *defective blood coagulation* and

8 Digestion and Absorption of Other Dietary Lipids

excessive bleeding; the serum of vitamin K-deficient patients contains inactive clotting factors lacking the Gla residues. Additional proteins in liver and in other tissues also undergo **vitamin K-dependent γ-carboxylation**, including proteins C and S, which are inhibitors of coagulation, and the bone proteins osteocalcin and matrix Gla-protein, which play a role in calcification of bone.

8.2 Dietary Sources of Vitamin K

Phylloquinone is the major dietary form of vitamin K. **Green leafy vegetables** are the best source, providing about 100–500 µg/100 g of food. Smaller amounts of vitamin K are present in milk and other dairy products, meats, eggs, cereals, other vegetables, and fruits, oils, and margarine. Human milk is relatively low in vitamin K, containing about 2 µg/L.

8.3 Intestinal Absorption of Phylloquinone

Phylloquinone is **highly insoluble in water**, and bile salts are therefore required for solubilization in mixed micelles. **Severe vitamin K malabsorption** occurs in patients with *cholestatic liver diseases*, and vitamin K deficiency is more common in this group than in other gastrointestinal diseases. Pancreatic secretions also facilitate phylloquine absorption, probably by generating fatty acids and β-monoglycerides that are components of mixed micelles.

Studies of phylloquinone absorption in the rat demonstrated that intestinal uptake was saturable and dependent on metabolic energy. These features suggest the involvement of a plasma membrane transporter or intracellular binding proteins, but these components have not been definitively identified to date. Other lipids such as fatty acids, monoglycerides, and phospholipids have been found to influence phylloquinone absorption in experimental animal models.

Most of the absorbed phylloquinone is incorporated into chylomicrons and transported via intestinal lymph to the circulation. Phylloquinone is taken up into the liver as a component of chylomicron remnants, and subsequently some is secreted from the liver in VLDL and HDL and distributed to other tissues. A portion of the absorbed phylloquinone is released from the intestine into the portal venous circulation.

8.4 Absorption and Utilization of Menaquinones

The extent to which menaquinones produced by bacteria in the GI tract are absorbed and utilized in the vitamin K dependent γ-carboxylase reaction is very controversial. Measurements of the vitamin K content of human liver have shown that 75–90 % of hepatic vitamin K is in the form of various menaquinones, with menaquinones-7, -9, and -11 being the predominant forms. Hepatic menaquinones could represent absorption of the small amount of menaquinones present in the diet or absorption of menaquinones derived from gut bacterial production, with accumulation in the liver due to a relatively slow turnover rate. Studies using various experimental preparations of rat small and large intestine have demonstrated limited uptake of menaquinones into all intestinal segments. **Menaquinone-4**, a minor bacterial form, was slowly absorbed from the jejunum and appeared in both mesenteric lymph and portal blood, whereas absorption from the colon occurred mainly into the portal circulation. In contrast, **menaquinone-9**, a typical bacterial form, was slowly absorbed from the jejunum and transferred mainly to lymph, whereas no transfer to lymph or to the portal circulation was seen when menaquinone-9 was administered in the colon. These data suggest therefore that menaquinones ingested in the diet, produced by small bowel bacteria, or synthesized by colonic bacteria can, at least to some extent, be absorbed and delivered via the mesenteric lymph and portal blood to the circulation. In the colon, however, there is only limited absorption of menaquinones into the portal circulation, and the absorption rate markedly declines with an increase in the number of isoprenoid units in the side chain. These differences likely reflect the low concentration of bile salts and the lack of formation of chylomicrons in the colon.

Studies in rats and in humans have demonstrated that consuming a vitamin K-deficient diet rapidly causes a fall in hepatic and plasma phylloquinone, with little change in liver menaquinones, and induces biochemical changes of vitamin K deficiency. In some studies, however, relatively mild vitamin K deficiency was produced, and in others the biochemical changes reverted toward normal in spite of continuing the vitamin K-deficient diet. These data can, therefore, be interpreted as indicating that **gut bacterial vitamin K production is not sufficient to ensure completely normal vitamin K nutritional status, but may contribute sufficiently to prevent severe deficiency**. Human studies and clinical experience indicate that clinically apparent vitamin K deficiency and bleeding rarely occurs due to dietary lack, but is regularly seen in patients with the combination of poor dietary intake and use of broad-spectrum antibiotics. Interpretation of this observation is complicated as some antibiotics have direct effects on coagulation independent of alteration of gut bacterial flora.

8.5 Hepatic Vitamin K Metabolism

The forms of vitamin K absorbed from the GI tract have a stable quinone structure. The **vitamin K-dependent carboxylase** that catalyzes the conversion of glutamate residues to γ-carboxyglutamate residues uses the quinol form of the vitamin (Fig. 8.12). This carboxylase also requires molecular oxygen, carbon dioxide, and the precursor of the vitamin K-dependent protein as a substrate. **Vitamin K-dependent clotting factors** contain a **carboxylation recognition site** in the proprotein region that designates an adjacent glutamic acid-rich domain for

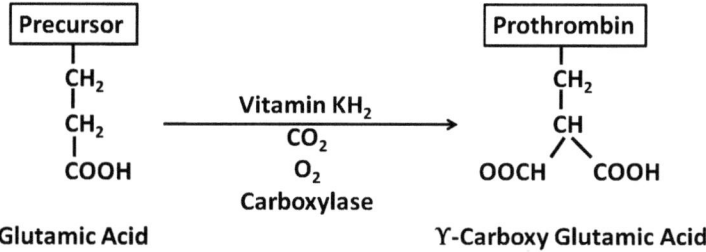

Fig. 8.12 Vitamin K-dependent post-translational modification of prothrombin to the ϒ-carboxylated form (Adapted from American Gastroenterological Association Teaching Slide Collection 13B ©, Bethesda, Maryland, slide 69; Used with permission)

γ-carboxylation. This γ-carboxylation recognition site probably directly binds to the vitamin K-dependent carboxylase, which is an integral endoplasmic reticulum membrane protein. Following γ-carboxylation, the substrate protein is released and transported to the Golgi apparatus for further processing and secretion. During **γ-glutamyl carboxylation**, the vitamin K quinol is simultaneously converted to the **vitamin K 2,3-epoxide** (Fig. 8.12). Evidence indicates that the same enzyme contains both the carboxylase and epoxidase activities. The epoxide is then recycled by two membrane-bound enzyme activities. First, the epoxide is reduced to vitamin K quinone by a **dithiol-dependent reductase**; and second, the quinone is reduced back to the active quinol form. Again, these two reactions may be catalyzed by the same enzyme. The **warfarin-type anticoagulants** are potent *inhibitors* of the activities of dithiol-dependent vitamin K-epoxide reductase and vitamin K-reductase. Vitamin K is therefore trapped in the epoxide form, resulting in deficiency of the active quinol, reduced γ-glutamyl carboxylation, and diminished production of active clotting factors. A second **NADPH-dependent pathway** that is relatively *insensitive* to warfarin exists for reduction of vitamin K to the quinol form. In the **treatment of warfarin overdose** with **large doses of vitamin K**, vitamin K is converted to the quinol using this alternative pathway, reversing the anticoagulant effect.

Under normal physiological conditions, 30–40 % of absorbed vitamin K is excreted into bile as partially degraded, conjugated, water-soluble metabolites, whereas 15 % is excreted as water-soluble forms in the urine.

Clinical Correlations

Case Study 1

A 42-year-old woman with a history of Crohn's disease for many years had a resection of 3 ft of terminal ileum 5 years ago. Since then, she has taken cholestyramine, a bile salt-binding resin, for chronic diarrhea. She now complains of diffuse bone pain, most severe in the lower back and upper legs, and is noted to be hypocalcemic. Her serum 25-hydroxyvitamin D_3 level is 5 ng/mL (normal: 10–50 ng/mL).

Questions

1. **Why would a patient with an ileal resection develop vitamin D deficiency?**
   Answer: Patients with a resection of 3 ft of ileum have **bile salt malabsorption**, diminished enterohepatic circulation, and a reduced bile salt pool. Because of intraluminal bile salt deficiency, they have **malabsorption of dietary lipids**, including **vitamin D**.
2. **How might the use of cholestyramine contribute to her vitamin D depletion?**
   Answer: Malabsorbed bile salts cause colonic fluid and electrolyte secretion and diarrhea. A **bile salt binding resin** can therefore be *useful* in the *treatment of watery diarrhea* in patients with **small ileal resections** (less than 3 ft), where increased hepatic bile synthesis maintains the bile salt pool size. In patients with **larger ileal resections**, increased bile salt synthesis cannot maintain a normal pool size, cholestyramine treatment will further reduce the enterohepatic circulation and the intraluminal bile salt concentration and will *worsen* lipid malabsorption.
3. **Why might treatment with $25(OH)D_3$ orally be preferable to supplementation with vitamin D_3?**
   Answer: $25(OH)D_3$ has been shown to be **more efficiently absorbed** than vitamin D_3 in patients with ileal resections. Absorption of $25(OH)D_3$ is less dependent on intraluminal bile salts, probably because it has somewhat greater solubility in an aqueous system than does vitamin D_3.

Case Study 2

A 14-year-old boy with cystic fibrosis has pancreatic insufficiency and is being treated with enzyme replacement. Because of worsening pulmonary symptoms, he is admitted to the hospital for intensive respiratory therapy and broad spectrum antibiotics. At the time of admission, he has a normal prothrombin time, a serum alkaline phosphatase that is five times normal, and serum transaminases that are twice normal. After 1 week of treatment in the hospital, he complains of a severe nosebleed and is now noted to have a prothrombin time of 21 s (normal, 10–13 s).

Questions

1. **What factors are responsible for the development of an abnormal prothrombin time during his hospitalization?**
   Answer: The patient would likely have some degree of **vitamin K malabsorption** due to his pancreatic insufficiency. Cystic fibrosis can also cause cholestatic liver disease. An **elevated prothrombin time**, however, occurs when the *level of active γ-carboxylated prothrombin falls to less than 30 % of normal*, a deficit that correlates with a clinically significant bleeding risk. During the hospitalization, he may have been eating more poorly than usual because of his worse pulmonary symptoms. In addition, the **broad-spectrum antibiotics** would alter his intestinal flora and decrease bacterial menaquinone synthesis. Although the contribution of these menaquinones to vitamin K nutritional status is still controversial, most data suggest that they are absorbed to some extent and may

provide enough vitamin K to prevent severe deficiency. Some antibiotics directly interfere with coagulation independent of effects on gut flora.
2. **How does pancreatic insufficiency cause vitamin K malabsorption?**
   Answer: Dietary vitamin K does not require intraluminal digestion. The vitamin K malabsorption seen in pancreatic insufficiency is likely due to **abnormal formation of mixed micelles** and to the **trapping of vitamin K within lipid droplets** composed of undigested triglyceride and other lipids.
3. **How might the patient's abnormal liver chemistries relate to his vitamin K depletion?**
   Answer: Some patients with cystic fibrosis develop *cholestasis* and *biliary cirrhosis* due to inspissation of bile in the biliary tract. Cholestasis leads to **severe vitamin K malabsorption**, as bile salts are required for solubilization of vitamin K in mixed micelles.

Case Study 3

A 70-year-old man has extensive diverticulosis of the small intestine. He complains of chronic diarrhea, bloating, and a 20-lb weight loss. On laboratory testing, he is noted to have 30 g/day of steatorrhea while ingesting a 100 g/day-fat diet. A bile acid breath test is performed, in which cholyl-^{14}C-glycine is given by mouth and breath samples are collected for measurement of $^{14}CO_2$. The patient is found to have a markedly elevated breath $^{14}CO_2$ excretion.

Questions

1. **How do you explain the results of the bile acid breath test?**
   Answer: In a normal individual, the administered cholyl-^{14}C-glycine is efficiently absorbed by the **ileal Na$^+$-bile salt cotransporter** and enters the enterohepatic circulation. In this patient with small bowel diverticuli, bacteria proliferate in the stagnant environment of the diverticuli. These bacteria deconjugate the ^{14}C-glycine and the ^{14}C-glycine is absorbed and metabolized, forming $^{14}CO_2$ that appears in the breath.
2. **Why would bile salt deconjugation result in fat malabsorption?**
   Answer: Unconjugated bile acids are, to some extent, passively absorbed by the small bowel. Deconjugation of bile acids by bacteria in the proximal intestine will, therefore, result in **absorption of the unconjugated bile acids in the proximal jejunum** and cause a decrease in the intraluminal bile salt concentration in the remaining small bowel. In addition, unconjugated bile acids have **limited solubility** in an aqueous system and will tend to precipitate in the luminal contents. These two factors can result in intraluminal bile salt deficiency and fat malabsorption.
3. **What other disease process would give a similar bile acid breath test?**
   Answer: Ileal dysfunction causes bile salt malabsorption. The bile salts are then deconjugated in the colon, and the $^{14}CO_2$ is excreted in breath. Thus, an abnormal bile acid breath test is also characteristic of patients with **ileal disease** or **resection**. Although on average, the $^{14}CO_2$ appears in the breath earlier in patients with small bowel bacterial overgrowth than in those with

ileal dysfunction, intestinal transit time is so variable that this cannot be used to reliably distinguish between these two causes of an abnormal bile salt breath test.

Case Study 4

A 55-year-old woman has advanced primary biliary cirrhosis. She is jaundiced with a serum bilirubin of 11.0 g/dl. Laboratory testing shows that she has a very low serum retinol level.

Questions

1. **Why would this woman have vitamin A malabsorption?**
   Answer: Patients with advanced primary biliary cirrhosis have cholestasis and intraluminal bile salt deficiency. Bile salts are required for solubilization of retinol and carotenes in mixed micelles, permitting efficient intestinal absorption. Pancreatic bile salt-activated lipase is also one of the enzymes that hydrolyze retinyl esters prior to intestinal absorption.
2. **What other factor would contribute to the very low serum retinol level in this patient?**
   Answer 2: Retinol is secreted from the liver into the circulation bound to serum retinol binding protein (RBP). Hepatic synthesis and secretion of RBP would also be impaired in advanced liver disease, and both serum RBP and serum retinol levels would be low in this patient.
3. **What symptoms might this patient experience?**
   Answer 3: The patient might have visual problems, particularly poor vision when in dim light (night blindness). She could also have a skin rash, damage to the conjunctivae leading to eye infections and eventually blindness, immune dysfunction, and bone disease.

Further Reading

1. Abumrad NA, Davidson NO (2012) Role of the gut in lipid homeostasis. Physiol Rev 92: 1061–1085
2. Brown JM, Yu L (2010) Protein mediators of sterol transport across intestinal brush border membrane. Subcell Biochem 51:337–380
3. Cohn JS, Wat E, Kamili A, Tandy S (2008) Dietary phospholipids, hepatic lipid metabolism and cardiovascular disease. Curr Opin Lipidol 19:257–262
4. D'Ambrosio D, Clugston RD, Blaner WS (2011) Vitamin A metabolism: an uptake. Nutrients 3:62–103
5. Harrison E (2012) Mechanisms involved in the intestinal absorption of dietary vitamin A and provitamin A carotenoids. Biochim Biophys Acta 182:70–77
6. Henry HL (2011) Regulation of vitamin D metabolism. Best Pract Res Clin Endocrinol Metab 25:531–541
7. Hofmann A (1994) Intestinal absorption of bile acid and biliary constituents. In: Johnson LR (ed) Physiology of the gastrointestinal tract, 3rd edn. Raven Press, New York, pp 1845–1865
8. Iqbal J, Hussain MM (2008) Intestinal lipid absorption. Am J Physiol Endocrinol Metab 296:E1183–E1194

9. Lemair-Ewing S, Desrumaux C, Neel D, Lagrost L (2010) Vitamin E transport, membrane incorporation and cell metabolism: is α-tocopherol in lipid rafts an oar in the lifeboat? Mol Nutr Food Res 54:631–640
10. Ong DE (1994) Absorption of vitamin A. In: Blomhoff R (ed) Vitamin A in health and disease. Marcel Dekker, New York, pp 37–72
11. Reboul E, Borel P (2011) Proteins involved in uptake, intracellular transport and basolateral secretion of fat-soluble vitamins and carotenoids by mammalian enterocytes. Prog Lipid Res 50:388–401
12. Reboul E, Goncalves A, Comera C, Bott R, Nowicki M, Landrier JF, Jourdheuil-Rahmani D, Dufour C, Collet X, Borel P (2011) Vitamin D intestinal absorption is not a simple passive diffusion: evidences for involvement of cholesterol transporters. Mol Nutr Food Res 55: 691–702
13. Shearer MJ, Fu X, Booth SL (2012) Vitamin K nutrition, metabolism, and requirements: current concepts and future research. Adv Nutr 3:182–185
14. Takada T, Suzuki H (2010) Molecular mechanisms of membrane transport of vitamin E. Mol Nutr Food Res 54:616–622
15. Wang DQH (2007) Regulation of intestinal cholesterol absorption. Annu Rev Physiol 69: 221–248
16. Wang LJ, Song BL (2012) Niemann-Pick C1-Like 1 and cholesterol uptake. Biochim Biophys Acta 1821:964–972
17. Wolf G (1991) The intracellular vitamin A-binding proteins: an overview of their functions. Nutr Rev 49:1–12

Chapter 9
Absorption of Water-Soluble Vitamins and Minerals

Michael D. Sitrin

1 Introduction

The **nine water-soluble vitamins**, *thiamine, riboflavin, niacin, pantothenic acid, folate, biotin*, and *vitamins B_6, B_{12}*, and *C* are a diverse group of organic compounds that are consumed in the daily diet in microgram to milligram amounts and are essential for normal growth, development, and maintenance of the human organism. These compounds are generally metabolized to forms that serve as **coenzymes** in various biochemical reactions; *vitamin C is an exception*, as it functions as an essential water-soluble antioxidant.

The mechanisms of intestinal absorption of the various water-soluble vitamins share some important **general characteristics**. The vitamins are usually *present in the diet as complex coenzyme forms* that must be digested intraluminally or at the brush-border membrane surface into simpler forms prior to transport across the intestinal epithelium. In addition, *dietary vitamins are often associated with proteins* (e.g. flavoproteins), and digestion of the protein component is needed to liberate the vitamin prior to absorption. At the **low concentrations** present in the diet (typically 10^{-9}–10^{-7} mol/L), transport of the vitamins across the brush-border membrane occurs by *specialized mechanisms*, such as membrane carriers, active transport systems, and membrane binding proteins and receptors, that are specific for a particular vitamin. At **higher intraluminal concentrations** attained with pharmacologic vitamin supplementation, uptake occurs via *passive diffusion*, either transcellularly or through the paracellular pathway. Extensive metabolism of water-soluble vitamins occurs within the enterocyte, and metabolism may be coupled to the rate of uptake. Mechanisms for extrusion of the vitamins from the enterocyte are

M.D. Sitrin, M.D. (✉)
Department of Medicine, University at Buffalo, The State University of New York, Buffalo, NY, USA
e-mail: mdsitrin@buffalo.edu

Table 9.1 Mechanisms of water-soluble vitamin absorption

Vitamin	Luminal Events	Transport
Thiamin	Hydrolysis of phosphorylated form	Na-independent, pH dependent and amiloride sensitive uptake; 2 carriers: one both apical and basolateral, one only apical
Folate	Hydrolysis of folate polyglutamates	Acid pH-dependent carrier; neutral pH carrier
Biotin	Digestion of protein-bound biotin	Na-dependent carrier
Vitamin C		Ascorbic acid - Na-dependent carriers, one apical and one basolateral; dehydro-L-ascorbic acid - glucose transporters, GLUT1,3,4
Vitamin B_6	Hydrolysis of phosphorylated form	Na-independent, acid pH dependent amiloride sensitive carrier
Riboflavin	Digestion of protein-bound riboflavin	Na-independent carrier
Niacin		Na-independent, acid pH-dependent carrier
Pantothenic acid	Digestion of protein-bound pantothenic acid	Same carrier as biotin
Vitamin B_{12}	Release of bound B_{12} by acid-peptic digestion; binding of B_{12} to haptocorrin; release of B_{12} from haptocorrin by pancreatic enzymes; complex with intrinsic factor	Membrane receptor (cubilin); endocytosis and processing via endosomal-lysosomal pathway; B_{12} enters transcobalamin-containing secretory vesicles, transports through basolateral membrane, and binds to transcobalamin II

less understood, but they often involve membrane carriers and specialized transport proteins. Table 9.1 summarizes some of the features of intestinal absorption of the water-soluble vitamins.

Essential mineral elements must also be absorbed to maintain normal physiological function and health. The typical diet contains several hundred milligrams per day of **macrominerals** such as *calcium, sodium, potassium, magnesium, chloride, phosphorus,* and *sulfur*. **Microminerals or trace elements** are present in smaller amounts, ranging from a few milligrams to micrograms per day. Essential trace elements including *chromium, cobalt, copper, fluoride, iodide, iron, manganese, molybdenum, selenium,* and *zinc* have well-established functions in human physiology. Other trace elements (*arsenic, boron, cadmium, nickel, silicone, tin, vanadium*) have physiological functions in some species, but an essential role in human metabolism had not been clearly defined. Minerals and trace elements are also present in various gastrointestinal (GI) secretions and are variably reabsorbed by the intestine. Minerals serve diverse physiological roles, including structural functions (bone minerals), components of metalloproteins (enzymes, transporters) and as ions involved in neurotransmission, muscle function, regulation of fluid and acid-base balance, energy gradients, and as second messengers.

Intraluminal factors substantially affect the efficiency of mineral and trace element absorption by altering the following processes: (**1**) intraluminal pH; (**2**) redox state of the metal; (**3**) formation of chelates that enhance solubility of the mineral; (**4**) formation of insoluble complexes that diminish absorption; and (**5**) digestion of proteins that are associated with dietary minerals. Kinetic studies of mineral uptake

Table 9.2 Common mineral interactions and antagonisms[a]

Mineral	Condition or state	Effect on net absorption
Iron	Fe deficiency	↑
	Fe excess	↓
	Mn excess	↓
	Co excess	↓
	Conditions favoring Fe^{2+}	↑
	Conditions favoring Fe^{3+}	↓
Copper	Cu deficiency	↑
	Cu excess	↓
	Zn excess	↓
	Cd excess	↓
	Ag excess	↓
	Conditions favoring Cu^{+}	↓
	Conditions favoring Cu^{2+}	↑
Zinc	Zn deficiency	↑
	Zn excess	↓
	Cu excess	↓
	Cd excess	↓
Manganese	Mn deficiency	?
	Mn excess	?
	Fe excess	↓
	Co excess	↓

From Rucker et al. [11]

[a]For given antagonists, a significant effect on absorption is often observed at intakes corresponding to 5 to ten times the normal requirements (i.e. Fe, Cu, Mn, Zn) or at concentrations >2–4 μmol/L (150–300 ng/g of intestinal content) in the case of Ag, Cd or Co when present

across the small intestine brush-border membrane generally have been consistent with *facilitated diffusion* or *active transport* when studied at **low physiological concentrations**, although many of these putative carriers have not been definitively identified. At **higher concentrations**, intestinal uptake via *passive diffusion*, either paracellularly or transcellularly, becomes quantitatively more important. Within the enterocyte, minerals often associate with intracellular ligands that play critical roles in regulating absorption and delivery of these elements into the circulation. Mechanisms for extrusion of minerals across the basolateral membrane into the portal circulation are incompletely characterized, but associations with transport proteins and other ligands are important. Effects of one mineral on the absorption of others are commonly observed, reflecting interactions in the luminal environment and shared mechanisms for absorption (Table 9.2). For example, high doses of zinc interfere with copper absorption. Individuals chronically taking zinc supplements for prevention of colds, etc. may develop copper deficiency with anemia, neutropenia, and bone disease. Zinc is used in the treatment of Wilson's disease, a genetic disorder characterized by diminished biliary copper excretion and copper overload. Zinc treatment decreases intestinal copper absorption, contributing to a reduction in body copper.

A detailed consideration of the absorption of each of the water-soluble vitamins, minerals, and trace elements is beyond the scope of this book. In this chapter, *two water-soluble vitamins*, **folate** and **vitamin B_{12}**, and *two minerals*, **iron** and **calcium**, will be considered in depth as illustrations of the important principles of intestinal absorption of these classes of nutrients.

2 Folate

2.1 Structure and Biochemical Function

The term folate denotes a family of compounds with nutritional properties and biochemical structures similar to the reference compound, **folic acid** or **pteroylglutamic acid** (Fig. 9.1). Folic acid comprises **three moieties**: a *pteridine* linked by a methylene bridge to *para-aminobenzoic acid* (*PABA*) and joined by a peptide bond to *glutamic acid*. The coenzyme forms of folate have reduced pteridine rings at positions 5, 6, 7, and 8 and one-carbon additions at N-5 or N-10. In addition, naturally occurring folates are mainly in the polyglutamate form, in which up to nine glutamate residues are conjugated via a unique γ-glutamyl bond forming a peptide chain (Fig. 9.1).

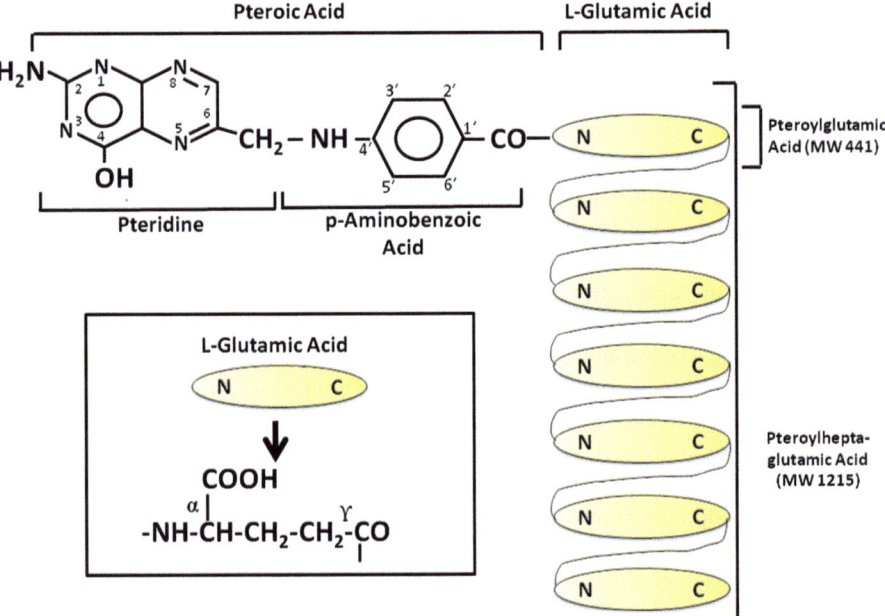

Fig. 9.1 Structure of conjugated folates (Adapted from Mason and Rosenberg [7])

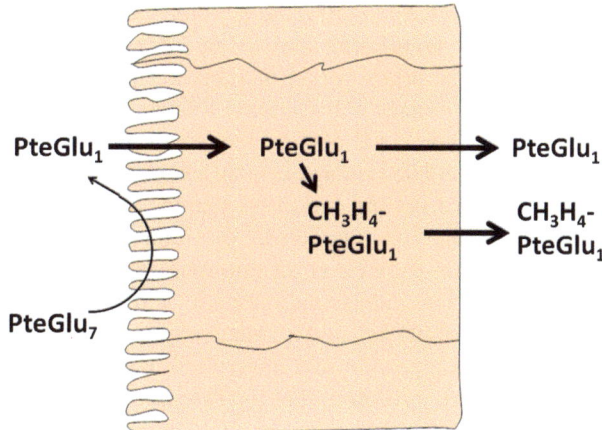

Fig. 9.2 Enterocyte metabolism and absorption of conjugated folates. $PteGlu_7$ heptaglutamyl folate, $PteGlu_1$ monoglutamyl folate, $CH_3H_4PteGlu_1$ 5-methyltetrahydrofolate (Adapted from Mason and Rosenberg [7])

Folate is widely distributed in foods, with liver, yeast, leafy vegetables, legumes and some fruits being especially rich sources. Only a few foods, such as milk and egg yolk, contain principally monoglutamyl forms, whereas organ meats contain mainly penta- and heptaglutamates. Folic acid is the most common form of the vitamin in pharmaceutical preparations and has been used in much of the research on intestinal folate absorption. A substantial amount of folate is also secreted in bile, mainly as **monoglutamyl 5-methyltetrahydrofolate**, and must be reabsorbed by the intestine to maintain the normal folate economy.

Folates function metabolically as **coenzymes** in biochemical reactions that transfer one-carbon units from one compound to another. These reactions are important in amino acid metabolism and in nucleic acid synthesis. **Folate deficiency** therefore impairs cell division and alters protein synthesis, with the most prominent effects noted in rapidly growing tissues. The most common clinical manifestation of human folate deficiency is a *macrocytic anemia*, characterized by abnormally large red blood cells (**macro-ovalocytes**). *Hypersegmentation of the chromatin of circulating neutrophils* is also observed, and with *severe* folate deficiency *neutropenia* and *thrombocytopenia* may also be present. **Bone marrow examination** demonstrates abnormal precursors of red blood cells (megaloblasts), neutrophils, and platelets resulting from defective DNA synthesis.

2.2 Hydrolysis of Polyglutamyl Folates

In the lumen of the small intestine, **polyglutamyl folates** must be hydrolyzed to the **monoglutamyl form** before transport through the enterocyte (Fig. 9.2). Because of the unique **γ-glutamyl linkage**, polyglutamyl folates are not hydrolyzed by typical pancreatic or intestinal proteases, and specific enzymes known as **folate conjugases** are required.

In humans, folate conjugase activities are present in both the *small intestinal brush-border membrane* and in an *intracellular* fraction composed mainly of lysosomes. The brush-border membrane enzyme appears to be responsible for the hydrolysis of dietary polyglutamyl folates. This enzyme has a pH optimum of pH 6.5–7.0, the typical pH of the upper small intestine, and is activated by Zn^{2+}. The brush-border folate conjugase is an **exopeptidase** that sequentially cleaves glutamate residues from the end of the peptide chain, eventually producing the monoglutamate form. The function of the intracellular conjugase enzyme is currently unknown. This enzyme is an **endopeptidase**, cleaving the polyglutamate chain between the first and second glutamic acid residues. Intracellular conjugase has a pH optimum of pH 4.5, and no metal requirement for this enzyme has been defined. During folate digestion and absorption, monoglutamyl folates accumulate in the intestinal fluid, indicating that the *rate-limiting step* is the **absorption of monoglutamyl folates** rather than deconjugation of polyglutamate forms. Studies have shown that alcohol decreases brush-border conjugase activity, a factor that may contribute to the high prevalence of folate deficiency in alcoholics. Several drugs have also been shown to inhibit folate deconjugation, including the anti-inflammatory medication *sulfasalazine* used in the treatment of inflammatory bowel disease.

2.3 Absorption of Monoglutamyl Folates

Studies of the intestinal uptake of monoglutamyl folates using different experimental preparations have consistently demonstrated the presence of both **saturable** and **nonsaturable uptake mechanisms**. Saturable uptake appears to reflect carrier-mediated transport, whereas the nonsaturable component is due to diffusion through the cell membrane and/or through the paracellular pathway. At physiological concentrations, carrier-mediated transport accounts for most of the folate absorption, whereas diffusion predominates at high folate concentrations.

Two carriers involved in intestinal folate absorption have been identified, **RFC** (**reduced folate carrier**, the product of the SLC19A1 gene) and **PCFT** (the product of the SLC46A1 gene). RFC is expressed in the *intestinal cell brush-border membrane* and functions at neutral pH. PCFT is expressed mainly in the *apical membrane of jejunal enterocytes* with low expression in the *ileum* and *colon*. Transport of folate through the PCFT system is proton coupled and occurs via a **folate-proton symport** using energy generated from the downhill movement of protons into the enterocyte. It is possible that PCFT is responsible for folate absorption in the proximal small bowel where the luminal pH is somewhat acidic, whereas RFC plays a role in absorption of folate in the distal small intestine and colon. **Mutation of the PCFT transporter** results in *hereditary folate malabsorption*. Folic acid, reduced folates such as 5-methyltetrahydrofolate, and the antimetabolite methotrexate (a competitive inhibitor of the enzyme dihydrofolate reductase) all appear to share the same brush-border membrane carrier, with similar affinities for the transporter. Folate uptake, however, is structure-specific, as degradation products such as PABA or glutamic acid and inactive diastereoisomers of 5-methyltetrahydrofolate do not interact with the brush-border membrane transporter. In addition to its effect on folate deconjugation, *sulfasalazine* is also a **competitive inhibitor** of intestinal monoglutamyl folate transport. The mechanism of transport of folate across the basolateral enterocyte membrane is not well understood, but some data suggest involvement of **MDR (multidrug resistance) proteins**.

Bacteria in the intestine synthesize folate, with a substantial portion in the monoglutamate form. Although the contribution of colonic absorption of bacterially-derived folate via the RFC transporter expressed in the colonocyte brush-border membrane to folate economy is uncertain, it is possible that it plays a significant role in human nutrition, particularly supplying folate for colonocytes.

Folate digestion and absorption are both regulated by the level of folate in the diet. Folate deficiency causes an increase in folate conjugase and in carrier-mediated folate transport with increased expression of PCFT and RFC. Carrier-mediated folate transport and expression of RFC and PCFT increase as enterocytes mature to villus cells.

Some data have suggested an alternative mechanism for intestinal folate absorption in the neonate. Folate in milk is largely bound to a **high-affinity folate-binding protein**. In contrast to free folate, this protein-bound folate is more avidly absorbed in the ileum than jejunum and is not inhibited by sulfasalazine.

2.4 Folate Metabolism in the Enterocyte

At physiological concentrations, folic acid is largely reduced and methylated or formylated within the enterocyte. Appearance of folate in the blood is faster after intraluminal administration of 5-methyltetrahydrofolate folate than after folic acid, suggesting that **folic acid reduction in the enterocyte via dihydrofolate reductase** may be *rate-limiting*. At pharmacological concentrations, unmodified folic acid appears in the portal blood.

3 Vitamin B_{12}

3.1 Structure and Biochemical Function

The basic structure of vitamin B_{12} is illustrated in Fig. 9.3. A **central cobalt atom** is surrounded by a **planar corrin nucleus** comprising **four reduced pyrrole rings** linked together. Below the corrin nucleus is a nucleotide moiety (1-α-D-ribofuranosyl-5,6-dimethylbenzimidazole-3-phosphate) that lies at a right angle to the corrin nucleus and is joined to the rest of the molecule at two points: (**1**) via a **phosphodiester bond** to a 1-amino-2 propanol group and (**2**) through coordination to the central cobalt via one of its nitrogens. The various forms of

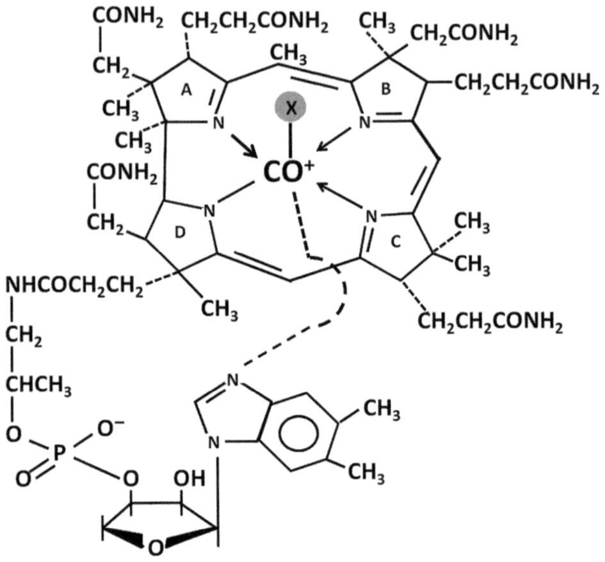

Fig. 9.3 Structure of vitamin B_{12}. X, axial ligand in coordinate linkage with cobalt

vitamin B_{12} have different anionic groups in coordinate linkage with the cobalt. The coenzyme forms of the vitamin are **methylcobalamin** and **adenosylcobalamin** (**5′-deoxyadenosylcobalmin**), formed via a unique carbon-cobalt bond. Other important forms of vitamin B_{12} are **hydroxocobalamin** and **cyanocobalamin**.

Vitamin B_{12} is synthesized *only* by microorganisms, and in the human diet it is almost entirely by animal products. Strict vegetarians are therefore at increased risk for vitamin B_{12} deficiency. The daily losses of vitamin B_{12} are, however, very small compared with the body pool size, and 10–20 years of a deficient diet is required to produce clinically significant depletion. Meat contains mainly adenosyl- and hydroxocobalamin, whereas dairy products have predominantly methyl and hydroxocobalamin. Cyanocobalamin is a stable form of the vitamin used in pharmaceutical preparations. Cyano- and hydroxocobalamin are readily converted to the coenzyme forms by enzyme systems found in the cytoplasmic and mitochondrial fractions. In **human plasma and tissue**, the predominant forms of vitamin B_{12} are *methylcobalamin*, *adenosylcobalamin*, and *hydroxocobalamin*. Bile contains a significant amount of vitamin B_{12} that is reabsorbed by the small intestine. The importance of this enterohepatic circulation in the maintenance of the vitamin B_{12} pool is indicated by the observation that patients with vitamin B_{12} malabsorption become deficient in only 2–3 years compared with the 10–20 years needed for deficiency to develop in individuals who lack dietary vitamin B_{12}, but normally conserve biliary cobalamins.

Vitamin B_{12} serves as a **coenzyme** for two important enzymatic reactions. Methylcobalamin is required for the conversion of homocysteine to methionine (Fig. 9.4). This reaction is catalyzed by the cytoplasmic enzyme **5-methyltetrahydrofolate-homocysteine methyltransferase**, which utilizes 5-methyltetrahydrofolate as a methyl donor. This pathway is therefore important to maintain the

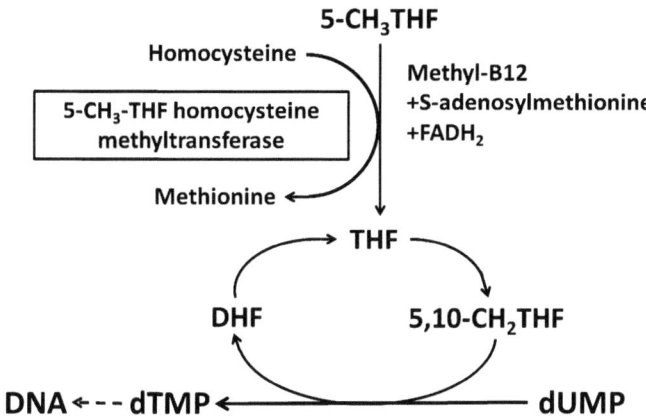

Fig. 9.4 Role of methylcobalamin in the conversion of homocysteine to methione. *5-CH₃THF* 5-methyltetrahydrofolate, *THF* tetrahydrofolate, *CH₂THF* methylenetetrahydrofolate, *DHF* dihydrofolate, *dUMP* uridylate, *dTMP* thymidylate (Adapted from Kano et al. [5])

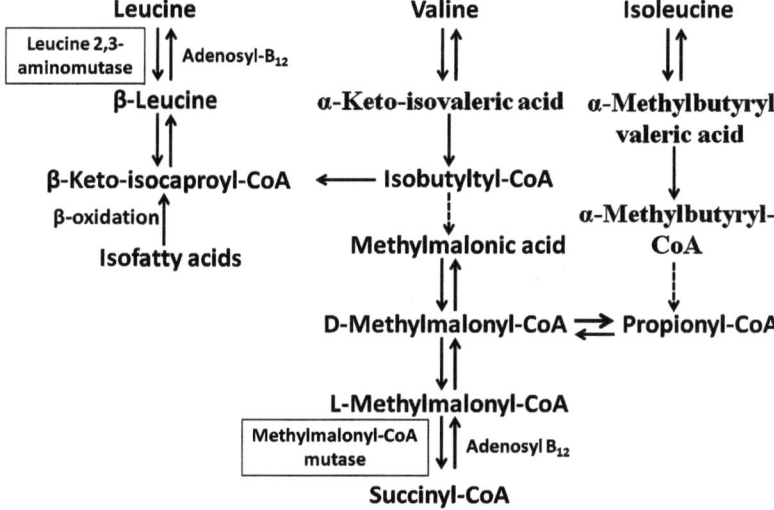

Fig. 9.5 Function of adenosylcobalamin in the mitochondrial isomerase reaction that converts methylmalonyl CoA to succinyl CoA (Adapted from Kano et al. [5])

supply of both methionine and tetrahydrofolate. Tetrahydrofolate is subsequently converted to 5,10-methylenetetrahydrofolate, which donates its one-carbon unit to deoxyuridylate, forming thymidylate and contributing to DNA synthesis. **Adenosylcobalamin** is required for the isomerase reaction in mitochondria that converts methylmalonyl coenzyme A (CoA) to succinyl CoA (Fig. 9.5). Methylmalonyl CoA is derived from propionate and amino acids such as valine, isoleucine, and threonine. In **vitamin B_{12} deficiency**, propionate and methylmalonate accumulate and result in impaired fatty acid synthesis.

The major clinical manifestations of vitamin B_{12} deficiency are **hematologic** and **neuropsychiatric abnormalities**. The hematologic changes are identical to those seen in folate deficiency (see above) and are thought to be due impaired generation of tetrahydrofolate and altered DNA synthesis. The neuropsychiatric abnormalities may be a consequence of alterations in brain and peripheral nerve fatty acid synthesis due to deficient methylmalonyl-CoA mutase activity, but disordered folate and methionine metabolism may also play important roles.

3.2 Intraluminal Events in Vitamin B_{12} Absorption

In the diet, vitamin B_{12} is predominately protein-bound, either to transport proteins or to the enzyme systems described above. Acid and pepsin play important roles in the digestion of these proteins and in the release of vitamin B_{12} into the gastric fluid. Individuals with reduced gastric acid and pepsin secretion are often able to absorb pure crystalline vitamin B_{12} normally, but have impaired absorption of vitamin B_{12} contained in food.

Gastric juice contains **two important vitamin B_{12} binding proteins**, *haptocorrin (R protein-type binder)* and *intrinsic factor (IF)*. Haptocorrin is a 60–66-kd glycoprotein present in many digestive secretions, although the haptocorrin in gastric juice is mainly derived from the salivary gland. Vitamin B_{12} has a **much higher affinity for haptocorrin** than for IF, particularly at low pH. The vitamin B_{12} liberated from food therefore binds preferentially to haptocorrin in gastric juice. In addition, bile contains a substantial amount of vitamin B_{12} bound to haptocorrin. Both haptocorrin and IF are unaffected by acid-peptic digestion. In contrast, pancreatic proteases do not alter IF, but modify haptocorrin to a smaller molecular weight form that has a markedly decreased affinity for vitamin B_{12}. In the small intestinal fluid, therefore, vitamin B_{12} is rapidly and essentially completely transferred from haptocorrin to IF (Fig. 9.6).

IF is a 48–50-kd glycoprotein produced by the **gastric parietal cells**, the same cell type that is responsible for acid secretion. Within the parietal cell, immunoreactive IF can be found in the endoplasmic reticulum, Golgi apparatus, and in tubulovesicles. IF secretion is stimulated by **gastrin**, **histamine**, and **cholinergic agonists**. Following stimulation, tubulovesicles containing IF migrate to the periphery of secretory canaliculi, and IF is then observed on the secretory microvilli. These findings indicate that IF secretion occurs via membrane-associated vesicular transport and fusion of the tubulovesicles with the secretory canalicular membrane. Although IF secretion is stimulated by the same agents that induce gastric acid secretion, these two events are regulated differently. Following stimulation, IF secretion is rapid, perhaps due in part to wash out of preformed IF, and subsequently declines to a much lower plateau value. The IF-mRNA level in the parietal cells is not affected by secretagogues. In contrast, gastric acid secretion has a slower

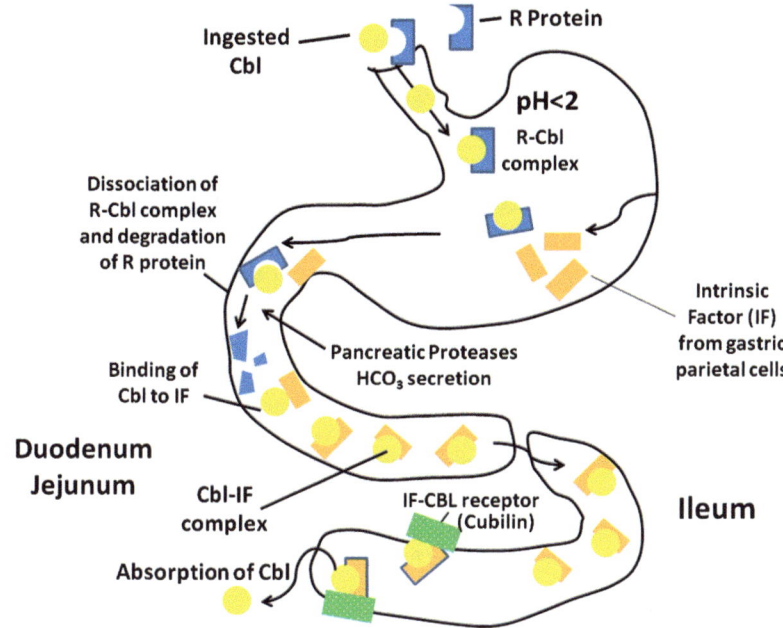

Fig. 9.6 Absorption of cobalamin. Vitamin B_{12} is associated with binding proteins in the gastric and intestinal lumen. *Cbl* vitamin B_{12}, *IF* intrinsic factor, *R* haptocorrin

time course and is sustained at a high level. Some data suggest that gastric acid inhibits secretion of IF, accounting for the different secretory patterns. **Histamine H_2-receptor blockers**, such as *cimetidine*, substantially decrease gastric acid but have only a slight inhibitory effect on IF secretion. The hormone **secretin** decreases acid secretion, but has no effect on IF. Agents, such as *omeprazole*, that block acid secretion by inhibiting the gastric H^+/K^+-ATPase do not alter IF secretion.

Vitamin B_{12} appears to fit into a hydrophobic pit of the IF protein, with the nucleotide portion in the interior of the pit and the anionic moiety coordinated to cobalt facing outward. Binding of vitamin B_{12} to IF exposes hydrophilic regions of IF that increase binding of the IF-vitamin B_{12} complex to brush-border membrane receptors.

3.3 Mucosal Events in Vitamin B_{12} Absorption

The **IF-vitamin B_{12} complex** binds to a specific receptor present in the brush-border membrane of ileal enterocytes, but not in the proximal intestine (Fig. 9.7). The receptor for the IF-vitamin B_{12} complex in the distal ileum is a 460 kDa protein cubilin. Binding of IF-vitamin B_{12} to cubilin is enhanced by Ca^{2+} and

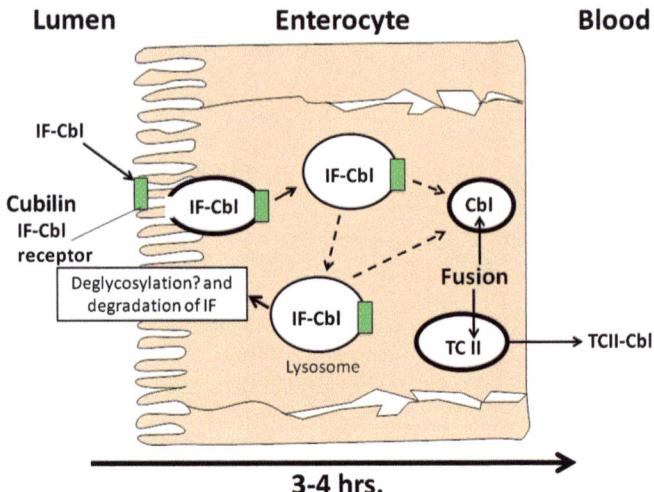

Fig. 9.7 Transport of IF- bound vitamin B_{12} across the ileal enterocyte. *IF* intrinsic factor, *Cbl* vitamin B_{12}, *TCII* transcobalamin II (Adapted from Seetharam [14])

possibly bile salts. **Cubilin** is associated with another protein **amnionless (AMN)** that is involved in the localization of cubilin to the apical membrane surface and in internalization by endocytosis of the IF-vitamin B_{12} complex. **Mutations in the genes coding for cubilin or amnionless** result in *congenital vitamin B_{12} malaborption (Imerslund-Grascbeck syndrome)*. Other proteins can associate with cubilin and may also be involved in vitamin B_{12} absorption. Within the enterocyte, the IF-vitamin B_{12} complex dissociates from cubilin in the endosome and reaches the lysosome were IF is degraded to forms that poorly binding vitamin B_{12}. The vitamin B_{12} leaves the lysosome and enters the cytoplasm probably mediated by a protein LMBD1.

Vitamin B_{12} is released from the ileum into the portal blood bound to another binding protein, **transcobalamin II (TC-II)**. The mechanisms for transfer of vitamin B_{12} to TC-II in the intestine are incompletely understood. TC-II is synthesized by enterocytes, and a complex of vitamin B_{12} with TC-II could traverse the enterocyte basolateral membrane. Evidence exists, however, for transport of free vitamin B_{12} across the basolateral membrane, perhaps mediated by **multidrug resistance protein1 (MDR1)**. The vitamin B_{12} would then associate with TC-II outside the enterocyte. The vitamin B_{12}-TC-II complex is taken up by peripheral tissues by receptor-mediated endocytosis.

Vitamin B_{12} absorption in neonates may occur through a different mechanism. Neonates have relatively low levels of IF, and in milk vitamin B_{12} is bound to haptocorrin. Neonatal small intestine expresses a receptor protein known as the **asialoglycoprotein receptor**. Haptocorrin-bound vitamin B_{12} is taken up into other cell types, such as hepatocytes, via this receptor, and it is possible that uptake of the haptocorrin-vitamin B_{12} complex in the neonatal intestine also occurs through this pathway. High oral doses of vitamin B_{12} (about 1–2 mg) can effectively treat patients with pernicious anemia, an auto-immune gastritis that causes IF deficiency and vitamin B_{12} malabsorption. In *pernicious anemia*, high dose vitamin B_{12} could be absorbed to a limited extent via the asialoglycoprotein receptor or an alternative paracellular route.

3.4 Schilling Test

The Schilling test is used clinically to **assess vitamin B_{12} absorption**. A tracer dose (0.5–1.0 mcg) of radioactive vitamin B_{12} is given orally, and 1–2 h later a large flushing dose (1,000 mcg) of unlabeled vitamin B_{12} is administered intramuscularly to saturate plasma binding sites so that most of the absorbed vitamin B_{12} is excreted in the urine. If the urinary excretion of radioactivity is low (less than 8 % of the administered dose in 24 h), a second-stage test is done in which IF is given orally along with the radioactive vitamin B_{12}. Patients with diseases causing diminished IF secretion, such as *pernicious anemia*, have **reduced radioactive vitamin B_{12} absorption that normalizes when given with IF**. In contrast, those with *ileal disease* or *resection* and reduced absorption of the IF-vitamin B_{12} complex have **diminished absorption in both parts of the Schilling test**. A singlestage absorption test has been developed in which the patient is given both ^{58}Cocobalamin and IF-bound ^{57}Co-cobalamin orally, and the urinary excretion of the two isotopes compared to distinguish between deficient IF secretion and ileal dysfunction. Tests utilizing radioactive cobalamin incorporated into food proteins have also been devised to more accurately assess the patient's capacity to absorb vitamin B_{12} from food sources.

4 Iron

4.1 Biochemical Function

Iron is an essential trace element because of its crucial roles in **cellular oxidative energy metabolism**. Iron is a component of the oxygen-transporting proteins *hemoglobin* and *myoglobin* and of specific redox enzymes. A **microcytic hypochromic anemia** (small red blood cells with reduced hemoglobin) is the most important clinical consequence of **iron deficiency**; however, diminished work and school performance may be observed in depleted individuals even before the development of anemia. **Cellular iron overload**, as occurs with the genetic disorder *hemochromatosis*, results in oxidant damage to many tissues.

4.2 Dietary Sources of Iron

Iron is present in the diet in **two forms**, as a component of *heme* (*heme iron*) and as various *nonheme iron compounds*. Heme iron represents about 40 % of the total iron in animal foods, whereas essentially all of the iron in plant foods is nonheme iron. In a typical mixed American diet, about 10 % of total iron is heme iron. The intestinal absorption of heme iron is considerably more efficient than that of nonheme iron, and uptake of these two forms into the enterocyte occurs via distinct pathways (see below). From a typical daily dietary intake of 10–20 mg of mixed heme and nonheme iron, about 10 % is absorbed by an iron sufficient individual, replacing the daily losses of 1–2 mg and maintaining the body iron content. In iron deficiency, the intestinal absorption of iron significantly increases. The regulation of body iron therefore occurs principally by adjusting intestinal absorption according to tissue requirements rather than by altering iron excretion.

4.3 Intraluminal Factors in Intestinal Iron Absorption

Nonheme iron is present in the diet mainly in the **ferric (Fe^{3+}) state**. Ferric iron is soluble at an acidic pH but precipitates above pH 3. Subjects with reduced gastric acid secretion may therefore develop *iron deficiency* because the ferric iron is not solubilized by an acidic pH within the stomach. **Ferrous (Fe^{2+}) iron salts** are used in pharmaceutical preparations, as ferrous iron is soluble at the nearly neutral pH of the small-intestinal luminal fluid and is therefore more efficiently absorbed than is ferric iron. Various compounds present in the diet or secreted into the intestine, such as certain sugars (e.g. fructose), amino acids (e.g. histidine), amines, and polyols, form unstable iron chelates by binding only a few of the six coordinating bonds of iron. These complexes help keep iron soluble in the intestinal luminal fluid. **Vitamin C (ascorbic acid)** is a well-known facilitator of iron absorption. Ascorbic acid forms an iron chelate that increases iron solubility but, more importantly, reduces iron to the more soluble Fe^{2+} state. The unstable iron chelates serve as iron donors to mucins produced by the upper GI tract. One molecule of macromolecular mucin can bind many iron atoms, and other transitional metals also bind mucins competitively with iron. The iron complexes with mucins, which keep the iron soluble and available in an acceptable form, and iron then undergoes uptake into the enterocyte. Bile increases intestinal nonheme iron absorption, probably by several mechanisms. Bile contains iron chelators and has reductants that convert Fe^{3+} to the more soluble Fe^{2+}. At concentrations below the critical micellar concentration, the bile salt **taurocholate** forms a soluble complex with Fe^{2+}. It has also been suggested that taurocholate may induce the formation of ion-permeable channels in the brush-border membrane, facilitating iron uptake.

Other components of the diet decrease nonheme iron absorption by precipitating iron or by forming stable chelates that interfere with the binding of iron to mucins. **Inhibitors of iron absorption** include *oxalates*, *phytates*, *tannates*, and *carbonates*. The efficiency of iron absorption from different foodstuffs may vary tenfold because of the presence of various enhancers and inhibitors. Fe^{3+} complexed with mucin is also reduced to Fe^{2+} prior to intestinal membrane transport. Reductants are present in the diet and in bile, and the small intestinal brush-border membrane also contributes to this process. Reductants of systemic or enterocyte origin (e.g. ascorbate, glutathione, etc.) are secreted into the luminal fluid, and the brush-border membrane enzyme reductase **duodenal cytochrome b** (**Dcytb**) acts to reduce Fe^{3+} to Fe^{2+} at the apical surface.

Heme iron is precipitated in an acid environment, but it is soluble at the more alkaline pH of the small intestinal fluid. Chelation is therefore not needed to facilitate solubility, and many of the substances that enhance or inhibit nonheme iron absorption have no effect on the absorption of heme iron. Iron is more poorly absorbed from a dose of heme, compared with an equivalent amount of hemoglobin, and it has been demonstrated that globin degradation products and certain amino acids, amines, and amides inhibit heme polymerization within the intestinal lumen and facilitate absorption.

4.4 Enterocyte Iron Transport

Iron is absorbed predominantly in the *duodenum* (Fig. 9.8). Fe^{2+} iron is transported across the brush-border membrane by the **divalent metal transporter DMPT1**, and loss of function mutations of this transporter results in very low rates of iron absorption and severe microcytic anemia. Following uptake into the enterocyte the iron can have **two fates**. (**1**) It can be sequestered in the iron storage protein **ferritin** that is expressed in intestinal epithelial cells. Iron bound to ferritin is lost back into the intestinal lumen as senescent epithelial cells are sloughed from the villus tips. (**2**) Alternatively, iron can be transported across the basolateral membrane. This occurs via the Fe transporter **ferroportin 1** (**FP1**). During the process of iron transport across the basolateral membrane, Fe^{2+} must be oxidized to Fe^{3+} prior to binding to the circulating iron transport protein transferrin. The oxidation of Fe^{2+} to Fe^{3+} at the basolateral membrane is facilitated by the protein **hephaestin** (**HP**). In a state of *iron deficiency*, duodenal levels of Dcytb, DMPT1, and FP1 are increased and the content of ferritin is reduced, favoring iron absorption. In contrast, *iron sufficiency* results in increased duodenal ferritin and decreased levels of Dcytb, DMPT1, and FP1 that would favor sequestration of iron in the mucosa and limit absorption.

Heme iron is taken up into the enterocyte as the intact **metalloporphyrin**. Evidence exists for uptake of heme both through a specific carrier and via diffusion through the lipid bilayer. Within the cell, iron is released from the porphyrin by the action of the enzyme **heme oxygenase** and enters the circulation as inorganic iron. Although there is no competition between heme and nonheme iron for uptake into the intestine, there is competitive inhibition in the overall process of intestinal absorption. This observation indicates that the inorganic iron released from heme intracellularly binds to the same cytosolic proteins and follows the same pathway through the cell and across the basolateral membrane as absorbed nonheme iron.

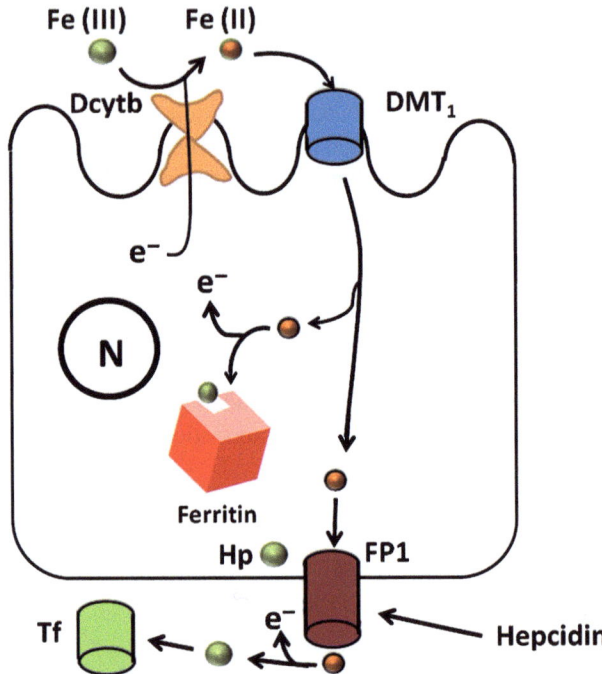

Fig. 9.8 Enterocyte iron transport. Fe^{3+} is reduced to Fe^{2+} by duodenal cytochrome b (Dcytb) in the brush-border membrane. Fe^{2+} is transported across the brush-border membrane by the divalent metal transporter DMPT1. Within the enterocyte, Fe^{2+} can bind to ferritin and be lost as cells are sloughed into the intestinal lumen, or transported across the basolateral membrane by the Fe transporter ferroportin 1 (FP1). During transport across the basolateral membrane, Fe^{2+} is oxidized to Fe^{3+} by hephaestin (HP) and binds to transferrin (Tf). Hepcidin secreted by the liver is the major systemic regulator of iron absorption. Hepcidin binds to FP1 and induces its degradation. Hepcidin secretion is inhibited by iron deficiency, favoring intestinal iron absorption

Several **iron-regulatory proteins** (**IRPs**) have been identified in duodenal enterocytes that interact with specific sequences in the mRNA transcripts of genes (or iron-responsive elements, **IREs**), coding for the proteins involved in intestinal iron transport. The IRP levels and binding to IREs are altered by cellular iron content. Interaction of IRPs with IREs alters the transcription or mRNA degradation rates of the different proteins involved in iron transport.

4.5 Systemic Regulation of Iron Absorption

The status of tissue iron stores in the most well-known systemic regulator of iron absorption. In addition to iron deficiency, however, iron absorption is also increased in *chronic hypoxia* and in diseases associated with *ineffective erythropoiesis* (**thalassemia, chronic hemolytic disease**). In contrast, various inflammatory states result in diminished iron absorption. **Hepcidin** is a protein secreted by the **liver** that is the **principal regulator of iron absorption**. Hepcidin secretion is inhibited by iron deficiency and increased by iron overload. Hypoxia and anemia also decrease hepatic hepcidin secretion. Hepcidin binds to FP1 in the duodenal enterocyte basolateral membrane, inducing endocytosis of FP1 and its proteolysis in lysosomes. The reduction of FP1 caused by hepcidin decreases iron export from the enterocyte and limits overall intestinal iron absorption.

The regulation of hepcidin secretion by iron is complex and incompletely understood. Extracellular iron is sensed by hepatocytes via binding of the circulating Fe^{3+}-transferrin complex to transferrin receptors in the hepatocyte membrane. The protein HFE is also a component of this complex, and **mutation of HFE** is the most common cause of the genetic disease *hemochromatosis*, which causes excessive iron absorption, iron overload and oxidative damage to many tissues. The regulation of hepcidin occurs at the transcriptional level. In addition to the extracellular iron-sensing mechanism, other proteins involved in the regulation of hepcidin mRNA synthesis by iron include **bone morphogenic protein 6** (**BMP6**) that responds to intracellular iron, BMP receptor, down-stream signaling elements, hemojuvelin, and others. Mutations in hepcidin, transferrin receptor 2, ferroportin, and ferritin are rare causes of iron overload.

The mediators of altered hepcidin secretion in hypoxia and anemia are not well-understood, but likely involve **signaling molecules** derived from the *bone marrow*. **Interleukin-6** and other inflammatory mediators have been demonstrated to induce hepcidin secretion.

Duodenal enterocytes express the **transferrin receptor** on the basolateral membrane. The circulating transferrin-Fe^{3+} may form a complex with HFE and the transferrin receptor that is taken up by endocytosis. At the acidic pH of the endocytic vesicle, iron is released, reduced to Fe^{2+}, and transported into the cytoplasm probably by DMPT1. The transferrin receptor and other proteins cycle back to the cell membrane. It is possible that iron sensing by this mechanism also regulates duodenal iron absorption.

5 Calcium

5.1 Functions and Dietary Sources of Calcium

The adult body contains approximately 1,200 g of calcium, with about 99 % present as a **structural element of bones and teeth**. The remaining 1 % of body calcium plays crucial **regulatory roles** in processes such a *nerve conduction, muscle contraction, blood clotting, membrane permeability, Ca^{2+}-dependent protein kinases*, and so on. Dairy products contribute more than 55 % of the average dietary calcium intake in the United States. Other sources include leafy green vegetables, soft fish bones, and calcium-fortified foods.

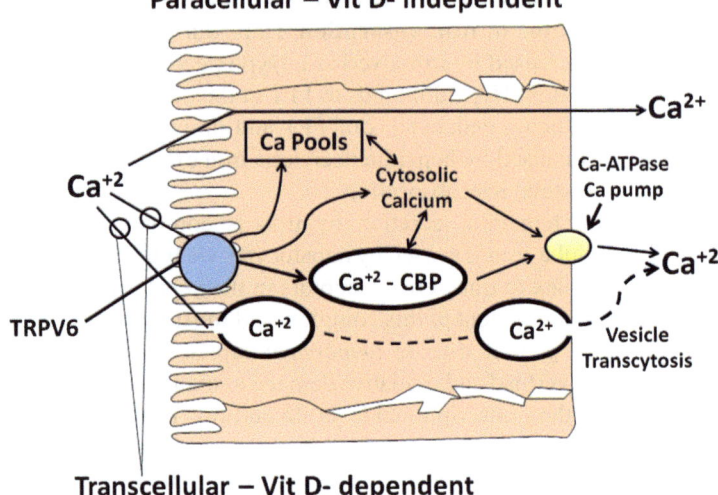

Fig. 9.9 Calcium absorption

5.2 Pathways of Intestinal Calcium Absorption

Calcium is absorbed throughout the small intestine and colon, but the **highest rate of absorption** is in the *duodenum*. At **high dietary intakes**, calcium is absorbed mainly via passive diffusion through a paracellular pathway. At **low levels of intake**, however, efficient calcium absorption is achieved by a saturable, energy-dependent transcellular pathway through the absorptive cells that is dependent on the hormonal form of vitamin D, 1,25-dihydroxyvitamin D_3 [1,25(OH)$_2$D$_3$] and is most active in the duodenum (see Chap. 8 for a brief description of vitamin D metabolism). Transcellular calcium transport involves **three separate steps**: (**1**) entry into the cell through the brush-border membrane down an electrochemical gradient; (**2**) translocation of calcium through the enterocyte; and (**3**) extrusion from the enterocyte at the basolateral membrane surface against an electrochemical gradient (Fig. 9.9). 1,25(OH)$_2$D$_3$ influences each of these processes, although the detailed mechanisms of these steps remain controversial.

5.3 Calcium Entry into Enterocytes

Intracellular Ca^{2+} concentration is approximately 100 nmol/L, whereas the luminal-fluid Ca^{2+} concentration is about 1–10 mmol/L. In addition, the electrical potential within the enterocyte is approximately −55 mV relative to the luminal fluid. Both **electrical** and **chemical gradient forces** therefore favor calcium entry into the cell, and no metabolic energy is needed for this step.

Much of the information concerning calcium uptake has come from studies using highly purified brush-border membrane vesicles prepared from experimental animal or human intestine. Calcium entry into these vesicles is not energy-dependent, but shows saturation kinetics consistent with uptake via calcium transporters or channels. The initial rate of entry of calcium into intestinal brush-border membrane vesicles prepared from vitamin D-deficient animals is reduced compared with vitamin D-sufficient animals; however, there is no effect on the final equilibrium concentration, indicating that vitamin D alters the rate of, but not the capacity for, uptake. **Vitamin D-inducible calcium selective ion channels *CaT1* (*TRPV6*)** and *CaT2* have been identified in the duodenal brush-border membrane. CaT1 knockout mice demonstrate reduced vitamin D-dependent intestinal calcium absorption, but it is not completely eliminated, indicating several pathways of calcium uptake. Evidence exists for calcium entry into the enterocyte via a mechanism involving endocytic vesicles. The relative importance of these different pathways for calcium entry into the enterocyte remains uncertain.

5.4 Transcellular Calcium Transport

Absorbed calcium must move from the brush-border membrane across the enterocyte to the basolateral surface for extrusion from the cell. The rate of free Ca^{2+} flux across the enterocyte has been estimated from the cell size, the presumed transcellular concentration gradient, and the diffusion constant of Ca^{2+} in water. The estimated value is two orders of magnitude slower than the observed absorption rate. **Three possible mechanisms** have been suggested for facilitating movement of calcium across the enterocyte: (**1**) *intracellular facilitated diffusion* involving binding of calcium to a calcium-binding protein, calbindin-D; (**2**) calcium transport in *fixed intracellular organelles*; and (**3**) *vesicular transport* involving endosomes, lysosomes, and the cytoskeleton.

In the intestine, $1,25(OH)2D_3$ induces the synthesis of the vitamin D-dependent calcium-binding protein **calbindin-D**, In mammals, calbindin-D is a 9-kd protein, whereas in avian intestine it is present as a 28-kd protein. In general, there is a linear relationship between intestinal calbindin-D content and the calcium absorption rate. In addition, calbindin-D is more abundant in the proximal small bowel than in the ileum, paralleling the calcium absorption rates. It is postulated that calbindin-D acts as an **intracellular carrier** to facilitate movement of calcium through the cell by **greatly increasing the transcellular calcium gradient**. Based on a calbindin-D concentration of 100 μmol/L, an intracellular Ca^{2+} concentration of 100 nmol/L, a diffusion coefficient for calbindin-bound calcium equal to 0.15 of the value for free calcium, and two calcium-binding sites on calbindin-D, it has been calculated that this binding protein would cause a 75-fold increase of transport beyond that of free Ca^{2+} ion, a value nearly identical to the experimentally measured rate. Recent studies of calbindin-D_{9K} knockout mice, however, have shown no significant reduction of calcium absorption compared to wild type controls, indicating that calbindin-D_{9K} is not essential for vitamin D-dependent calcium absorption.

Some studies reported vitamin D-stimulated binding of calcium to intracellular organelles such as Golgi apparatus, rough endoplasmic reticulum, or mitochondria, but these remain controversial. Vitamin D-dependent calcium transport has been observed in isolated Golgi membranes.

Recent research has suggested that much of vitamin D-dependent calcium absorption occurs via a **vesicular pathway** involving sequestration of calcium in endocytic vesicles, fusion of these endosomes with lysosomes, movement of vesicles and lysosomes along microtubules to the basolateral cell surface, and exocytosis from the enterocyte. Calbindin-D has been found within vesicles in the enterocytes, perhaps explaining their avidity for calcium. During calcium absorption, calbindin-D has been reported to decrease in the enterocyte and to appear in the core of the villus, suggesting extrusion from the cell along with the calcium. Some researchers have suggested that this vesicular transport pathway is initiated by a $1,25(OH)_2D_3$-induced rise in intracellular Ca^{2+} concentration and activation of cellular protein kinases.

5.5 Calcium Extrusion

Calcium is extruded from the basolateral enterocyte surface against a steep electrochemical gradient requiring metabolic energy. As in many cell types, enterocytes and colonocytes contain an *ATP-dependent calcium pump* in the basolateral membrane called the **calcium-transporting ATPase**. This transport activity is correlated with intestinal calcium absorption, as it is greater in the proximal than distal bowel and greater in villus than crypt cells, and declines with aging. The calcium-binding affinity of the calcium-transporting ATPase is about 2.5 times that of calbindin-D, providing a gradient of binding affinities from brush-border membrane to basolateral membrane that favors vectorial transport. Studies with basolateral membrane vesicles have shown that calcium pump activity is **stimulated by $1,25(OH)_2D_3$**. Kinetic analyses showed that this was due to an *increase in transport capacity, with no change in affinity*. In **vitamin D-replete animals**, the calcium-pumping rate is greater than the calcium absorption rate, indicating that pump activity can accommodate the active calcium transport process. In **vitamin D deficiency**, however, pump activity could be rate-limiting, and treatment of vitamin D-deficient animals with $1,25(OH)_2D_3$ has been shown to increase immunodetectable pumps in the basolateral membrane.

At least four genes coding for plasma membrane calcium-transporting ATPases have been identified, and the diversity of pump proteins is further increased by alternative splicing of these gene transcripts. Different isoforms of the calcium-transporting ATPase are found in the small intestine and colon. Enterocytes contain a **basolateral Na^+/Ca^{2+} exchange system**, but this system is not regulated by vitamin D, does not vary along the length of the small intestine, and has much less activity than the calcium-transporting ATPase. The possibility of vesicular calcium transport has been discussed above.

5.6 Other Factors Affecting Calcium Absorption

In addition to vitamin D, a number of other dietary constituents have been suggested to alter intestinal calcium absorption. **Phosphate, oxalate, phytate,** and **fatty acids** can precipitate soluble calcium, and certain types of **dietary fiber** can bind calcium. In some studies, **lactose** has been shown to increase calcium absorption, probably by influencing the paracellular pathway. Overall, at intakes of the typical American diet, these factors appear to influence calcium absorption only modestly. **Bile salts** increase calcium absorption by similar mechanisms to those previously described for iron absorption.

Clinical Correlations

Case Study 1

A 40-year-old woman with long-standing Crohn's disease had a resection of 150 cm of terminal ileum. Two years later she develops recurrent disease and is placed on the treatment of sulfasalazine. After 1-year treatment, she presents with complaints of fatigue and numbness and tingling in her feet. She reports eating a normal, unrestricted diet and does not take vitamin supplements. Laboratory evaluation shows a megaloblastic anemia and reduced serum folate and vitamin B_{12} levels.

Questions

1. **What is the pathogenesis of vitamin B_{12} deficiency in this patient, and what treatment would you recommend?**
   Answer: Vitamin B_{12} bound to intrinsic factor (IF) is absorbed via binding to cubilin and then taken up by endocytosis in the ileum. This patient with a **large ileal resection** would have *vitamin B_{12} malabsorption*. Large doses of oral vitamin B_{12} may be absorbed in the proximal small intestine by passive diffusion or other mechanisms and could be tried as replacement therapy; however, **intramuscular or intranasal vitamin B_{12}** will correct the deficiency.
2. **What is the pathogenesis of this patient's folate deficiency?**
   Answer: The patient reports eating an unrestricted diet thus inadequate folate intake is unlikely. Many Crohn's disease patients, however, avoid eating fruits and vegetables, which are important dietary sources of folate. Folate is absorbed mainly in the proximal intestine, which is likely to be functioning well in this patient. She is, however, taking the medication sulfasalazine as treatment for her Crohn's disease, and **sulfasalazine** is an **inhibitor of folate conjugase**

(the enzyme that deconjugates polyglutamyl folates) and **of monoglutamyl folate uptake**. It is likely that her folate deficiency is due to this drug-nutrient interaction.

3. **What would be the results of a Schilling test in this patient?**
   Answer: The Schilling test would demonstrate **reduced urinary excretion** of **orally administered radioactive vitamin B_{12}** given either alone or with IF.

Case Study 2

A 65-year-old man undergoes a total gastrectomy for a gastric adenocarcinoma. After surgery, he is not given any vitamin or mineral supplements. Four years later, he presents to his doctor with marked fatigue. He is found to be severely anemic. The average size of his red blood cells is normal, but examination of a blood smear shows that some cells are abnormally large, whereas others are very small.

Questions

1. **Which nutritional deficiencies are likely to be causing this patient's anemia?**
   Answer: The patient is likely to be **vitamin B_{12}-deficient**, since total gastrectomy eliminates secretion of IF needed to bind vitamin B_{12} prior to receptor-mediated ileal uptake. He is also likely to be **iron-deficient**, as gastric acid is needed to solubilize dietary ferric iron and he has not received iron supplementation. **Folate absorption** could be somewhat impaired since folate is optimally absorbed in the proximal intestine at a slightly acidic pH, and the lack of gastric acid will cause the intraluminal pH to rise. **Patients who have had total gastrectomy, however, are *not* commonly folate-deficient**, probably because the lack of gastric acid permits the proliferation of ingested folate-producing bacteria in the upper intestine and that folate is absorbed.

2. **How do you explain the findings regarding the size of his red blood cells?**
   Answer: Vitamin B_{12} and **folate deficiency** cause a *macrocytic anemia*, whereas **iron deficiency** causes a *microcytic anemia*. In this patient with both vitamin B_{12} and iron deficiency, the average red blood cell size may be normal, but individual erythrocytes may be either macrocytic or microcytic.

3. **Why did this patient become vitamin B_{12}-deficient in only 4 years?**
   Answer: The daily turnover rate of vitamin B_{12} is normally very low compared with body stores. Vitamin B_{12} depletion will therefore not occur until 10–20 years of dietary lack. This patient, however, will have malabsorption of both dietary vitamin B_{12} and vitamin B_{12} present in bile. The **lack of reabsorption** and **conservation of biliary vitamin B_{12}** results in *accelerated depletion*.

Case Study 3

A 45-year-old woman has a 15-year history of primary biliary cirrhosis, a disease characterized by progressive destruction of intrahepatic bile ducts and reduction of bile flow. She presents with a fractured humerus after lifting a small bag of groceries. She is found to have markedly reduced bone density. Laboratory testing

9 Absorption of Water-Soluble Vitamins and Minerals

reveals reduced serum calcium, phosphorus, and 25-hydroxyvitamin D_3 levels, markedly reduced urinary calcium excretion, and 30 g/day of steatorrhea (normal, less than 7 g/day). She is noted to eat few dairy products because they produce gas and diarrhea.

Questions

1. **What factors contribute to calcium malabsorption in this patient?**
   Answer: The most important cause of calcium malabsorption in this patient is **vitamin D deficiency**, as vitamin D is necessary for efficient calcium transport through the transcellular pathway (see Chap. 8 for a discussion of vitamin D absorption). In addition, bile salts directly enhance intestinal calcium absorption, and **reduced bile flow** in this patient would diminish the intraluminal bile salt concentration. The patient has steatorrhea, and calcium complexes with unabsorbed fatty acids, forming soaps that are excreted in the stool.

2. **What nutritional therapy would you prescribe for her metabolic bone disease?**
   Answer 2: The patient should receive **supplementation** to correct her vitamin D deficiency. **25-Hydroxyvitamin D_3** is absorbed better than vitamin itself in patients with severe cholestatic liver disease and may be the better form of therapy (Chap. 8). She should be placed on a **low fat diet** to decrease her steatorrhea and be given **calcium supplementation** in the form of low-lactose, low-fat dairy foods or pharmaceutical preparations.

3. **How does her avoidance of dairy products contribute to her bone disease?**
   Answer 3: The patient has symptoms of lactose intolerance and avoids dairy products (see Chap. 6 for a discussion of lactose absorption). **Dairy foods** are the *major dietary sources* of both **vitamin D** and **calcium**. In addition, **lactose** may *enhance the efficiency of calcium absorption* through the *paracellular pathway*.

Case Study 4

A 55-year-old man is found on routine testing to have abnormal liver function tests (elevated serum transaminases and alkaline phosphatase). His father died of cirrhosis and congestive heart failure. Laboratory testing shows that his serum iron is 300 mcg/dl, serum iron-binding capacity is 350 mcg/dl, and serum ferritin is 4,500 ng/ml.

Questions

1. **What additional tests would you order?**
   Answer: The patient likely has hereditary hemochromatosis. Genetic testing can be performed for certain mutations causing this disorder. The most common mutation is in the HFE gene (mutation C282Y). The HFE protein is involved in the regulation of hepcidin secretion from the liver by iron, which in turn controls intestinal iron absorption. The patient should also have a liver biopsy to assess the degree of liver damage from hepatic iron overload.

2. **What other gene mutations could cause hemochromatosis?**
   Answer: Mutations in the genes coding for hepcidin, transferrin receptor 2, ferroportin, and ferritin, all genes involved in the regulation of iron absorption, can also cause hemochromatosis.
3. **What is the appropriate treatment for this patient?**
   Answer: The patient needs to be placed on a regular schedule of blood donation until the body iron burden is reduced, assessed by a fall in the serum ferritin level to 50–100 ng/ml.

Further Reading

1. Bronner F (2003) Mechanisms of intestinal calcium absorption. J Cell Biochem 88:387–393
2. Bronner F (2009) Recent developments in intestinal calcium absorption. Nutr Rev 67:109–113
3. Ganz T (2011) Hepcidin and iron regulation, 10 years later. Blood 117:4425–4433
4. Hentze MW, Muckenthaler MU, Galy B, Camaschella C (2009) Two to tango: regulation of mammalian iron metabolism. Cell 142:24–38
5. Kano Y et al (1985) Disorders of cobalamin metabolism. Crit Rev Oncol Hematol 3:1–34
6. Kozyraki R, Cases O (2013) Vitamin B12 absorption: mammalian physiology and acquired and inherited disorders. Biochimie 95:1002–1007
7. Mason JB, Rosenberg IH (1994) Intestinal absorption of folate. In: Johnson LR (ed) Physiology of the gastrointestinal tract, 3rd edn. Raven Press, New York, pp 1979–1996
8. Peitrangelo A (2002) Physiology of iron transport and the hemochromatosis gene. Am J Physiol Gastrointest Liver Physiol 282:G403–G414
9. Philpott CC (2001) Molecular aspects of iron absorption: insights into the role of HFE in hemochromatosis. Hepatology 35:993–1001
10. Quadros EV (2009) Advances in the understanding of cobalamin assimilation and metabolism. Br J Haematol 148:195–2004
11. Rucker RB, Lonnerdal B, Keen CL (1994) Intestinal absorption of nutritionally important trace elements. In: Johnson LR (ed) Physiology of the gastrointestinal tract, 3rd edn. Raven Press, New York, pp 2183–2204
12. Said HM (2011) Intestinal absorption of water-soluble vitamins in health and disease. Biochem J 437:357–372
13. Said HM, Mohammed ZM (2006) Intestinal absorption of water-soluble vitamins: an update. Curr Opin Gastroenterol 22:140–146
14. Seetharam B (1994) Gastrointestinal absorption and transport of cobalamin (vitamin B_{12}). In: Johnson LR (ed) Physiology of the gastrointestinal tract, 3rd edn. Raven Press, New York, pp 1997–2026

Part III
Hepatobiliary Physiology

Chapter 10
Structure, Functional Assessment, and Blood Flow of the Liver

Dennis D. Black

1 Introduction

The liver is a truly remarkable organ when one considers the diversity of its many functions. This organ has a central role in **nutrient** (carbohydrate, protein and lipid) and **vitamin metabolism** and serves as an intermediary between dietary sources of energy and the extrahepatic tissues, which utilize such energy. Another major function is the **detoxification** and elimination of xenobiotic toxins and drugs, as well as endogenous metabolites, such as bilirubin. The liver is responsible for the **synthesis of biologically important proteins** (i.e., albumin, transferrin, clotting factors, complement factors, apolipoproteins, etc.). The liver synthesizes and secretes the components of **bile**, including bile acids, which are important for the intestinal solubilization and absorption of the products of lipid digestion. Finally, the liver has an important role in **immune function,** due primarily to its resident macrophage, the **Kupffer cell.**

2 Liver Anatomy and Blood Supply

In order to efficiently carry out its myriad functions, the liver has unique requirements with regard to anatomic location, blood supply, and structural organization. The liver is the largest solid organ in the body, accounting for 3–5 % of body mass. The liver is subdivided into **two main lobes** (right and left) and **two accessory lobes** (quadrate and caudate) (Fig. 10.1). These lobes are supplied with blood by the right and left branches of the *portal vein* and *hepatic artery,* and bile is drained by the

---

D.D. Black, M.D. (✉)
Department of Pediatrics, University of Tennessee, Memphis, TN, USA
e-mail: dblack@uthsc.edu

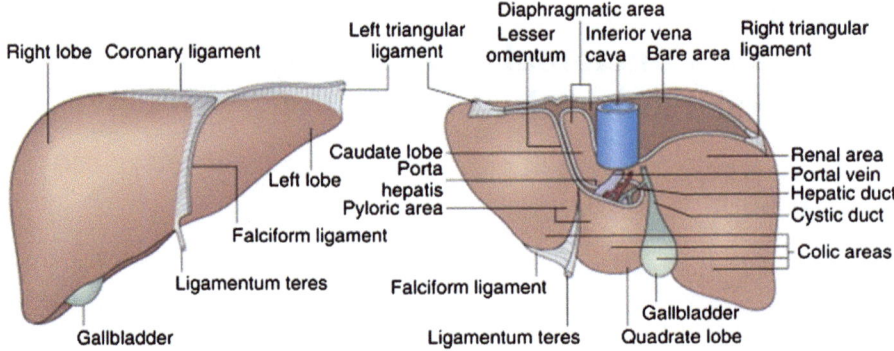

Fig. 10.1 Anterior (*left*) and posteroinferior (*right*) views of the liver (Reproduced with permission from Elsevier [14])

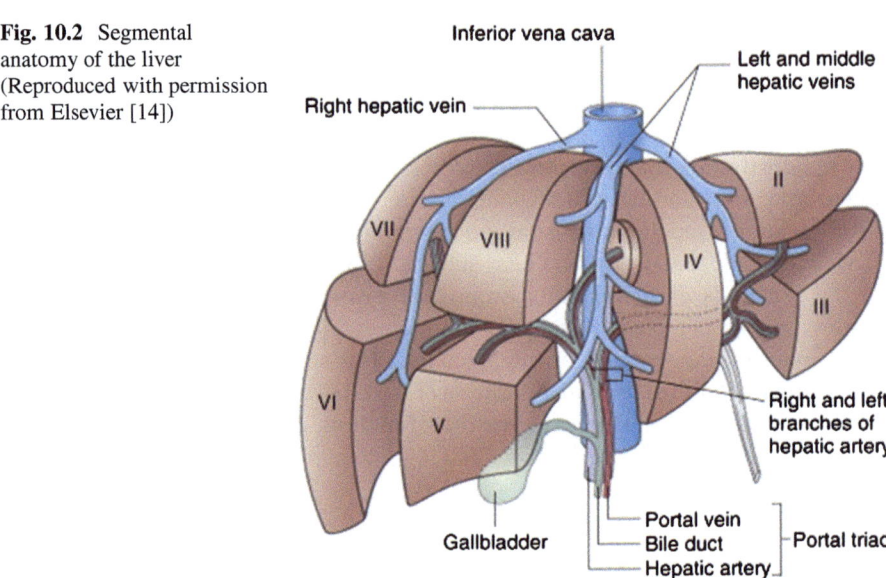

Fig. 10.2 Segmental anatomy of the liver (Reproduced with permission from Elsevier [14])

right and left hepatic bile ducts. Of more significance from a functional standpoint is the *segmental anatomy,* whereby the liver is divided into **eight segments**, each with distinct blood supplies and biliary drainage without collateral circulation between segments (Fig. 10.2). The liver receives approximately 28 % of the body's total blood flow and consumes 20 % of the total oxygen used by the body. The liver is situated as a "sentinel" for the interception and processing of nutrients, toxins and bile acids from the intestine because of its unique **dual blood supply**. The liver receives all of the venous blood draining from the splanchnic vascular bed by way of the *portal vein* (Fig. 10.3). Seventy-five percent of the incoming blood volume to the liver is supplied from the portal vein. Normally, the portal system is a **low-pressure (5–10 mmHg) system** designed to optimally deliver the venous blood for flow through the hepatic sinusoids to allow the most efficient exposure

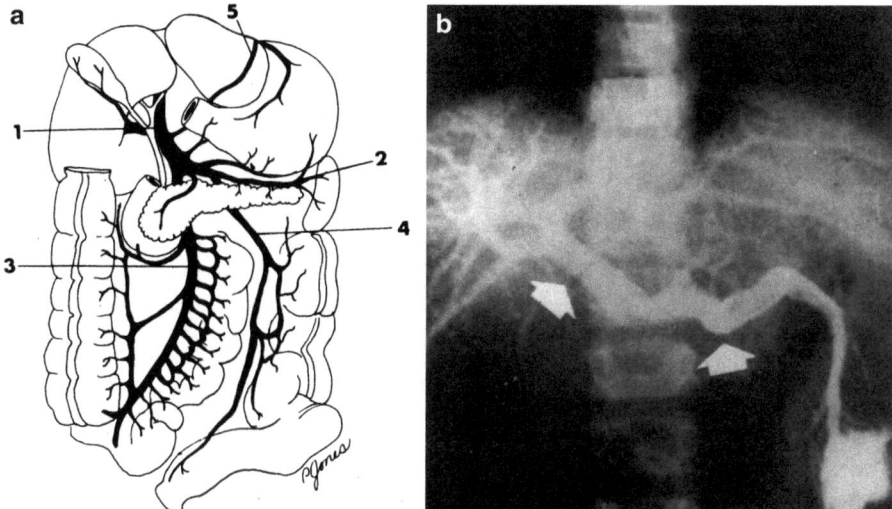

Fig. 10.3 (**a**) Prehepatic portal venous system. *1* portal vein, *2* splenic vein, *3* superior mesenteric vein, *4* inferior mesenteric vein, *5* coronary vein. (**b**) Normal portal venous system shown by splenoportography. The contrast material immediately opacifies the splenic and portal veins (*arrows*). The intrahepatic branches of the portal system are also well visualized (Reproduced with permission from Silverman and Roy [19])

to the hepatocyte plasma membrane. The liver receives oxygenated arterial blood from the hepatic artery. A portion of the blood from the terminal branches of the hepatic arterial system flows through plexuses surrounding the interlobular bile ducts before entering the sinusoids. Individuals who have received liver transplants may survive with normal liver function after complete thrombosis of the hepatic artery because oxygen delivered by the portal venous blood is adequate. However, the dependence of the bile ducts on hepatic arterial blood is underscored by the fact that these individuals often experience degeneration and loss of bile ducts after arterial thrombosis.

2.1 Microanatomy of the Liver

The liver comprises several cells types in addition to hepatocytes. The **hepatocytes** make up approximately 80 % of the volume of the liver and approximately 60 % of the total hepatic cell population. **Mesenchymal cells,** including *Kupffer cells, stellate (fat-storing or Ito) cells,* and *sinusoidal endothelial cells*, comprise 30–35 % of the total population, and **bile ductular epithelial cells** contribute 3–5 %. Liver cells are arranged in a *polarized* fashion on a basement membrane in the liver lobule forming cords of cells (Fig. 10.4). As shown in Fig. 10.5, hepatocytes are complex cells containing the necessary subcellular machinery to function in synthesis, processing and secretion, as well as degradation and oxidation

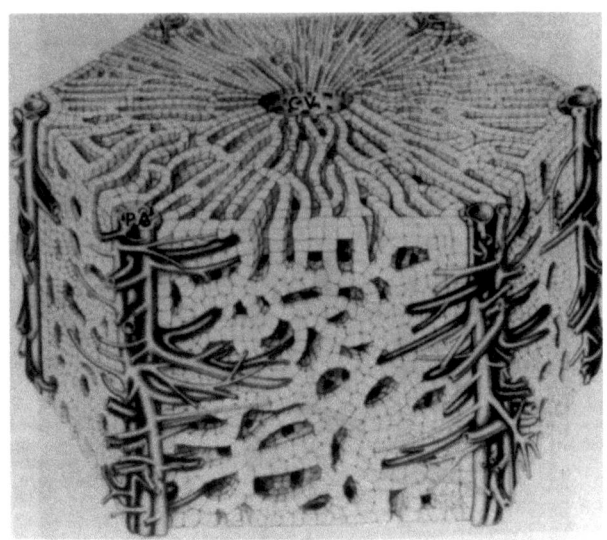

Fig. 10.4 Liver lobule. The central vein (*CV*) lies in the center of the figure, surrounded by anastomosing cords of block-like hepatocytes. Around the periphery are six portal areas (*PA*) consisting of branches of the portal vein, the hepatic artery, and the bile duct (Reproduced with permission from Jones and Spring-Mills [12])

of a variety of substrates. This is reflected by the presence of **highly developed endoplasmic reticulum, mitochondria, peroxisomes, lysosomes, Golgi complex** and **cytoskeleton.** Hepatocytes have a specialized cell membrane domain called the **canaliculus** (Fig. 10.5), which functions in the secretion of bile constituents from the hepatocyte. Newly formed bile coalesces in bile ductules (<20 μm diameter), which empty into progressively larger biliary channels to finally empty into the common bile duct.

Other **non-hepatocyte (non-parenchymal) cells** in the liver include *stellate (fat-storing or Ito) cells, Kupffer cells, pit cells* and *sinusoidal endothelial cells.* **Stellate cells** are located in the *perisinusoidal space,* usually in intimate contact with endothelial cells via cytoplasmic processes and are concentrated in periportal areas. As will subsequently be discussed, these cells function in the storage of vitamin A. Stellate cells also produce extracellular matrix components, as well as matrix degradative enzymes, and appear to play a role in the remodeling and maintenance of the extracellular matrix. However, when stimulated by cytokines and other factors, these cells are also responsible for hepatic fibrosis leading to *cirrhosis.* Stellate cells also are capable of Ca^{2+}-dependent contractility and may function in regulation of sinusoidal blood flow. **Kupffer cells** are the liver's resident macrophages and are located within the *sinusoidal lumen.* These cells phagocytize dead cells, as well as debris and microorganisms brought in by the portal blood from the intestine. When activated by substances such as bacterial endotoxin, Kupffer cells secrete cytokines, such as *tumor necrosis factor-α, interleukin-1,* and *interleukin-6,* which can modulate several hepatocyte functions, including synthesis of **acute phase reactants** (fibrinogen, α-1-antitrypsin, and serum amyloid A), **albumin,** and **apolipoproteins,** as well as lipid metabolism. These soluble factors also affect the metabolism of sinusoidal endothelial cells. **Pit cells** are *intrasinusoidal* lymphocytes with *natural killer (NK) cell activity* and play a role in

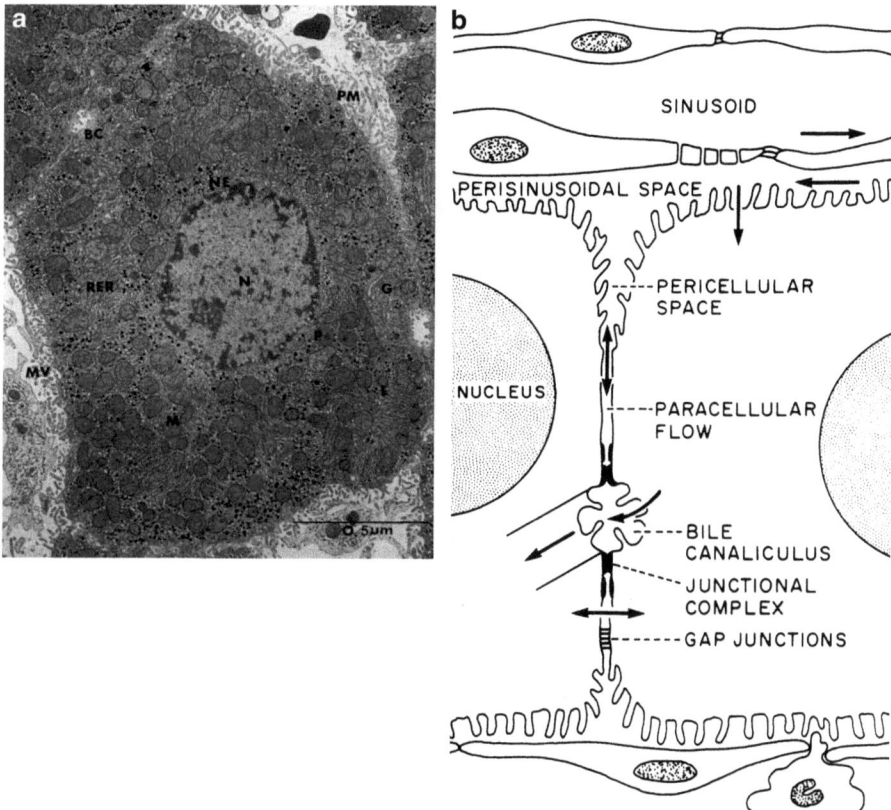

Fig. 10.5 Hepatocyte cellular (**a**) ultrastructure and (**b**) organization. *BC* bile canaliculus, *G* Golgi apparatus, *L* lysosome, *M* mitochondrion, *MV* microvilli, *N* nucleus, *NE* nuclear envelope, *P* peroxisome, *PM* plasma membrane, *RER* rough endoplasmic reticulum. (**a**: Reproduced with permission from Hanover [9]; **b**: Reproduced with permission [10])

surveillance for viral infection and malignancy. **Sinusoidal endothelial cells** line the *hepatic sinusoids* and are *fenestrated* to allow macromolecular permeation (Fig. 10.6). These cells can respond to humoral factors to change the size of these fenestrations to regulate permeability. These endothelial cells also are very metabolically active and have numerous other functions, including active transport, participation in coagulation, fibrinolysis, and immune responses. Sinusoidal endothelial cells secrete *nitric oxide (NO)* (vasodilation) and *endothelin* (vasoconstriction) and **function in tandem with the stellate cell** to regulate sinusoidal vascular tone and pressure.

The **liver lobule** is a histologic and functional unit of the liver (Fig. 10.4). A typical lobule is *hexagonal* in cross section with **interlobular portal triads** (each containing branches of the *portal vein, hepatic artery*, and *bile duct*) at the angles of the hexagon. In the center of the lobule is the **central vein** or **terminal hepatic venule**. Hepatocytes are arranged as single cell plates radiating from the central vein to the portal triads in the periphery of the lobule. These plates are located between

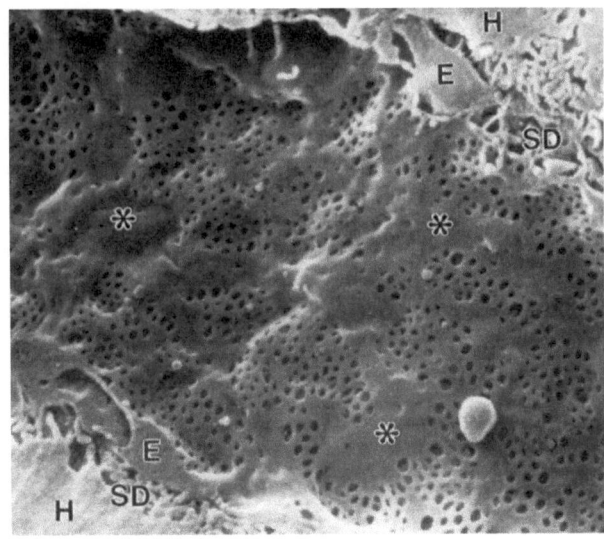

Fig. 10.6 Scanning electron photomicrograph of a sinusoid illustrating sieve plates comprising clusters of endothelial fenestrae. Areas of endothelium lacking fenestrae (*asterisks*) overlay processes of stellate cells. *E* endothelial cell, *SD* space of Disse containing microvilli extending from hepatic parenchymal cells (*H*) (Reproduced with permission from McCuskey [13])

the **endothelial-lined sinusoids**, which serve as vascular channels to carry blood from the vessels in the portal triad at the periphery of the lobule to the central vein in the center of the lobule. As discussed previously, the endothelial cells lining the sinusoids are ***fenestrated*** and ***highly permeable*** to macromolecules. This allows optimal exposure of hepatocyte plasma membranes to the mixed arterial-portal venous blood, as well as allowing the unimpeded movement of larger particles, such as lipoproteins, to and from the hepatocytes.

Another conceptual unit of liver microanatomy defined by Rappaport is the **hepatic acinus** (Fig. 10.7). In this structural model the **portal triad** is at the center of the acinus, and the **terminal hepatic venules** are at the periphery. Cells are grouped into zones based upon their distance from the blood vessels in the center of the acinus (hepatic artery and portal vein branches). The closest, **zone 1 cells**, are the first to receive oxygenated, nutrient-rich blood and to undergo regeneration, and are the last to develop hypoxic injury. The opposite is true of **zone 3 cells. Zone 2 cells** are intermediate.

There is an anastomosing network of **hepatocyte canaliculi** in the hepatic lobule, which coalesces to drain bile into the interlobular bile duct in the portal triad through the **canal of Hering** (Fig. 10.8). Bile flows into larger intrahepatic ducts and finally into the right and left hepatic ducts, which exit the liver, merge in the porta hepatis, and form the **common hepatic duct**. This duct is joined by the cystic duct from the gallbladder to form the **common bile duct**, which empties into the duodenum.

An important structural-functional concept in the liver is that of **hepatocyte heterogeneity** and **metabolic zonation** (Fig. 10.9). For example, **zone 1 hepatocytes** are most active in glucose synthesis and release, oxidative energy metabolism, amino acid utilization, and bile acid and bilirubin excretion. **Zone 3 hepatocytes**

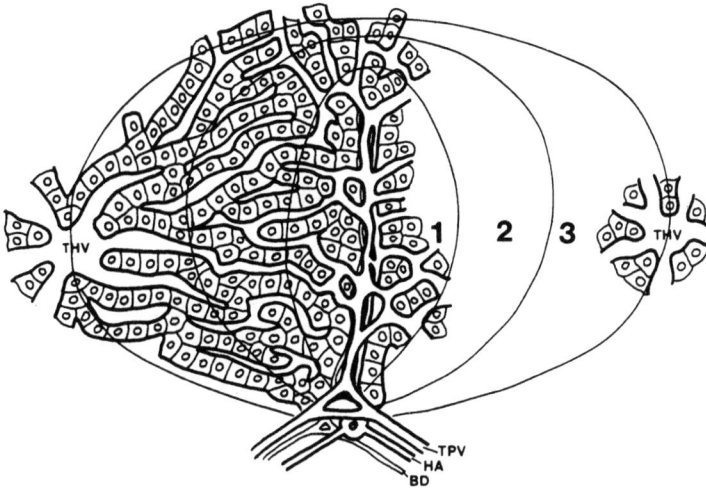

Fig. 10.7 Hepatic acinus. This three-dimensional structure is the microvascular unit of the liver parenchyma. The acinar axis is formed by the terminal hepatic venules (*TPV*), the hepatic arteriole (*HA*), and the bile ductule (*BD*). The perfusion of this unit is unidirectional from the acinar axis to the acinar periphery, where two or more terminal hepatic venules (*THV*) empty the acinus. The arbitrary division of the acinus into functional zones is represented by *1*, *2* and *3* (Reproduced with permission from Gumucio and Chianale [8])

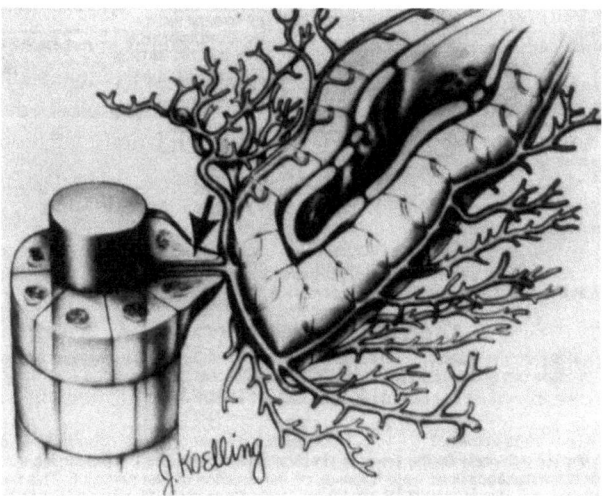

Fig. 10.8 An anastomosing network of hepatocyte canaliculi in the hepatic lobule. The terminal portion of a rat biliary tree shows a bile ductule, a canal of Hering (*arrow*), and the fine anastomosing network of bile canaliculi (Reproduced with permission from Jones [11])

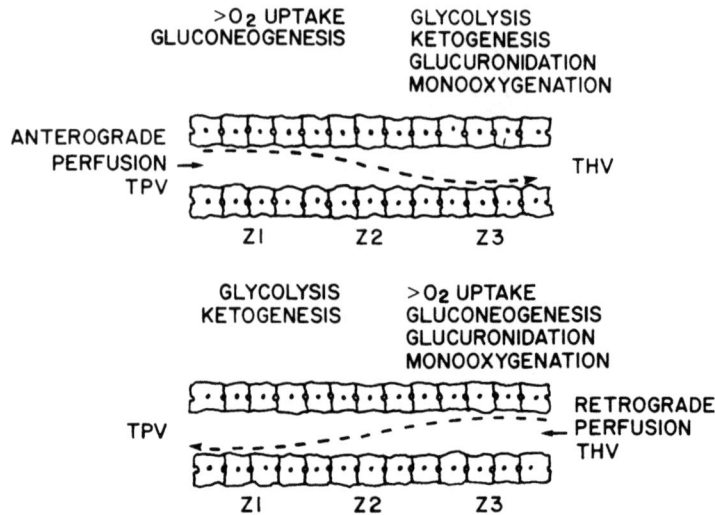

Fig. 10.9 Metabolic zonation of the liver and direction of liver perfusion. The predominant zonal location of some metabolic pathways depends on the direction of liver perfusion. This has been demonstrated for gluconeogenesis, glycolysis, and ketogenesis. Monooxygenation and glucuronidation were not affected by the direction of perfusion. *TPV* terminal portal venule, *THV* terminal hepatic venule (Reproduced with permission from Gumucio and Chianale [8])

are most active in glucose uptake and utilization, biotransformation, and ammonia detoxification. Although lobular blood flow patterns may be responsible for at least part of this zonation, in experiments in which flow patterns in rat liver were reversed, some, but not all, of these zonal metabolic patterns were reversed. For example, gluconeogenesis shifted from zone 1 to zone 3, and glycolysis and ketogenesis shifted from zone 3 to zone 1. However, glucuronidation and monooxygenation remained localized to zone 3.

3 Evaluation of Liver Function

Clinicians routinely rely on the performance of a battery of "**liver function tests**" to assess the degree of liver injury and functional impairment in various disease states. However, many of these tests really do *not* measure function at all, but rather indirectly reflect degree of hepatocellular or bile ductular injury. It is also important to be aware of all of the non-hepatic factors, as well as inherent limitations of the tests themselves, which may affect the reliability of these tests. Table 10.1 lists many of these tests and their uses. One test, the **serum lactate dehydrogenase (LDH) level,** has such *poor* specificity for liver disease that it is not included.

Table 10.1 Liver evaluation tests

Test	Mechanism	Significance
Hepatic function		
Albumin	Decreased synthesis	Decreased hepatic synthesis, increased loss by protein-losing enteropathy or nephrotic syndrome, increased volume of distribution (edema, ascites)
Prothrombin time (indirect assessment of coagulation factors)	Decreased synthesis of factors I, II, V, VII, and X	Hepatic dysfunction-unresponsive to vitamin K, Malabsorption-responsive to vitamin K
Coagulation factors (direct measurement)	Decreased production of liver-dependent factors I, II, V, VII, IX and X	Hepatic synthetic impairment
Ammonia	Decreased ureagenesis, increased portosystemic shunting, increased ammonia load (GI bleeding)	Increased ammonia associated with hepatic encephalopathy
Indocyanine green clearance	Low dose – indicator of liver blood flow; High dose – indicator of hepatic processing	May be particularly sensitive for detecting cirrhosis
Galactose elimination test	Hepatic uptake and metabolism	Reflective of functional liver cell mass – useful for following declining hepatic function
Aminopyrine breath test	Hepatic uptake, demethylation by cytochrome P_{450}, measurement of breath excretion of ^{13}C label from methyl group	Index of hepatic excretion – useful for detection of cirrhosis, as well as prognostic indicator in patients with cirrhosis
Lidocaine clearance test	Measurement of appearance of MEGX, major hepatic metabolite	Useful to evaluate liver dysfunction, especially liver transplantation (pre- and post-)
Cellular injury		
Aspartate aminotransferase (AST/SGOT)	Released from damaged hepatocytes	Also in heart, skeletal muscle, kidney, pancreas, RBCs, and brain
Alanine aminotransferase (ALT/SGPT)	Released from damaged hepatocytes	More specific for liver injury than AST

(continued)

Table 10.1 (continued)

Test	Mechanism	Significance
Cholestasis		
Alkaline phosphatase	Greatest elevation usually in extrahepatic biliary obstruction	Also in bone, intestine, placenta, kidney – may need to be fractionated for liver-specific isozyme
γ-glutamyltranspeptidase (GGTP)	Sensitive, but not specific, index of biliary tract obstruction/disease	Inducible by ethanol, phenobarbital, and other drugs; elevated in diabetes, myocardial infarction, renal failure, and other conditions
5′-Nucleotidase	More specific for liver disease than alkaline phosphatase or GGTP	Biliary disease
Conjugated bilirubin	Decreased hepatic excretion in cholestasis	May also be elevated with hepatocellular injury (drug-induced, viral hepatitis); may be normal to slightly elevated with significant cholestasis (primary biliary cirrhosis, chronic cholestatic syndromes)
Bile acids	Decreased extraction from portal blood, decreased hepatic excretion in cholestasis	Do not discriminate hepatocellular versus obstructive cholestasis
Tumor marker		
α-fetoprotein	Derepressed fetal gene	Hepatic tumor
Hepatic fibrosis		
Biomarkers of hepatic fibrosis (FibroTest[R], FibroSure[R], others)	Panel of serum markers, including components of extracellular matrix, and clinical data predictive of degree of liver fibrosis	Indirect measure of liver fibrosis
Elastography	Measures hepatic stiffness as an indicator of liver fibrosis	Invalid measurements due to improper probe positioning and obesity
Imaging		
Ultrasonography	Gross anomalies and blood flow using Doppler probe	Gallstones, larger masses; can assess hepatic vein and portal vein thrombosis/obstruction
Computed axial tomography (CT scan)	Greater resolution than ultrasound	Particularly useful for detecting abscesses and tumors

(continued)

Table 10.1 (continued)

Test	Mechanism	Significance
Angiography	Definition of hepatic vessels by contrast injection	Particularly useful for better definition of hepatic vein/portal vein thrombosis/obstruction; defining blood supply to tumors
Cholangiography	Definition of biliary tree by contrast injection	Defining biliary tract anatomy for source/site of obstruction
Endoscopic retrograde cholangiopancreatography (ERCP)	Endoscopic injection of contrast via the ampulla of Vater into the pancreatic ducts and biliary tree	Allows visualization of ductal anatomy, as well as interventions such as stone retrieval and stent placement
99mTc-sulfur colloid	Kupffer cell uptake	Large focal lesions
99mTc-DISIDA	Hepatic uptake and biliary excretion	Discriminating hepatocellular from obstructive cholestasis; confirming presence of choledochal cyst
Magnetic resonance imaging (MRI)	Greater resolution than CT	Evaluation of focal lesions (especially hemangiomas), biliary tree (MRCP) and detecting excess hepatic iron
Liver biopsy		
Percutaneous Transjugular Laparoscopic Surgical	Obtaining liver tissue for histology, electron microscopy, immunostaining, measurement of copper or iron, and specific enzyme assays	May provide specific diagnosis and prognosis (degree of necrosis/fibrosis) for liver diseases

3.1 Measures of Cellular Injury

One of the most common groups of tests is the **serum aminotransaminases, alanine aminotransferase** (ALT; serum glutamic pyruvic transaminase or SGPT) and **aspartate aminotransferase** (AST; serum glutamic oxaloacetate transaminase or SGOT). These enzymes catalyze the reversible transfer of the alpha amino group of the amino acids **alanine** and **aspartic acid,** respectively, to the α-keto group of **α-ketoglutaric acid.** These enzymes are released into the circulation whenever there is hepatocyte plasma membrane damage or necrosis. Whereas ALT is localized exclusively in the cytosol, 20 % of AST is in the cytosol and 80 % is found in the mitochondria. It is also important to note that *AST is present in a variety of tissues other than liver,* including *heart, skeletal muscle, kidney, pancreas, red blood cells* and *brain*. However, **ALT is relatively specific for liver.** Although serum levels of ALT and AST are elevated to some degree in almost all liver diseases, the **highest levels** are seen with **acute hepatocellular injury,** such as *viral hepatitis, toxin-induced hepatic necrosis,* and *hypoxic liver injury.* The decline in serum levels after such an insult, particularly viral hepatitis, may *not* necessarily reflect recovery, but may indicate massive necrosis with loss of viable hepatocyte mass available to release transaminases. The same is true in the "burned out" cirrhotic patient with **chronic liver disease** who also has little remaining viable liver cell mass. In the majority of other liver diseases, including cholestatic and chronic liver diseases and hepatic malignancy, serum transaminase levels are only slightly to moderately elevated. A **ratio of serum AST to ALT of more than 2** is suggestive of *alcoholic liver disease,* if the ALT level is normal or only minimally elevated and the AST level is only modestly elevated. This observation may be due to relatively selective mitochondrial injury in alcoholic liver disease. If the AST is elevated significantly out of proportion to a minimally elevated ALT, one should look for *extrahepatic reasons* for the elevation, such as *hemolysis, myocardial infarction* or *myopathy*. Concomitant elevation of LDH may suggest *hemolysis,* and elevation of creatine phosphokinase (CPK) and aldolase would suggest *muscle disease.* Both ALT and AST levels may be low in pyridoxine deficiency. Serum AST levels are depressed in patients on **long-term hemodialysis.**

3.2 Measures of Cholestasis

Another group of tests is considered useful in the evaluation of **cholestasis (bile excretory failure). Alkaline phosphatase** is actually a broad family of enzymes found in a variety of tissues, including *liver, bone, intestine, placenta, kidney* and *leukocytes,* which hydrolyze organic phosphate esters. In the liver, the enzyme is associated with both *sinusoidal* and *canalicular membranes,* as well as the *cytosol.* Normally, liver and bone isozymes are found in the circulation. Normal serum levels are **highly dependent on age,** with high levels in *childhood and puberty* (due to active bone growth), a plateau in *middle age,* and another increase in *old age.* The various isozymes in serum can be biochemically fractionated to determine their source. The *most striking elevations of alkaline phosphatase* are seen in **cholestatic liver disease,** particularly *extrahepatic biliary obstruction,* and *hepatic carcinoma*. With regard to the mechanism of elevation in cholestasis, certain bile acids may induce synthesis of alkaline phosphatase at the translational level. Obviously, many types of bone disease, including *biliary rickets* seen in patients with chronic cholestasis who are vitamin D deficient, may elevate the bone component of the serum alkaline phosphatase level. The placental isozyme may account for serum elevation during pregnancy. Serum alkaline phosphatase may be spuriously *low* in the face of **zinc deficiency.**

Another enzyme used to evaluate cholestasis, **γ-glutamyl transpeptidase (GGTP),** is present throughout the *liver* and *biliary tract,* as well as other organs, including *small intestine, kidney, testes, pancreas, spleen, heart* and *brain*. This enzyme catalyzes the transfer of γ-glutamyl groups from glutathione and other peptides to other amino acids. The serum level of GGTP is a sensitive, though not specific, test for detecting biliary tract disease and generally correlates well with alkaline phosphatase levels. GGTP is *inducible by drugs* such as *ethanol, phenytoin, phenobarbital,* and *warfarin*. In addition, levels may be elevated in a number of conditions, including *diabetes, chronic ethanol consumption, myocardial infarction, renal failure, pancreatic disease* and *chronic obstructive pulmonary disease*. Another enzyme, **5′-nucleotidase**, is found in *liver, intestine, brain, heart,* and *pancreas*. In liver it is primarily associated with *canalicular* and *sinusoidal membranes,* and elevation of the serum level is more specific for liver disease than either alkaline phosphatase or GGTP. However, this assay is often not available on a timely basis in many hospitals.

Although hepatic metabolism of bilirubin and bile acids will be discussed subsequently in detail, they are discussed here in the context of their diagnostic usefulness. **Serum conjugated bilirubin levels** are elevated in a number of disease states, including *hepatitis, drug-induced liver injury,* and *obstructive and hepatocellular cholestasis.* Although conjugated bilirubin elevation in cholestasis is to be expected in the face of *biliary excretory failure,* its elevation in *hepatocellular injury* is due to the fact that, except in cases such as massive hepatocellular necrosis, hepatic bilirubin conjugating activity is relatively preserved in the face of hepatocellular injury and impaired excretion with release of the conjugated bilirubin into the plasma compartment. Unlike unconjugated bilirubin, conjugated bilirubin is readily filtered by the kidney and excreted in the urine. **Urobilinogen** is formed by the action of intestinal bacteria on conjugated bilirubin. Up to 20 % of urobilinogen is absorbed and undergoes an enterohepatic circulation, and a small fraction is excreted in the urine. With *liver dysfunction,* a smaller fraction of absorbed urobilinogen is cleared by the liver and an increased amount is excreted in the urine. In the case of *biliary obstruction,* with decreased entry of bilirubin into the intestine, urinary urobilinogen levels fall.

Measurement of **bile acids** in serum can be a **useful marker for cholestasis,** which is more specific and sensitive than the bilirubin level. In fact, elevation of serum bile acid levels is implicit in the diagnosis of cholestasis. Measurements may include *total bile acids* or specific ones such as *cholylglycine.* Unfortunately, bile acid measurement does not reliably differentiate hepatocellular and obstructive cholestasis. **Fasting and post-prandial bile acid** levels may be used to evaluate the integrity of the enterohepatic circulation, because a slight elevation is expected after a meal despite efficient first-pass clearance by the liver under normal conditions. The post-prandial rise is even more pronounced in the face of mild liver disease, with sub-optimal first-pass clearance. However, it has not been confirmed that post-prandial measurements are more sensitive than those obtained with fasting for the diagnosis of liver disease. Analysis of bile acid metabolites in urine is used to diagnose *inherited defects of bile acid metabolism,* which may cause severe liver disease in the neonatal period.

3.3 Measures of Liver Synthetic Function

Measurement of **serum proteins** produced by the liver can provide an index of liver synthetic function. The liver synthesizes **six coagulation factors,** including *factors I (fibrinogen), II (prothrombin), V, VII, IX,* and *X.* Although these factors may be measured individually, the prothrombin time determination can indirectly assess these factors. This assay evaluates the **extrinsic coagulation pathway** and is *prolonged* when factors I, II, V, VII, and X are deficient, singly or in combination. In addition to *liver synthetic failure,* other causes for prolongation of the prothrombin time include *congenital coagulation factor deficiency, consumptive coagulopathy with disseminated intravascular coagulation (DIC), drugs* such as *warfarin,* and *vitamin K deficiency* (factors II, VII, IX, and X). In fact, one should always check for **correction of the prothrombin time** by parenterally administered vitamin K. Correction is rapid, usually within 24 h. In the case of liver disease, there is generally a large reserve for coagulation factor synthesis, such that prolongation occurs only when a large portion of the liver's synthetic functional reserve is impaired, usually in the setting of acute hepatocellular necrosis or chronic liver disease with cirrhosis.

Albumin is the most abundant plasma protein synthesized by the liver, and the serum level can serve as an index of liver synthetic function. The half-life of serum albumin is 20 days, which limits its usefulness as a synthetic marker for acute liver disease. In order to use the serum albumin level as an indicator of liver synthetic function, one must be aware of the factors regulating serum levels, including synthesis, distribution, and catabolism. Factors regulating albumin synthesis include nutrition, plasma oncotic pressure, and liver dysfunction. **Nutrition** is an extremely important regulator, particularly the availability of amino acids, especially *tryptophan.* **Abnormal loss of serum albumin** through protein-losing enteropathy or nephrotic syndrome should always be considered. In fact, some degree of protein-losing enteropathy may accompany portal hypertension. In the cirrhotic patient with edema and ascites, the volume of distribution for albumin may be expanded, leading to reduced serum concentrations.

3.4 Quantitative Liver Function Tests

Several so-called "quantitative" liver function tests have been developed over the years that generally rely on the plasma clearance of various substances by the liver. **Bromosulfophthalein (BSP)** is an organic ion that was extensively used in the past. However, it is presently not available in the United States for intravenous injection. **Indocyanine green** is an inert dye that is exclusively eliminated by the liver, unchanged, into bile. Administered intravenously in low doses, it is a *good indicator of liver blood flow.* In higher doses, clearance is more dependent on processing of the dye by the liver. This test is particularly sensitive for detecting *cirrhosis.* The **galactose elimination test** is reflective of *functional liver cell mass.* It is particularly useful in following individual patients with progressive liver disease and declining hepatic function. Another quantitative test is the **aminopyrine breath test.** After hepatic uptake, this compound is demethylated by cytochrome P_{450}. The labeled methyl group is eventually excreted in expired air as CO_2. Measurement of the labeled CO_2 provides an index of hepatic excretion. Non-radioactive ^{13}C has replaced the formerly used radioactive tracer. This test has utility for the detection of *cirrhosis,* as well as a *prognostic indicator for the survival of patients with cirrhosis.* Another quantitative test is the **lidocaine clearance test.** This test is performed by injecting a standardized dose of lidocaine intravenously and measuring the appearance of its main metabolite, **monoethylglycinexylidide (MEGX)** in serum. MEGX is formed in the hepatocyte by cytochrome P_{450}-dependent demethylation. Clinical studies indicate that the MEGX test may be useful in evaluating liver dysfunction, particularly in the area of liver transplantation, as an *indicator for the need for transplantation, evaluation of graft function,* and *determination of outcome after transplantation.* Currently, all of these tests are expensive, difficult to perform and not readily available in most hospitals for routine use.

3.5 Imaging Studies

Several imaging studies are used for the evaluation of liver disease. **Plain films** of the liver rarely provide useful information, except in the occasional case with an *intrahepatic calcification* or *calcified gallstones.* **Ultrasonography** can provide detailed anatomic data, as well as evaluate blood flow using a Doppler probe. Ultrasonography can detect *gallstones, bile duct dilatation, abscesses* and *masses,* such as tumors, down to 1–2 cm in size, and diagnose *portal vein* and *hepatic vein (Budd-Chiari syndrome) thrombosis/obstruction.* Ultrasonography is less reliable for evaluation of diffuse parenchymal lesions, such as cirrhosis and fatty liver. **Contrast-enhanced ultrasound (CEUS)** uses an intravenous microbubble contrast agent. This imaging modality is most useful in evaluating *focal lesions,* including *hemangiomas* and *focal nodular hyperplasia.* However, CEUS currently has limited availability.

Greater resolution is possible with **computed tomography (CT) scanning**, which can be performed with and without intravenous and intestinal contrast administration. CT scanning is particularly useful in detection of *abscesses* and *masses* such as tumors. **Angiography** has largely been supplanted by newer imaging methods, but still may be useful for *studying the hepatic vessels and blood supply,* particularly in identifying the blood supply to tumors and evaluating the patient with *portal hypertension.*

Magnetic resonance imaging (MRI) is gaining widespread use for liver imaging. It is especially useful for detecting *focal hepatic lesions,* particularly in distinguishing hemangiomas from other lesions, and detecting excessive hepatic iron deposition in *hemochromatosis.* Imaging by MRI is also the *most* accurate method for assessing *hepatic steatosis,* as occurs in *non-alcoholic fatty liver*

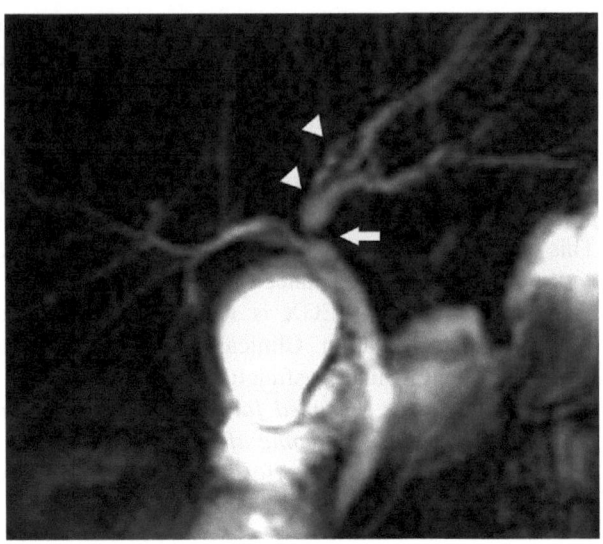

Fig. 10.10 Magnetic resonance cholangiopancreatography from a patient with primary sclerosing cholangitis showing stenosis at the estuary of the left hepatic duct in the common hepatic duct (*arrow*). Stenoses are also present along the branches of the left hepatic duct (*arrowheads*) with mild prestenotic dilations (Reproduced with permission from Alexopoulou et al. [1])

disease (NAFLD) associated with obesity and insulin resistance. **Intraluminal injection of contrast** into the biliary tree, either percutaneously (percutaneous transhepatic cholangiography) or endoscopically directly through the ampulla of Vater (endoscopic retrograde cholangiopancreatography or ERCP) is useful for evaluating the source *of biliary tract obstruction*. **Biliary imaging by MRI (magnetic resonance cholangiopancreatography or MRCP)** is useful for the diagnosis of biliary tract disease, such as primary sclerosing cholangitis (PSC), and is especially useful in children, because it carries less risk than ERCP and avoids radiation exposure. An MRCP image of the bile duct lesions in PSC is shown in Fig. 10.10. **Intravenous cholangiography** is rarely used because of its relatively poor imaging. **Radionuclide imaging** of the liver is useful in the detection of *filling defects* and *diffuse hepatocellular disease.* Hepatobiliary scans utilizing radionuclides, such as 99mTc-DISIDA, taken up by the liver and excreted into bile, are used to evaluate biliary excretory function. When used in conjunction with ultrasonography, these scans can often reliably *differentiate intrahepatic from extrahepatic biliary obstruction.*

3.6 Elastography

A promising technology for the non-invasive assessment of liver fibrosis is **transient elastography.** This technique can be performed with either ultrasonography or magnetic resonance and is based on the observation that the *propagation of a waveform through a homogeneous tissue,* such as liver, *is proportional to its stiffness,* which in the case of liver is most often due to fibrosis or cirrhosis.

Increased stiffness causes the wave to propagate more quickly, which can be measured. Invalid measurements may result from improper positioning of the transducer over rib or lung or in obese patients. Elastography is a rapidly evolving technology, and its routine clinical utility remains to be proven.

3.7 Liver Biopsy

Examination of liver histology obtained by liver biopsy usually provides useful information for both *diagnosis* and *prognosis*. In addition to providing sections for **light microscopy,** tissue may be sent for **special studies** such as *electron microscopy* for the diagnosis of mitochondrial or peroxisomal disorders, *immunostaining* for antigens such as hepatitis B or α-1-antitrypsin, *measurement of iron or copper content* for diagnosis of hemochromatosis or Wilson disease, respectively, and *specific enzyme assays* for the diagnosis of metabolic disorders such as glycogen storage disease. **Percutaneous needle biopsy** is generally safe and provides adequate tissue. However, other approaches include the *transjugular route* when coagulopathy is present, *laparoscopic biopsy* to allow direct visualization of the liver, and *open surgical wedge biopsy*.

3.8 Other Testing

Additional disease-specific tests include **detection of viral infections,** such as hepatitis A, B or C and other hepatotropic viruses by **polymerase chain reaction (PCR)** and **serology.** Detection of specific **autoantibodies** is useful in the diagnosis of autoimmune liver diseases, including *autoimmune hepatitis* and *primary sclerosing cholangitis*. **Genetic testing** is widely available for inherited diseases such as alpha-1-antitrypsin deficiency, hemochromatosis, and tyrosinemia. **Tumor markers,** such as alpha-fetoprotein (AFP) and carbohydrate antigen 19-9 (CA19-9), are helpful in the diagnosis of hepatocellular carcinoma and cholangiocarcinoma, but are not specific for liver cancer.

A new class of biomarkers for liver disease undergoing intense scrutiny is **microribonucleic acid (miRNA).** MicroRNAs are small (19–25 nucleotide) noncoding single-stranded RNA molecules that regulate protein-coding genes and are involved in modulation of many cell functions, both normal and aberrant. They also function in complex gene regulatory networks. Their clinical utility resides in their usefulness as biomarkers for disease and as potential therapeutic targets. They may be measured in serum where they are protected from degradation by RNases. Specific patterns of changes in serum levels of miRNA species are observed in a variety of liver diseases, including *viral hepatitis, malignancy, biliary atresia* (a serious liver disease in infants) and in *liver graft rejection* after transplantation.

4 Hepatic Blood Flow, Cirrhosis, and Portal Hypertension

Under normal conditions, an important feature of hepatic blood flow is a **very low outflow resistance from the sinusoids.** The sinusoidal pressure is generally constant despite changes in blood flow. Maintenance of a low sinusoidal pressure prevents an imbalance of Starling forces, which would result in excessive movement of fluid, solutes and proteins through the highly permeable sinusoidal endothelial lining into the space of Disse. Another feature of the hepatic circulation is the **regulation of hepatic arterial blood flow in response to changes in portal venous blood flow and sinusoidal pressure.** This results in a reflex increase in hepatic arterial blood flow in the face of a decrease in portal venous flow or sinusoidal pressure, and a decrease in hepatic arterial flow in the face of an increase in sinusoidal pressure. Hepatic arterial vascular resistance is responsive to intrinsic, hormonal, and neural factors, all of which may play a regulatory role.

The hepatic circulation is also very sensitive to changes in **systemic pCO_2 levels,** much more so than to changes in pO_2 levels. In animal models, higher systemic pCO_2 levels result in increased hepatic arterial and portal venous flow due to decreased arterial resistance. **Portal blood acidosis** increases portal resistance and decreases arterial resistance in the liver, and **portal alkalosis** has the opposite effects. During digestion there is an increased oxygen extraction by the intestine with an accompanying increase in portal venous pCO_2 and decrease in pH. These changes would be expected to increase hepatic arterial blood flow during a period of enhanced hepatic metabolic activity. Increased portal flow also occurs after a meal secondary to the accompanying splanchnic hyperemic response.

The normal portal venous pressure is 5–10 mmHg. Because the direct measurement of portal venous pressure *in vivo* is not generally practical, a **wedged hepatic venous pressure** is often used as an indirect measurement. **Portal hypertension** is defined as a wedged hepatic venous pressure that is more than 5 mmHg greater than the pressure in the inferior vena cava. Portal hypertension is also present when the splenic pulp pressure is greater than 15 mmHg, or the pressure in the portal vein directly measured in surgery is greater than 30 cm H_2O. **Doppler ultrasonography** allows non-invasive measurement of portal venous flow. Portal hypertension results from either an increase in portal blood flow or any process that increases the resistance to portal blood flow at the level of the portal vein, liver, hepatic veins, or heart. Therefore, portal hypertension may result from increased portal blood flow (**arteriovenous fistula**), pre-hepatic obstruction (**portal vein thrombosis** or **compression by tumor**), increased vascular resistance at the hepatic level (**cirrhosis** or **congenital hepatic fibrosis**), or post-hepatic venous obstruction/congestion (**Budd-Chiari syndrome** or **right heart failure**). Different types of portal hypertension may also be classified as **pre-sinusoidal, sinusoidal**, or **post-sinusoidal**, based upon information obtained from measurement of wedged hepatic venous pressure. However, technical limitations of this measurement and the fact that the pattern may change over time with the same disease process limit the usefulness of this system of classification. Table 10.2 lists causes of

Table 10.2 Causes of portal hypertension

Prehepatic
Splenic vein thrombosis
Portal vein thrombosis
Congenital stenosis of the portal vein
Extrinsic compression of the portal vein
Arteriovenous fistula (splenic, aortomesenteric, aortoportal, and hepatic artery-portal vein)

Intrahepatic
Partial nodular transformation
Nodular regenerative hyperplasia
Congenital hepatic fibrosis
Peliosishepatis
Polycystic disease
Idiopathic portal hypertension
Hypervitaminosis A
Arsenic, copper sulfate, and vinyl chloride poisoning
Sarcoidosis
Tuberculosis
Primary biliary cirrhosis
Schistosomiasis
Amyloidosis
Mastocytosis
Rendu-Osler disease
Liver infiltration in hematologic diseases
Acute fatty liver of pregnancy
Severe acute viral hepatitis
Chronic active hepatitis (viral and autoimmune)
Hepatocellular carcinoma
Hemochromatosis
Wilson disease
Hepatic porphyrias
Tyrosinemia
α-1-antitrypsin deficiency
Cyanamid toxicity
Chronic biliary obstruction
Alcoholic cirrhosis
Sinusoidal obstruction syndrome (veno-occlusive disease)

Posthepatic
Budd-Chiari syndrome
Congenital malformations and thrombosis of the inferior vena cava
Constrictive pericarditis
Tricuspid valvular disease

Adapted from Arias [2]

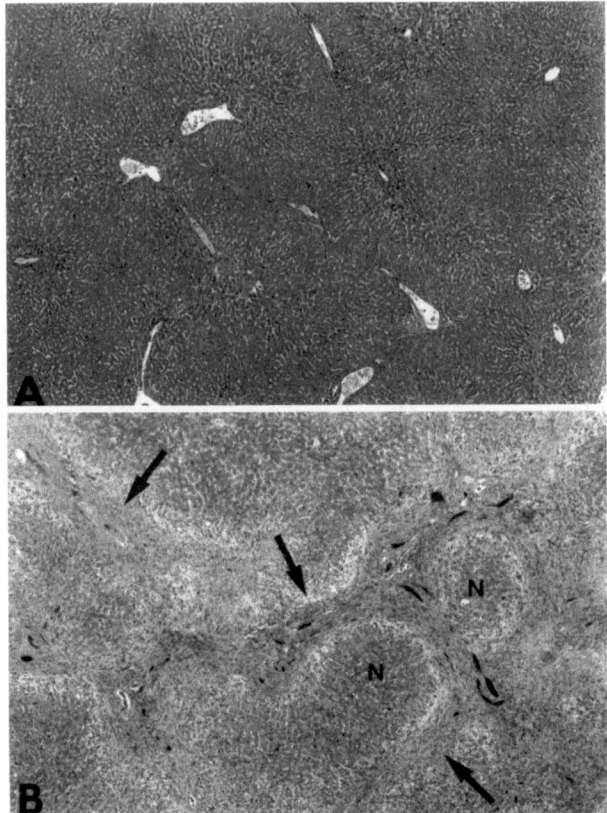

Fig. 10.11 Normal (**a**) and cirrhotic (**b**) liver. Note the fibrous bands (*arrows*) and regenerative nodules (*N*), which lack central veins (hematoxylin-eosin, original magnification 100×; Courtesy of Dr. John Hart)

portal hypertension. The following discussion will generally be limited to portal hypertension resulting from hepatic cirrhosis.

A number of insults to the liver, including **hepatitis** (viral or autoimmune), **toxins** (alcohol, drugs, and environmental toxins), **metabolic disease** (Wilson disease, α-1-antitrypsin deficiency), and **extrahepatic biliary obstruction** (stones, tumor) may result in hepatic fibrosis and marked alteration of the normal hepatic microanatomy. Figure 10.11 shows a photomicrograph of a liver biopsy from a patient with advanced cirrhosis. Normal lobular architecture is abolished. Present are regenerative nodules surrounded by fibrous bands. Although vessels are identified in the fibrous bands, no central veins are present in the nodules. These changes result in inefficient perfusion of hepatocytes with intrahepatic shunting of blood and increased resistance to blood flow through the liver, contributing to the development of portal hypertension.

The interplay of hepatic sinusoidal endothelial cells and stellate cells is a major determinant of the vascular changes that occur in the liver in cirrhosis and associated portal hypertension. There are likely two components to the increase in resistance to flow through the hepatic sinusoids. One is **structural** and related to the *deposition of collagen in the sinusoids,* as well as *distortion by cirrhotic nodules,* resulting in a largely irreversible increase in resistance to portal flow and increased portal pressure. The stellate cell becomes activated in response to exposure to inflammation and chemokines and is responsible for collagen and matrix molecule deposition. The other component is **dynamic** and related to *dysregulation of vascular tone.* Stellate cells and stellate cell-derived myofibroblasts also are capable of Ca^{2+}-dependent contractility around the sinusoids to regulate vascular tone. These cells respond to endothelial cell secretion of NO mediating vasodilation and endothelins and other molecules modulating vasoconstriction. In disease states with stellate cell proliferation and activation, this balance is tipped in favor of vasoconstriction, contributing to increased resistance to blood flow through the liver. Future therapeutic targets may be induction of **NO production** by *endothelial nitric oxide synthase (eNOS)* and inhibition of **endothelins** in the liver.

The increase in intrahepatic vascular resistance in portal hypertension is accompanied by **increased splanchnic blood flow.** In response to the rise in portal pressure due to increased intrahepatic resistance, porto-systemic collateral blood vessels, such as esophageal and gastric varices, develop in response to increased release of **vascular endothelial growth factor** (**VEGF**) and other factors, and shunting contributes to a fall in splanchnic resistance and development of a hyperdynamic circulation. Another mechanism for the drop in splanchnic resistance is an **increase in NO production** in this vascular bed, in contrast to the local decrease in NO in the hepatic sinusoidal bed. The resultant increase in flow into the portal circulation in the face of increased resistance in the liver results in further elevation of the portal pressure.

In the patient with cirrhosis, **collateral circulatory pathways** develop to allow portosystemic shunting of blood. These collaterals may include prominent dilated *superficial abdominal veins, caput medusae* (tortuous periumbilical veins), *recanalization of the umbilical vein remnant,* and *rectal varices,* as shown in Fig. 10.12. Of the greatest clinical significance is the formation of esophageal varices, thin-walled superficial veins in the distal esophagus, which can result in **life-threatening hemorrhage** (Fig. 10.13a). Varices and portal hypertensive gastropathy may develop in the stomach, which are also often a source of bleeding. The factors that predispose varices to bleed are not well known. Irritation from gastroesophageal reflux, a popular theory in the past, is probably not a common event. The initiation of hemorrhage is probably a complex event dependent on factors such as **wall tension (T), transmural pressure [intravariceal pressure (P_1) – esophageal lumen pressure (P_2)], vessel radius (r),** and **wall thickness (W)** as related to each other by the following equation involving **Laplace's law:**

$$T = (P_1 - P_2)(r/W)$$

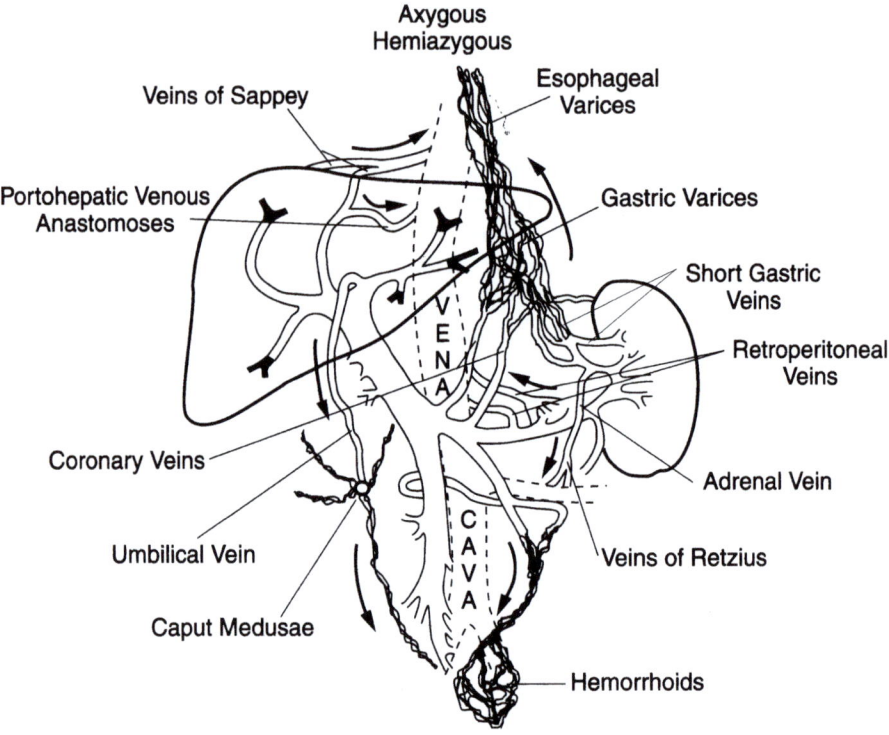

Fig. 10.12 Collateral blood flow in portal hypertension (Reproduced with permission from Black [4])

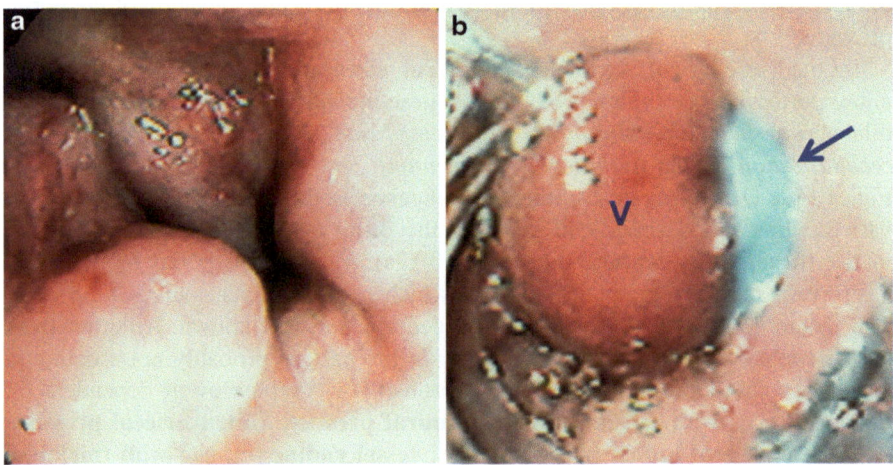

Fig. 10.13 Esophageal varices. (**a**) Fiberoptic esophagoscopy reveals esophageal varices in the distal esophagus of a patient with cirrhosis and portal hypertension. (**b**) Esophageal varix (*V*) after endoscopic ligation with a rubber band (*arrow*)

Varices probably burst open and bleed when the wall tension can no longer contain the forces expanding the varices. Large varices with thin walls would be expected to have a greater tendency to bleed, compared to smaller varices with thinner walls at the same transmural pressure.

Bleeding esophageal varices may be treated by endoscopic injection of a sclerosing solution (**sclerotherapy**) or placement of bands around the varices (**banding,** Fig. 10.13b). **Acute reduction in portal hypertension** is possible by the intravenous infusion of *vasopressin* or *octreotide* (a somatostatin analogue), which results in splanchnic arteriolar constriction and temporary reduction in portal venous pressure. **Chronic reduction of portal hypertension** is possible with administration of *β-adrenergic antagonists*, particularly *propranolol,* although not all patients are responsive. This therapy appears to work predominantly by enhancing splanchnic arteriolar vasoconstriction by blocking $β_1$-receptor-mediated vasodilatation. These therapeutic modalities have replaced balloon tamponade using a Sengstaken-Blakemore tube. Another approach to treat the underlying portal hypertension is surgical creation of a **portosystemic shunt** between the portal system and systemic venous system to reduce portal blood pressure. The technique of **transjugular intrahepatic portosystemic shunting (TIPS)** via interventional radiology is widely used as an emergency procedure to reduce portal pressure to avoid surgery and as a bridge to liver transplantation.

Portal hypertension also results in **splenic congestion** and **splenomegaly,** often causing **functional hypersplenism.** Hypersplenism results in sequestration of red blood cells, white blood cells, and platelets with resultant depression of counts of one to all of these cell types in peripheral blood. Occasionally, the magnitude of this reduction in circulating cell number is great enough to result in significant **anemia (red blood cells),** risk for **infection (leukocytes),** or risk of **bleeding (platelets).**

Chronic portosystemic shunting can also result in **chronic intermittent encephalopathy** resulting from the entry of ammonia and other toxins, which would normally be cleared by the liver, from the intestine into the systemic circulation via collateral circulation and subsequently into the central nervous system. As might be expected, surgical shunting to treat portal hypertension often leads to worsening of encephalopathy.

Another consequence of cirrhosis and portal hypertension is the **development of fluid in the peritoneal cavity (ascites)** and **peripheral edema,** as well as **impaired renal function.** The major factors contributing to these complications of cirrhosis and portal hypertension are summarized in Fig. 10.14. A discussion of the pathophysiology of these conditions provides an excellent insight into the important role of the liver in the maintenance of normal circulatory homeostasis. The increased lymph formation due to increased hydrostatic pressure in the splanchnic vascular bed as a result of portal hypertension, as well as increased hepatic lymph formation from locally increased vascular resistance, exceeds the capacity of the thoracic duct to return the lymph to the systemic circulation. This event, along with venous pooling in the splanchnic bed, results in a redistribution of the blood volume and ascites formation. One of the major factors resulting in the development of ascites is avid **renal sodium retention** in the patient with cirrhosis. A concept

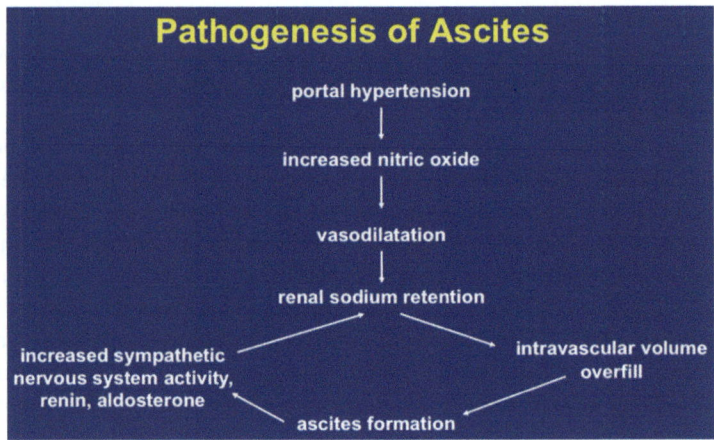

Fig. 10.14 Schematic representation of the major pathophysiological mechanisms leading to renal sodium retention and ascites in cirrhosis

important to understanding this phenomenon is diminished effective blood volume. **Effective plasma volume** refers to that fraction of the total circulating volume that is effective in stimulating volume receptors. **Decreased effective plasma volume** is the result of several factors. *Hypoalbuminemia,* which often is present in cirrhosis, results in decreased plasma oncotic pressure and an imbalance in Starling forces favoring an extravascular shift of fluid. Other factors include a *decrease in systemic vascular resistance* caused by peripheral arteriovenous shunts, as well as circulating vasodilators, such *NO, carbon monoxide,* and others. The net result of these processes is a decreased effective plasma volume, despite the fact that the actual plasma volume is increased. Lack of appropriate stimulation of volume receptors sets in motion several compensatory mechanisms to correct what is inappropriately perceived as a reduced vascular volume. These mechanisms include activation of the renin-angiotensin system with resultant hyperaldosteronism and renal sodium retention. Decreased aldosterone clearance by the cirrhotic liver also contributes to the hyperaldosteronism. In addition, increased renin-angiotensin and sympathetic nervous system activity leads to increased renal vascular resistance and a decreased glomerular filtration rate (GFR). Other humoral factors may also contribute to decreased renal blood flow and increased tubular reabsorption of sodium, including *renal prostaglandins,* the *kallikrein-kinin system, circulating natriuretic factor,* and *atrial natriuretic factors.*

All of these factors in the patient with cirrhosis and portal hypertension, particularly those affecting renal blood flow, can result in renal failure, which is termed the **hepatorenal syndrome.** This syndrome usually occurs in the setting of end-stage liver disease and may be precipitated by overzealous **diuresis** or **paracentesis** for ascites or variceal bleeding. It is characterized by the development of *oliguria* with *decreased GFR* in the face of preservation of tubular function, as reflected by a normal or increased fractional excretion of sodium and increased water reabsorption. The most striking pathological event in the kidney is **marked vasoconstriction** and ***diminution in arterial blood flow***, particularly that supplying the cortex. Kidneys from individuals with hepatorenal syndrome dying from liver failure demonstrate a normal arterial tree on angiogram and have histologically normal blood vessels when removed from the patient. Furthermore, these kidneys function normally when transplanted into normal individuals with renal failure. This suggests that the renal abnormalities in the hepatorenal syndrome are functional and due to local and/or circulating factors.

Complications of ascites in the patient with cirrhosis include ***spontaneous bacterial peritonitis*** and ***impaired ventilation*** in extreme cases. Treatment of ascites in the cirrhotic patient should be dictated by knowledge of the pathophysiology. The cornerstone of treatment is **sodium restriction.** In fact, in many patients, this one maneuver, if complied with, may result in a striking diuresis. Since hyperaldosteronism is a prominent abnormality, treatment with the **aldosterone antagonist,** *spironolactone,* is often effective and can be combined with a **loop diuretic** if needed. However, one should be cautious, because aggressive diuresis may precipitate hepatorenal syndrome or encephalopathy. Complete **elimination of ascitic fluid** should never be the goal, but reducing it to make the patient more comfortable and to eliminate any respiratory compromise should be the objectives. **Intravenous infusion of albumin** in the patient with hypoalbuminemia may facilitate a transient diuresis. Refractory cases may require **paracentesis,** particularly in the face of respiratory compromise. The placement of **shunts** to drain ascitic fluid back into the systemic circulation has been used, but is fraught with complications, especially infection.

5 Hepatic Sex Hormone Metabolism in Cirrhosis

The role of the liver in sex hormone metabolism is underscored in the cirrhotic male patient, particularly in those with alcoholic cirrhosis. **Hypogonadism** is frequently present, with **testicular atrophy** and **decreased libido**. The testicular atrophy reflects a decrease in seminiferous tubule mass, most likely the result of chronic ethanol toxicity. **Leydig cell dysfunction** is also present and reflected by decreased production of testosterone. Cirrhosis results in an ***increase in plasma levels of sex hormone-binding globulin***, and the combination of reduced total levels of testosterone and increased protein binding results in significantly reduced levels of free, biologically active hormone. Pre-menopausal women with alcoholic cirrhosis have **reduced levels of estradiol and progesterone** and accompanying **loss of breast and pelvic fat** and **anovulation.** Pathologic changes in the ovaries may be at least partially due to ***ethanol toxicity,*** as is postulated in the testes in males. At the hypothalamic-pituitary level, there is **reduced basal gonadotropin secretion** and a **reduced luteinizing hormone (LH) response to LH-releasing factor.** Males with cirrhosis frequently exhibit **feminization** manifested by *gynecomastia, palmar* *erythema, cutaneous spider angiomata,* and a *female escutcheon*. Although the mechanisms of this estrogenization are not completely known, hepatic clearance of estrogens is reduced and may contribute to elevated circulating levels. Another mechanism may involve the **reabsorption of estrogens and androgens excreted in bile** with accompanying **inefficient first-pass clearance by the liver** in the face of portosystemic shunting and hepatocellular dysfunction. This allows exposure of estrogen-sensitive peripheral tissues to these estrogens. There also appears to be **increased peripheral conversion of reabsorbed androgens to estrogens.** Finally, there may be **increased sensitivity of peripheral tissues to estrogens** in these individuals.

Clinical Correlations

Case Study 1

A 55-year-old white male is brought to the emergency room by family members with a history of passing three large, foul-smelling, black, tarry stools over the past 12 h. He has also been "acting strangely" over the past 2 days and has been verbally abusive and very agitated, more so than usual when intoxicated. Family members report that the patient has a history of "drinking heavily" for the past 25 years. He often will consume a pint of whiskey per day. There has also been an inversion of sleep patterns. Today, he has been very sleepy and difficult to arouse. They also report that they have noticed a large increase in the size of his abdomen over the past several weeks, and his pants no longer fit. There is no past history of blood transfusion, intravenous drug use, or hepatotoxic drug exposure.

Physical examination: Respirations 22, temperature 37.2 °C rectally, pulse 96, and blood pressure 122/81. When tilted up, the pulse rate goes up to 116, and the blood pressure drops to 106/66. The patient is sleepy, but arousable with vigorous stimulation. However, he is very combative and uncooperative when aroused. Pupils are equal and reactive to light. The chest is clear to auscultation. There is bilateral breast enlargement. Cardiac examination reveals a regular rate and rhythm and a grade II/VI systolic ejection murmur. The abdomen is distended and exhibits shifting dullness, indicating ascites. There are dilated superficial abdominal veins. The liver edge cannot be palpated but has a span of 12 cm by percussion. The spleen is palpable 8 cm below the left costal margin. There is reduced pubic hair in a female distribution. Rectal examination reveals black stool strongly positive for occult blood. Skin examination reveals several spider angiomata.

Laboratory studies: Hematocrit 28 %; hemoglobin 9.2 g/dL; white blood cell count 5,400 cells/mm^3 (4,500–11,000); 136,000 platelets/mm^3 (150,000–350,000); prothrombin time 18 s (control 12.4 s); albumin 2.8 g/dL (3.7–5.6); total bilirubin 2.8 mg/dL (0.1–1.0); direct bilirubin 2.4 mg/dL (0–0.2); ALT 200 IU/L (10–30); AST 280 IU/L (15–30); GGTP 180 IU/L (14–25); alkaline phosphatase 50 IU/L (30–94); blood ammonia 277 μmol/L (29–57); sodium 130 mEq/L (135–145), potassium 2.9 mEq/L (3.5–5.0), chloride 91 mEq/L (94–106); carbon dioxide 33 mEq/L (22–29); blood urea nitrogen (BUN) 77 mg/dL (5–25); and creatinine 0.2 mg/dL (0.6–1.2).

Questions:

1. **What is the most likely source of bleeding (melena) in this patient?**
   Answer: With the history of long-term heavy ethanol abuse, **bleeding esophageal varices due to portal hypertension** from alcoholic cirrhosis would be a likely source. The physical findings discussed below would also support this diagnosis. It should be remembered that **ethanol** itself is an irritant and can produce gastritis with bleeding. Also, there is a higher than usual incidence of *peptic ulcer disease* in individuals with cirrhosis.

2. **What is the cause of the altered mental state?**
   Answer: The history and present findings are certainly compatible with **hepatic encephalopathy**, particularly the inversion of the sleep/wake cycle, somnolence, and concomitant gastrointestinal bleeding, which often exacerbates hepatic encephalopathy because of the increased intestinal ammonia load. However, ethanol abuse alone can produce an **organic brain syndrome.** Electrolyte derangements can cause altered mentation and should always be checked. **Wernicke-Korsakoff psychosis** may develop in alcoholics in the setting of thiamine deficiency. Also, **abuse of other psychoactive drugs** should be considered, often on an unintentional basis because of confusion as to proper dosing of prescribed medication. Remember that the patient must be conscious and cooperative to allow elicitation of asterixis and that asterixis is not specific for hepatic encephalopathy.

3. **What findings support a diagnosis of cirrhosis?**
   Answer: The suspicion of **esophageal varices** and the presence of **dilated superficial abdominal veins** as evidence of portosystemic collaterals and the presence of **ascites** and **splenomegaly** indicate portal hypertension. There are signs of hyperestrogenism and hypogonadism (gynecomastia, spider angiomata, and a female escutcheon). Palmar erythema may not be obvious in the face of blood loss and hypovolemia. In the setting of heavy ethanol abuse, alcoholic cirrhosis would be most likely. However, other considerations would **include chronic active viral (HBV or HCV) hepatitis with cirrhosis** and **primary hemochromatosis**. Much less likely would be etiologies such as **Wilson disease** or **α-1-antitrypsin deficiency**. There is no history of drug or toxin exposure, which might result in chronic liver disease and development of cirrhosis, although the medical history in such a patient may not be reliable. The serum GGTP elevation is most likely due to induction by chronic ethanol ingestion.

4. **How would you interpret the laboratory tests reflective of renal function in this patient?**
   Answer: The **low serum sodium** may reflect a relative *dilutional hyponatremia* in the face of total body sodium overload due to the renal sodium retention present in cirrhosis. **Hypokalemia** is often present in the cirrhotic patient due to *respiratory and/or metabolic alkalosis* driving potassium into cells and chronic renal loss caused by *hyperaldosteronism.* These individuals are almost always total body potassium-depleted. The presence of the hypokalemia may further exacerbate the alkalosis and have adverse effects on renal and muscle function. It is very difficult to interpret the BUN and creatinine as indices of

renal function in this patient. The **BUN may be elevated** because of *reduced vascular volume (pre-renal azotemia), increased urea production from blood in the gastrointestinal (GI) tract,* or *renal failure (acute tubular necrosis or hepatorenal syndrome).* The **creatinine level** may be high, reflecting *renal failure*, or factitiously low in the face of a markedly reduced muscle mass in the cirrhotic patient. Close monitoring of urine output during replenishment of blood volume and evaluation of renal tubular function using indices such as the fractional excretion of sodium and urine osmolality and examination of the urine sediment may be useful.

5. **How should one proceed with the treatment of this patient?**
   Answer: Because the patient has significant hypovolemia, as demonstrated by the positive "tilt" test, adequate venous access should be established (preferably a **central line**) and blood should be sent for type and crossmatch immediately. **Volume expansion** should be started immediately; 5 % albumin would be a good choice. Also, **vitamin K** should be administered parenterally, and **fresh frozen plasma infusion** would provide volume as well as correction of coagulopathy. Active variceal bleeding may be treated with a **vasopressin** or **octreotide** infusion followed by **upper GI endoscopy with variceal banding.** Encephalopathy may be treated by **removing the source of GI bleeding** (esophageal varices) and administering **lactulose** or a **non-absorbable antibiotic**, such as rifaximin, by nasogastric tube.

Case Study 2

A 5-year-old female is brought to the emergency room with a history of vomiting blood twice and passing three large, tarry, foul-smelling stools in the last 6 h. She has remained alert and complained only of nausea. She has been healthy up to this time with no complaints of abdominal pain or jaundice. She was born prematurely at 32 weeks gestation and spent 2 weeks in the neonatal ICU with respiratory distress and sepsis. However, since then she has been healthy with no other significant illness.

Physical examination: Respirations 22, temperature 37.6 °C rectally, pulse 112, blood pressure 96/55, weight 18.2 kg (50th percentile), and height 108 cm (50th percentile). This is an anxious, pale female child who is alert and cooperative. There is no jaundice. Her physical examination is generally unremarkable without stigmata of chronic liver disease or cirrhosis except for the abdominal examination, which reveals an enlarged spleen palpable 4–5 cm below the left costal margin. There is no hepatomegaly or ascites. Rectal examination yields guaiac-positive stool. The neurologic examination is normal, and asterixis cannot be elicited.

Laboratory studies: Hematocrit 24 %; hemoglobin 8.1 g/dL; white blood cell count 3,600 cells/mm^3 (5,000–15,500); 78, 000 platelets/mm^3 (150,000–350,000); prothrombin time 13.0 (control 12.4 s); albumin 4.5 g/dL (3.7–5.6); total bilirubin 1.2 mg/dL (0.1–1.0); direct bilirubin 0.4 mg/dL (0–0.2); ALT 26 IU/L (10–30), AST 22 IU/L (15–30); GGTP 21 IU/L (14–25); alkaline phosphatase 296 IU/L

(100–300); blood ammonia 77 μmol/L (29–57); sodium 139 mEq/L (135–145), potassium 3.9 mEq/L (3.5–5.0), chloride 101 mEq/L (94–106); carbon dioxide 26 mEq/L (22–29); blood urea nitrogen (BUN) 39 mg/dL (5–25); and creatinine 0.6 mg/dL (0.3–0.7).

Questions:

1. **What is the most likely source of this girl's bleeding?**
   Answer: In a child with upper GI bleeding of this magnitude without significant liver dysfunction and with a past history of neonatal intensive care and splenomegaly, **portal hypertension** and **bleeding esophageal varices secondary to cavernous transformation of the portal vein** would be very likely. This form of post-hepatic portal hypertension often is a sequela of portal vein thrombosis from placement of an umbilical vein catheter and/or omphalitis in the neonatal period. The term **"cavernous transformation"** refers to the characteristic appearance of this lesion on angiogram demonstrating collaterals and recanalization in the region of the obstruction. This diagnosis can often be made by ultrasonography with Doppler flow analysis. Although these patients often have significant portal hypertension, they rarely have ascites, and they are rarely encephalopathic because of their normal liver function. This finding contrasts with hepatic (cirrhosis) or pre-hepatic (Budd-Chiari syndrome) portal hypertension, which is frequently complicated by ascites and liver dysfunction. Other causes of portal vein obstruction would include thrombosis due to a disorder in blood coagulation such as anti-phospholipid syndrome or protein C deficiency, extrinsic compression by tumor or other mass, or abdominal trauma. The majority of cases in otherwise healthy children are idiopathic. Indeed, in this patient esophageal varices were documented and sclerosed by endoscopy, and cavernous transformation of the portal vein was diagnosed by ultrasonography. Cavernous transformation of the portal vein demonstrated by venous phase superior mesenteric arteriography in a child is shown in Fig. 10.15.

2. **Why does this patient have elevated blood ammonia?**
   Answer: This patient has modest hyperammonemia because of the **increased intestinal ammonia load from the variceal bleeding** along with **decreased hepatic first-pass portal blood clearance** secondary to the portal vein obstruction and collateral circulation. The preservation of normal hepatic function prevents higher ammonia levels and clinical encephalopathy.

3. **What is the significance of the elevated BUN in this patient?**
   Answer: In the face of normal hepatic function and pre-existing normal renal function, the BUN is most likely elevated secondary to **increased BUN production** from **GI bleeding** and **volume depletion** leading to pre-renal azotemia. Although variceal bleeding and hypovolemia can lead to acute tubular necrosis, the normal creatinine level would not be consistent with renal failure. However, during hospitalization urine output and renal function should be monitored closely, particularly in the face of continued bleeding.

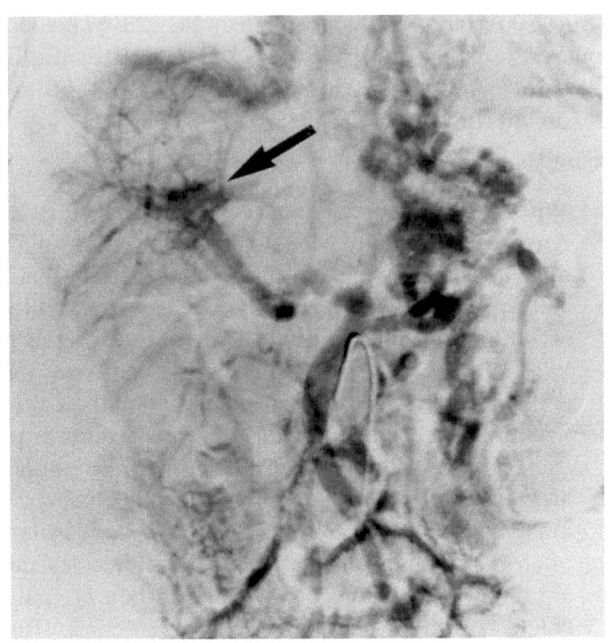

Fig. 10.15 Venous phase of a superior mesenteric arteriogram showing cavernous transformation of the portal vein (*arrow*) (Reproduced with permission from Brady et al. [5])

4. **Why are the white blood cell count and platelet count depressed?**
   Answer: The presence of splenomegaly in an individual with portal hypertension often leads to **hypersplenism,** with sequestration of red blood cells, leukocytes and platelets. In this patient it is difficult to determine how much of the present anemia was pre-existing from hypersplenism and how much is due to blood loss from variceal bleeding.
5. **How would you treat this patient?**
   Answer: The same measures to manage variceal bleeding discussed for the previous case would be used. However, treatment of coagulopathy and encephalopathy would not be necessary. These patients can often be managed by **multiple sessions of endoscopic banding** to obliterate their varices until they reach adolescence, when they will often have developed enough collateral circulation to prevent recurrence of varices. Occasionally, a **surgical portosystemic shunt procedure** is needed, but tends to be well tolerated without portosystemic encephalopathy because of the presence of normal liver function. Rarely is hypersplenism severe enough to be the sole indication for a surgical shunt procedure.

Case Study 3

A 19-year-old African-American male is referred to your clinic because of recent worsening of recurrent right upper quadrant pain that started 3 months ago. He has also noticed deepening yellowish discoloration of his sclerae. He relates that he has hemoglobin SS disease (sickle cell disease) that was diagnosed in childhood.

He has had recurrent painful crises throughout his life, and his present symptoms were initially thought to represent a component of one of these crises. However, the pain is now persistent, and the jaundice is unrelenting. He cannot relate the pain to eating or a specific activity. His urine is now dark colored. He has had multiple hospitalizations for his sickle cell disease, but no other serious illnesses. He has two younger siblings with the same disease.

Physical examination: Respirations 20, temperature 37.3 °C rectally, pulse 72, and blood pressure 122/81. The patient is clearly uncomfortable, but in no acute distress. There is bilateral scleral icterus. The chest is clear to auscultation. Cardiac examination reveals a regular rate and rhythm and a grade I/VI systolic ejection murmur. The abdomen is soft and non-distended. There is mild right upper quadrant tenderness. The liver edge cannot be palpated but has a span of 12 cm by percussion. The spleen is not palpable. There is no abdominal mass or ascites. Further examination reveals no spider angiomata or other stigmata of chronic liver disease or cirrhosis.

Laboratory studies: Hematocrit 25 %; hemoglobin 8.3 g/dL; white blood cell count 10,200 cells/mm^3 (4,500–11,000); 222,000 platelets/mm^3 (150,000–350,000); reticulocyte count 6 %; prothrombin time 11.4 s (control 12.4 s); albumin 3.9 g/dL (3.7–5.6); total bilirubin 12.2 mg/dL (0.1–1.0); direct bilirubin 10.1 mg/dL (0–0.2); ALT 202 IU/L (10–30), AST 180 IU/L (15–30); GGTP 905 IU/L (14–25); alkaline phosphatase 1,140 IU/L (30–94).

Questions:

1. **To what hepatic complications are patients with sickle cell disease susceptible?**
   Answer: These individuals were prone to **hepatitis C infection** in the era before accurate testing was available for donated blood. They are also at increased risk for autoimmune liver disease. They may also develop **liver disease secondary to sickled red blood cells** lodging in hepatic sinusoids and producing ischemic changes in the hepatocytes. This may range across a spectrum from acute hepatic crisis with fever, right upper quadrant pain, tender hepatomegaly and jaundice with mild transaminase elevation and total bilirubin levels usually less than 13 mg/dL that is usually readily reversible, to a sequestration crisis with massive sequestration of red blood cells in the liver and severe anemia, to sickle cell intrahepatic cholestasis with severe liver impairment, often leading to liver failure. Multiple blood transfusions may lead to iron overload and liver injury. They are also at risk for thrombotic complications, such as hepatic vein thrombosis and Budd-Chiari syndrome. Finally, the increased production of bilirubin due to enhanced turnover of the abnormal cells containing hemoglobin SS, as reflected by the anemia and elevated reticulocyte count indicating increased red cell production by the bone marrow, frequently leads to formation of **pigment gallstones.** See the discussion of bilirubin metabolism in Chap. 12.

2. **What is the likely diagnosis in this patient?**
   Answer: The elevated alkaline phosphatase and GGTP suggest **biliary obstruction.** An *ultrasound study* is quickly obtained that shows multiple gallstones in

the gallbladder, as well as a single stone in the common bile duct with significant dilatation of the duct proximal to the stone, indicating a high-grade obstruction.
3. **What is the recommended treatment?**
  Answer: After hydration and starting antibiotics to prevent bacterial cholangitis due to the obstruction, the patient undergoes an **ERCP,** and the endoscopist verifies the presence of a single stone in the common bile duct and is able to retrieve the stone with a basket. A few days later with further stabilization and improvement in the cholestasis, the patient undergoes a **laparoscopic cholecystectomy** to prevent any future stone formation and episodes of stones migrating from the gallbladder into the bile duct to cause obstruction.

Further Reading

1. Alexopoulou E, Xenophontos PE, Economomopoulos N et al (2012) Investigative MR cholangiopancreatography for primary sclerosing cholangitis-type lesions in children with IBD. J Pediatr Gastroenterol Nutr 55:308–313
2. Arias IM (1994) The liver: biology and pathobiology, 3rd edn. Raven Press, New York, p 1344
3. Berzigotti A, Seijo S, Reverter E, Bosch J (2013) Assessing portal hypertension in liver diseases. Expert Rev Gastroenterol Hepatol 7(2):141–155
4. Black DD (1996) Liver physiology I: structure, functional assessment, and blood flow. In: Chang EB, Sitrin MD, Black DD (eds) Gastrointestinal, hepatobiliary, and nutritional physiology, 1st edn. Lippincott-Raven, New York
5. Brady L, Magilavy D, Black DD (1996) Portal vein thrombosis associated with antiphospholipid antibodies in a child. J Pediatr Gastroenterol Nutr 23:470–473
6. DeLeve LD (2009) The hepatic sinusoidal endothelial cell: morphology, function and pathobiology. In: Arias IM (ed) The liver: biology and pathobiology, 5th edn. Wiley-Blackwell, West Sussex, pp 373–388
7. Duarte-Rojo A, Altamirano JT, Feld JJ (2012) Noninvasive markers of fibrosis: key concepts for improving accuracy in daily clinical practice. Ann Hepatol 11(4):426–439
8. Gumucio JJ, Chianale J (1988) Liver cell heterogeneity and liver function. In: Arias IM, Jacoby WB, Popper H, Schachter D, Shafritz DA (eds) The liver: biology and pathobiology, 2nd edn. Raven Press, New York
9. Hanover JA (1988) Molecular signals controlling membrane traffic. In: Arias IM, Jacoby WB, Popper H, Schachter D, Shafritz DA (eds) The liver: biology and pathobiology, 2nd edn. Raven Press, New York
10. Hanover JA (1988) Introduction: organizational principles. In: Arias IM, Jacoby WB, Popper H, Schachter D, Shafritz DA (eds) The liver: biology and pathobiology, 2nd edn. Raven Press, New York
11. Jones AL (1990) Anatomy of the normal liver. In: Zakim D, Boyer TD (eds) Hepatology: a textbook of liver disease, 2nd edn. W. B. Saunders Co., Ltd, Philadelphia
12. Jones AL, Spring-Mills E (1977) In: Weiss L, Greep RO (eds) Histology, 4th edn. New York, McGraw-Hill Book Co
13. McCuskey RS (1994) Hepatic microvascular system. In: Arias IM, Boyer JL, Fausto N, Jakoby WB, Schachter DA, Shafritz DA (eds) The liver: biology and pathobiology, 3rd edn. Raven Press, New York
14. McCuskey R (2012) Anatomy of the liver. In: Boyer TD, Manns MP, Sanyal AJ (eds) Hepatology: a textbook of liver disease, 6th edn. Elsevier Saunders, Philadelphia
15. Moore CM, Van Thiel DH (2013) Cirrhotic ascites review: pathophysiology, diagnosis and management. World J Hepatol 5(5):251–263

16. Poynard T, Imbert-Bismut F (2012) Laboratory testing for liver disease. In: Boyer TD, Manns MP, Sanyal AJ (eds) Zakim and Boyer's hepatology, 6th edn. Elsevier Saunders, Philadelphia, pp 201–215
17. Rockey DC, Friedman SL (2012) Hepatic fibrosis and cirrhosis. In: Boyer TD, Manns MP, Sanyal AJ (eds) Zakim and Boyer's hepatology, 6th edn. Elsevier Saunders, Philadelphia, pp 64–85
18. Rojkind M, Reyes-Gordillo K (2009) Hepatic stellate cells. In: Arias IM (ed) The liver: biology and pathobiology, 5th edn. Wiley-Blackwell, West Sussex, pp 407–432
19. Silverman A, Roy CC (1983) Pediatric clinical gastroenterology, 3rd edn. C. V. Mosby Co., St. Louis
20. Wang XW, Heegaard NH, Orum H (2012) MicroRNAs in liver disease. Gastroenterology 142(7):1431–1443

Chapter 11
Protein Synthesis and Nutrient Metabolism

Dennis D. Black

1 Introduction

The liver **produces and secretes most of the circulating proteins** in the body that function in an amazingly complex array of regulatory and metabolic processes. The liver also functions as a central metabolic way station for the **processing, partitioning and trafficking of nutrients**, including *lipids* and *carbohydrates*, as well as *vitamins*, at the intersection of the intestine and the rest of the body. It is remarkable that the regulatory mechanisms that have evolved to control these processes generally function in a highly responsive and coordinated manner. However, when dysregulation occurs due to genetic or environmental factors, such as in glycogen storage diseases or non-alcoholic fatty liver disease, significant morbidity may result. This chapter will highlight these functions of the liver and some of the disease processes that may occur.

2 Role of the Liver in Synthesis of Biologically Important Proteins

The majority of circulating plasma proteins is synthesized by the liver. A list of plasma proteins secreted by the liver and their characteristics and functions is shown in Table 11.1. To accomplish this task, the hepatocyte has a well-developed endoplasmic reticulum, Golgi system, and cellular cytoskeleton, all of which function in the synthesis, processing, and secretion of proteins. The most abundant plasma protein produced by the liver is **albumin**, which comprises

---

D.D. Black, M.D. (✉)
Department of Pediatrics, University of Tennessee, Memphis, TN, USA
e-mail: dblack@uthsc.edu

Table 11.1 Partial list of plasma proteins synthesized and secreted by the liver

Protein	Mol. Wt. (kDa)	Function	Ligand binding
Albumin	66	Binding and carrier protein, osmotic regulator	Hormones, amino acids, steroids, vitamins, fatty acids
α-1-acid glycoprotein (orosomucoid)	40	Uncertain, may have role in inflammation. Acute phase reactant	
α-1-antitrypsin	54	General protease inhibitor	Proteases in serum and tissue secretions
α-fetoprotein	72	Uncertain, expressed in fetus and malignancy. Tumor marker	Possibly the same as albumin
α-2-macroglobulin	720	Serum endoprotease inhibitor	Proteases
Antithrombin III	65	Protease inhibitor of intrinsic coagulation system	Proteases
Apolipoprotein A-I	26	Lecithin:cholesterolacyl-transferase activator. Esterifies cholesterol in HDL	Major apolipoprotein of plasma HDL
Apolipoprotein B-100	500	Lipoprotein assembly and secretion, LDL receptor ligand	Component of plasma VLDL and LDL, binds to LDL receptor
Ceruloplasmin	134	Transport of copper	6 atoms copper/mol
C-reactive protein	105	Uncertain, may have role in inflammation. Acute phase reactant	Complement C1q
Fibrinogen	340	Fibrin precursor in hemostasis	
Haptoglobin	100	Binding and transport of cell-free hemoglobin	Hemoglobin
Hemopexin	57	Binds to porphyrins, particularly heme for recycling	Porphyrins
Transferrin	80	Iron transport	2 atoms iron/mol

Adapted from Zakim [16, p. 125]

55–60 % of the total plasma protein pool. Albumin serves as a ***binding and carrier protein*** for hormones, amino acids, steroids, vitamins, and fatty acids. It is also an important ***osmotic regulator*** in the maintenance of normal plasma oncotic pressure. Albumin synthesis is exquisitely sensitive to nutritional status and availability of amino acids, particularly *tryptophan*. Also, it has been demonstrated that albumin gene transcription is regulated by changes in **serum colloid osmotic pressure** to facilitate maintenance of osmotic homeostasis. This regulation appears to be mediated by the interaction of transcription factor HNF-1α with a specific site in the albumin gene promoter region. Furthermore, it is interesting that expression of other genes with promoters containing the HNF-1α recognition site, such as α-1-antitrypsin and certain apolipoproteins, can be regulated by the serum colloid osmotic pressure. This is but one example of the complex regulatory mechanisms involved in hepatocyte gene expression. The other proteins synthesized and secreted by the liver are for the most part **glycosylated proteins (glycoproteins)**, which function in hemostasis, protease inhibition, transport, and ligand binding.

One especially interesting glycoprotein produced by the liver is important in both health and disease. **Alpha-1-antitrypsin (A-1-AT)** is a neutrophil protease inhibitor that is secreted and protects tissues, particularly *lung*, from damage by endogenous proteases. In **A-1-AT deficiency**, a point mutation produces an abnormal protein that accumulates in the ER and is susceptible to misfolding, polymerization and aggregation, resulting in toxicity to the hepatocyte. This disorder is a common cause of liver disease in infants and adults, as well as chronic lung disease in adults. Interestingly, only about 10 % of homozygotes develop clinically significant liver disease that may require liver transplantation. Genetic and environmental modifiers that impact on the adaptation to or degradation of the accumulated abnormal protein in the hepatocyte ER, through proteosomal and autophagic pathways, appear to play crucial roles in determining the severity of the liver disease. Development of new therapies has focused on targeting and enhancing these pathways.

3 Role of the Liver in Carbohydrate Metabolism

The liver plays a key role in the utilization of the **major monosaccharides**, *glucose*, *fructose*, and *galactose* (Fig. 11.1). The initial step in hepatic glucose metabolism is phosphorylation by **glucokinase**. Glucokinase synthesis is increased by high glucose diets and high insulin concentrations. Depending upon the metabolic need, the glucose may be utilized for energy production, synthesis of other substrates (i.e. amino acids, fatty acids), or stored as glycogen. **Glycogen** is a polymer of units of glucose (Fig. 11.2) with *linear 1-4 linkages* and *branch points with 1-6 linkages*. Glycogen serves as a storage depot for glucose in the liver, which can be readily mobilized when glucose is in immediate demand. Because of its polymeric structure, glycogen has a low osmolality and is more easily stored in hepatocytes than the monomeric glucose. The liver can store up to 65 g of glycogen per kilogram of liver tissue.

Fructose is taken up by the hepatocyte and phosphorylated by **fuctokinase** (Fig. 11.1). It may then be stored as glycogen by conversion to **glucose-6-phosphate**. When used for energy production, phosphorylated fructose may actually traverse the glycolytic pathway more readily than glucose. Also, fructose is a better substrate for lipogenesis than glucose in the liver. Hepatic metabolism of fructose follows the **Leloir pathway** to form either *glucose-6-phosphate* to enter the glycolytic pathway or *UDP-glucose* to enter the glycogenesis pathway. **Galactose** can also be taken up and used by the liver after phosphorylation by **galactokinase** (Fig. 11.1).

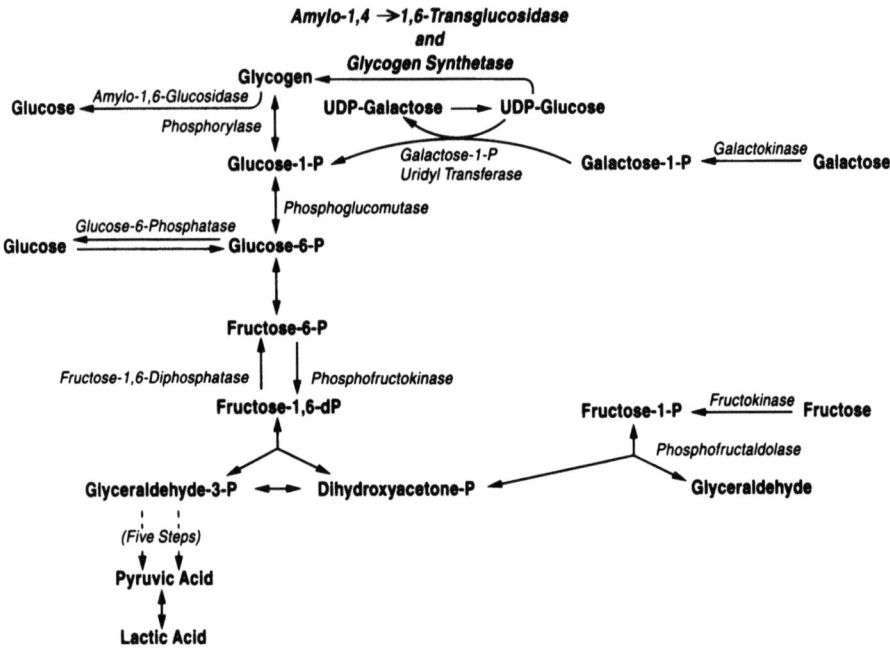

Fig. 11.1 Metabolic pathways for hepatic utilization of glucose, fructose, and galactose

Glycolysis is the only pathway by which glucose can be oxidized anaerobically with production of ATP. Under aerobic conditions the liver uses mainly fatty acids as substrates for oxidation, and the glycolytic rate is low. Higher rates of glycolysis result in lipogenesis from carbohydrate. Regulation of glycolysis in the liver is highly integrated with that of gluconeogenesis, lipogenesis, glycogen synthesis, and glycogenolysis.

Gluconeogenesis is the production of glucose from amino acids and lactate and is carried out solely in liver and renal cortex. **Two cycles** exist for hepatic gluconeogenesis from non-hepatic substrates (Fig. 11.3). These are the *lactic acid (Cori) cycle* and the *glucose-alanine cycle*. In the Cori cycle lactic acid produced by working muscle is taken up by the liver and provides a substrate for gluconeogenesis to produce glucose. Alanine is also released from muscle and serves as a substrate for hepatic glucose synthesis. Regulation of gluconeogenesis is dependent on substrate availability and hormonal factors, particularly insulin and glucagon levels. Overall, **glucagon** and **epinephrine** stimulate and insulin inhibits gluconeogenesis.

As mentioned previously, glycogen is a storage form of glucose consisting of a polymeric form of glucose. The pathway for glycogen metabolism is shown in Fig. 11.4. Most substrates enter the glycogen synthetic pathway by conversion to **UDP-glucose**. The glucosyl units are linked linearly by 1-4 linkages by the enzyme **UDPG-glycogen synthetase**. Branch points with 1-6 linkages are formed by the branching enzyme **amylo-1-4, 1-6-transglucosidase**. Glycogen synthesis is

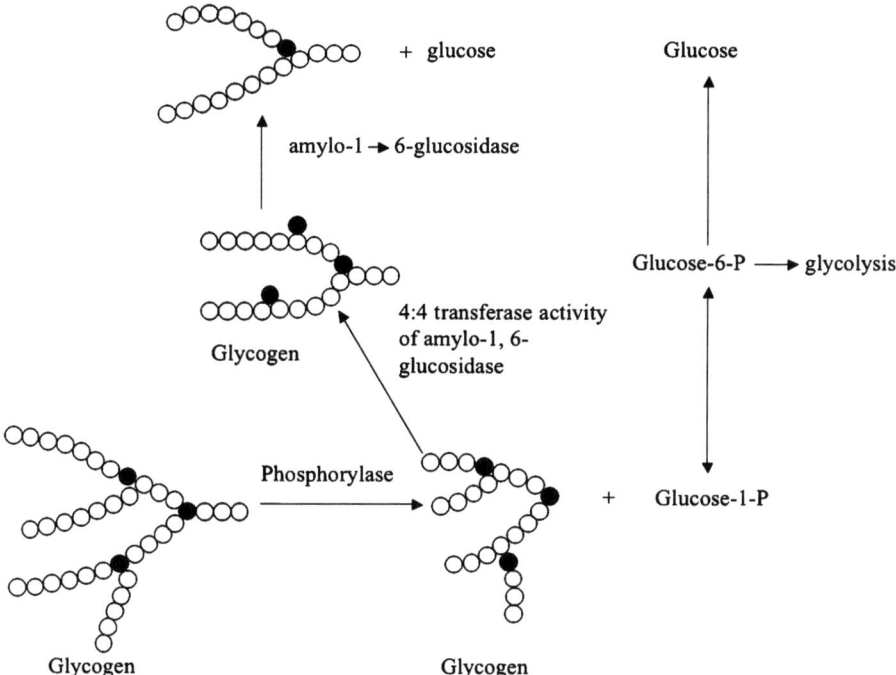

Fig. 11.2 Pathway for production of glucose from glycogen (glycogenolysis). Notice that the product of the phosphorylase-catalyzed hydrolysis of glycogen is glucose-1-phosphate. A small amount of glucose (about 6 % of the total produced) is released via action of the enzyme amylo-1,6-glucosidase. This last reaction removes the glucose residues that form branch points (*closed circles*) and occurs after the shortened oligosaccharide chain attached to a branch point (three glucose residues) is shifted to create a longer oligosaccharide. These reactions are required because phosphorylase will not catalyze hydrolysis of short oligosaccharide chains (Reproduced with permission from Zakim [16, p. 69])

promoted by insulin and glucocorticoids and by increased glucose concentrations. **Glycogen synthetase** is the *rate-limiting step* in glycogen synthesis and is converted from active to inactive form by phosphorylation by a **cAMP-dependent protein kinase**. **Glycogenolysis**, the breakdown of glycogen to release glucose units, is promoted by glucagon, epinephrine, vasopressin, angiotensin II, and oxytocin. Under conditions promoting glycogenolysis, **phosphorylase** catalyzes glycogenolysis by breakage of the linear 1-4 linkages. Phosphorylase is converted from the inactive to active form by phosphorylation by a cAMP-dependent protein kinase. Glucose molecules are removed from the 1-6 linkage branch points by the debrancher enzyme **amylo-1-6-glucosidase** (Fig. 11.2). Several inherited defects in various steps of glycogen metabolism have been identified, as shown in Fig. 11.4, and result in a clinical spectrum of glycogen storage diseases with the abnormal accumulation of glycogen in the liver.

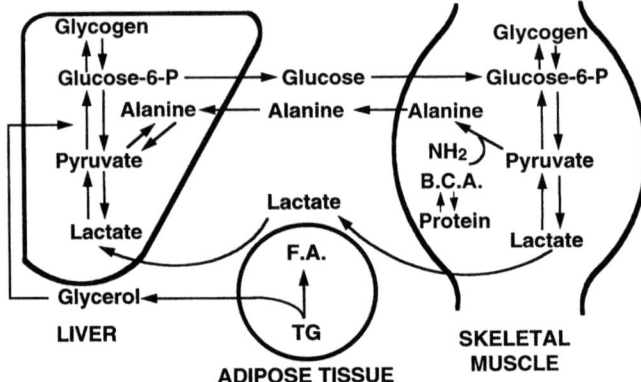

Fig. 11.3 The lactic acid (Cori) and glucose-alanine cycles. Lactic acid produced by working muscle is taken up by the liver and provides a substrate for gluconeogenesis to produce glucose. Alanine produced in muscle as a result of protein breakdown, is released and deaminated in the liver to form pyruvate. Glucose is formed from pyruvate and is released to be metabolized by muscle, where it is reconverted to pyruvate and alanine, respectively, completing the cycle. *BCA* branched-chain amino acids, *FA* fatty acid, *TG* triglyceride (Reproduced with permission from Van Thiel [13])

4 Role of the Liver in Lipid Metabolism

Fatty acids are synthesized in the liver from carbohydrate precursors by conversion of these precursors to **acetyl-CoA** by the **cytosolic fatty acid synthase complex**. Hepatic fatty acid synthesis is stimulated by carbohydrate feeding and insulin. Fatty acids are generally stored in the liver as **triglycerides**, consisting of three fatty acids esterified to a glycerol backbone. Fasting, starvation, and diabetes mellitus with insulin deficiency cause increased hepatic fatty acid oxidation and production of **acetoacetate** and **D-3-hydroxybutyrate** (also known as ketones), which can be used as an energy source by muscle and brain. This process is known as **ketogenesis.** Lipolysis of adipose tissue triglycerides provides a major source of substrate fatty acids to the liver for production of ketones. The **plasma glucagon to insulin ratio** is probably the main regulator of ketogenesis.

The liver is a major site of **fatty acid β-oxidation**, which results in the production of energy using fatty acids as substrates. There are **two postulated regulatory mechanisms for β-oxidation** when carbohydrate is in short supply: regulation at the level of *pyruvate formation* and regulation at the level of *fatty acid entry into the mitochondria*, the site of most fatty acid oxidation, by way of **carnitine acyltransferase I**. Another site of fatty acid oxidation in the hepatocyte, particularly for saturated long chain fatty acids, is the **peroxisome**. Peroxisomal fatty acid oxidation is regulated only by fatty acid substrate concentration. This pathway may provide a mechanism for the production of acetyl CoA outside the mitochondria without the participation of citrate formed in the mitochondria.

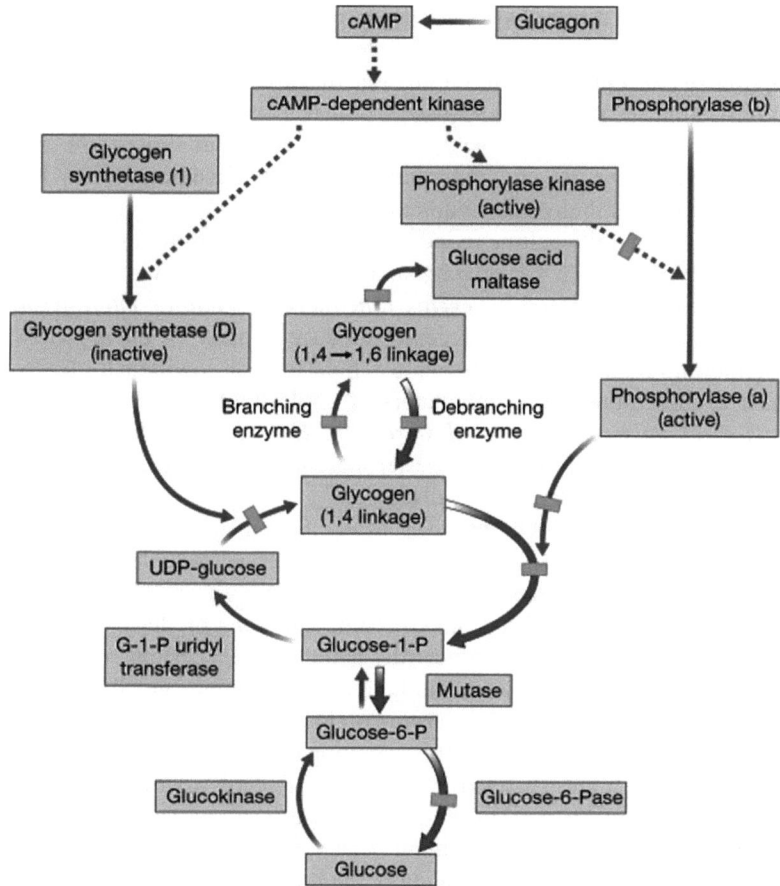

Fig. 11.4 Pathways for glycogen formation and glycogenolysis. *Broken lines* indicate enzymatic activation after glucagon stimulation; *thick arrows* indicate glycogen degradation from glucagon infusion, and *hatched boxes* indicate points in the metabolic sequence where enzymatic defects have been identified (Reproduced with permission from Elsevier [6])

Peroxisomal β-oxidation is inducible in rats by feeding a high-fat diet and treatment with certain hypolipidemic drugs that cause peroxisome proliferation.

The liver is a major site of **cholesterol synthesis** from **acetyl CoA**. In fact, all of the carbon atoms in cholesterol are derived from acetate. Cholesterol is ubiquitous throughout the body as a structural component of cell membranes and is a substrate for synthesis of several steroid hormones. Cholesterol is important in the liver as the precursor for bile acid synthesis. The **rate-limiting step** in cholesterol synthesis is the *cleavage of CoA from the synthetic intermediate β-hydroxy-β-methylglutaryl (HMG) CoA* with the *simultaneous reduction to mevalonate*. The enzyme which catalyzes this reaction is **HMG-CoA reductase**. Although the regulation of hepatic cholesterol synthesis is complex, several regulatory factors have been defined. In general, synthesis is **up-regulated** by *any event that depletes the hepatocyte of cholesterol*. For example, depletion of the bile acid pool by interruption of the enterohepatic circulation by biliary diversion, ileal resection or administration of a bile acid sequestrant such as *cholestyramine* will increase cholesterol synthesis, since bile acids are synthesized from cholesterol. Increased secretion of very low density lipoproteins (VLDL) stimulated by an influx of free fatty acids or carbohydrate requires cholesterol for packaging and depletes intracellular cholesterol resulting in increased synthesis. *Hormones* such as *thyroid hormone, corticosteroids,* and *glucagon* also up-regulate synthesis. Cholesterol synthesis is **down-regulated** by *events that increase cellular cholesterol*, such as the influx of cholesterol from receptor-mediated uptake of dietary cholesterol in chylomicron remnants and endogenous cholesterol in low density lipoproteins (LDL). The most potent down-regulators of cholesterol synthesis are *oxidation products of cholesterol, 7-ketocholesterol* and *25-hydroxycholesterol*. Bile acid synthesis and biliary secretion of cholesterol are the major secretory pathways for cholesterol and will be subsequently discussed.

Triglycerides synthesized by the liver are secreted as particles called **very low-density lipoproteins (VLDL)**. Lipoproteins are spherical particles, which serve as thermodynamically stable circulating packages for the transport of lipids through the aqueous environment of the bloodstream. The *core* of the particle contains hydrophobic, non-polar lipids, triglycerides and cholesteryl esters, and the *surface* coating consists of hydrophilic, polar lipids, phospholipid and free cholesterol, and lipid-binding peptides called **apolipoproteins**. Apolipoproteins generally serve important functions in the assembly, secretion, and peripheral metabolism of lipoprotein particles. The secreted hepatic VLDL serves to transport lipid from the liver to peripheral tissues. In humans, a large portion of the VLDL secreted is converted to **low density lipoproteins (LDL)**, which are the major transporters of cholesterol to various tissues throughout the body. This pathway is summarized in Fig. 11.5.

Carbohydrate feeding stimulates hepatic fatty acid production; thereby driving increased triglyceride and VLDL production and secretion. Secreted VLDL triglyceride is hydrolyzed by the enzyme **lipoprotein lipase** to liberate fatty acids for storage by adipose tissue and energy production by muscle. Lipoprotein lipase is bound to the *endothelium of capillary beds* in primarily adipose tissue and muscle and requires the apolipoprotein **apo C-II**, produced by the liver, as a cofactor. A related lipase produced by the liver, **hepatic lipase**, resides on *sinusoidal endothelial cells*. Hepatic lipase is important in HDL metabolism and in the hydrolysis of lipid in chylomicrons, VLDL and intermediate density lipoproteins (IDL).

The liver not only functions in the secretion of lipoprotein particles, but also is an essential organ for the **uptake and metabolism of lipoproteins. Chylomicrons** of intestinal origin, which carry mainly exogenous fatty acids and cholesterol, are metabolized by **lipoprotein lipase** to produce triglyceride-depleted, relatively cholesteryl ester-enriched remnants which are cleared by a receptor-mediated mechanism in the liver. This receptor-mediated clearance involves the **LDL receptor**

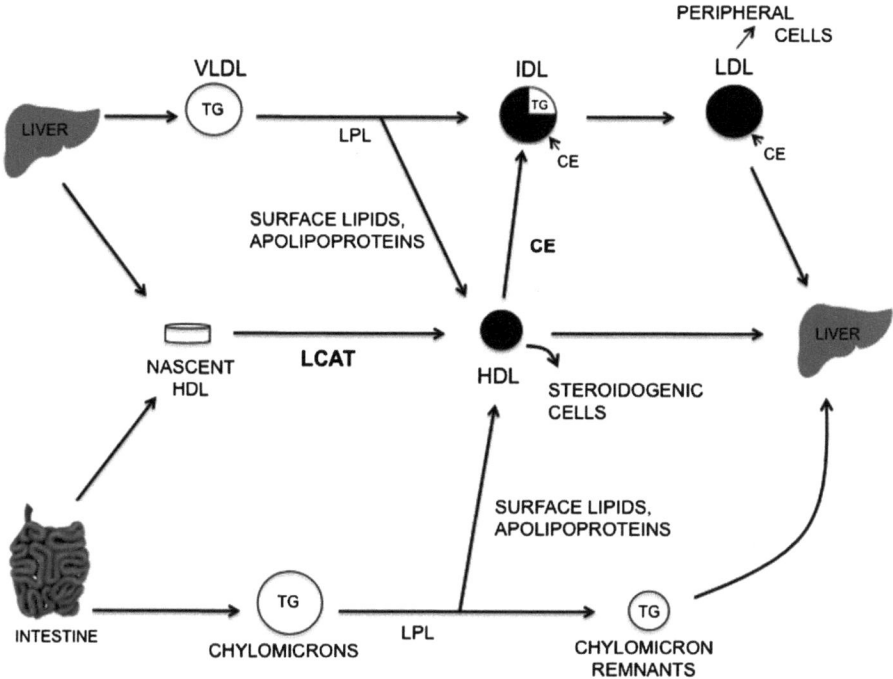

Fig. 11.5 Simplified scheme of plasma lipoprotein metabolism. Key events are the hydrolysis of triglyceride-rich lipoproteins by lipoprotein lipase (*LPL*), the conversion of VLDL to IDL, the movement of surface lipids and apolipoproteins released during lipolysis by LPL to HDL, the progressive enrichment of IDL and LDL with cholesteryl esters (*CE*) transferred from HDL as a result of the LCAT reaction, and the removal of LDL and chylomicron remnants by peripheral cells and the liver (Reproduced with permission from Glickman and Sabesin [7])

and the **LDL receptor-related protein**, a membrane receptor, which binds and internalizes a wide variety of ligands. A significant proportion of plasma LDL particles are taken up by the liver via the LDL receptor, as the liver contains over half of the body's total LDL receptors. Receptor-mediated LDL uptake results in an increase in esterification of the incoming cholesterol for storage, down-regulation of the number of LDL receptors, and suppression of cholesterol synthesis by the cell (Fig. 11.6). Finally, it has been demonstrated that the liver takes up a portion of VLDL particles shortly after they are secreted.

Recently, another key protein that regulates cellular cholesterol homeostasis in the liver through interaction with the LDL receptor was identified, **proprotein convertase subtilisin kexin type 9 (PCSK9)**. Originally, PCSK9 was identified as a protein up-regulated during apoptosis of nerve cells. Subsequently, **mutations of the PCSK9 gene** associated with hypercholesterolemia were noted leading to identification of its role in cholesterol metabolism. **Gain-of-function mutations** result in *increased plasma LDL cholesterol levels*, and **loss-of-function mutations** cause the *opposite*. The normal function of PCSK9 appears to be interaction with

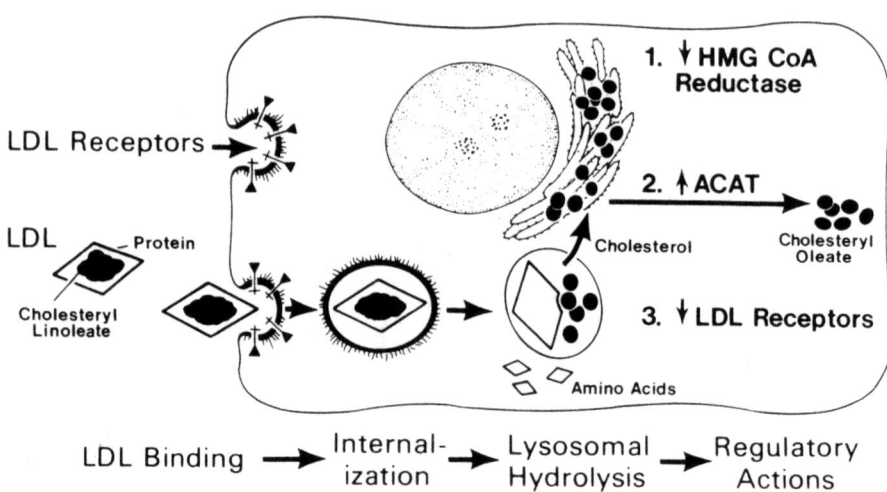

Fig. 11.6 Sequential steps in the LDL pathway. The numbers indicate the regulatory actions of LDL: *1* decrease in 3-hydroxy-3-methylbutyryl coenzyme A reductase (HMG CoA reductase) and cholesterol synthesis, *2* increase in acyl coenzyme A:cholesterol acyltransferase (ACAT) and cholesterol esterification, *3* decrease in the number of cell surface receptors for LDL (Reproduced with permission from Brown and Goldstein [4])

the LDL receptor protein on the cell surface to interfere with normal recycling of the protein and direct it to a degradative pathway. This role of PCSK9 in cholesterol metabolism makes it an attractive **therapeutic target for the treatment of hypercholesterolemia**.

Another lipoprotein produced by the liver is **high-density lipoprotein (HDL)**. This particle is secreted as a discoidal bilayer consisting primarily of phospholipid and free cholesterol. The major apolipoprotein is **apo A-I**, a cofactor for **lecithin:cholesterol acyltransferase (LCAT)**, which is produced by the liver and catalyzes the transfer of a fatty acid from lecithin to free cholesterol to produce cholesteryl ester. These HDL particles acquire free cholesterol from peripheral tissues, which is then esterified and thereby moves into the core of the particle to change the shape from discoidal to spherical. This esterification is a driving force to effect net movement of cholesterol from tissue membranes to HDL. A portion of the cholesteryl ester in the HDL core may be transferred to a chylomicron, VLDL, or LDL particle through the action of a plasma **cholesteryl ester transfer protein (CETP)** to allow transport of the cholesterol to a variety of tissues throughout the body. The cholesterol from mature HDL particles is ultimately taken up by the liver via **scavenger receptor-BI (SR-BI)**. Thus, HDL function in "**reverse cholesterol transport**" to effect cholesterol movement from peripheral tissues to the liver. Serum levels of HDL cholesterol correlate *inversely* with risk for atherosclerotic coronary artery disease.

4.1 Regulation of Hepatic Lipid Metabolism

Many of the regulatory processes described above for hepatic lipid metabolism are mediated by transcription factors called **nuclear receptors** that regulate the transcription of key proteins and enzymes in response to physiological cues, such as cellular lipid and substrate levels. One family of such receptors is that of the **sterol regulatory element binding proteins (SREBPs)**. Key enzymes involved in *de novo* **lipogenesis**, such as *acetyl-CoA carboxylase, fatty acid synthase,* and *stearoyl-CoA desaturase,* are regulated by **SREBP-1c**, and genes involved in cholesterol metabolism, including those of *HMG-CoA synthase* and *reductase,* the *LDL receptor* and *PCSK9,* are regulated by **SREBP-2**. Another regulatory transcription factor, **carbohydrate responsive element binding protein (ChREBP)**, mediates activation of several glycolytic and lipogenic regulatory enzymes, including *fatty acid synthase, acetyl CoA carboxylase* and *pyruvate kinase*. It is translocated to the nucleus and binds to the carbohydrate response element in the promoter of its target genes. Activation of ChREBP occurs by dephosphorylation in the presence of high levels of glucose-6-phosphate (glucose-responsive), leading to nuclear translocation.

The **lipid-sensing property of SREBP-2** occurs via a unique cellular mechanism called **regulated intramembrane proteolysis.** For example, in the case of cholesterol, SREBP-2 precursor protein is anchored to the membrane of the ER and complexed to the cholesterol sensing **SREBP cleavage activating protein (Scap)**, which in turn is bound to ER proteins, Insig-1 and -2. When cellular cholesterol falls below a certain level, Scap and the Insigs no longer bind to each other. The **Scap/SREBP precursor protein complex** then moves to the Golgi, where a **site-1 protease (S1P),** followed by a **site-2 protease (S2P),** cleave the SREBP precursor protein to produce two N-terminal fragments that dimerize to form the active transcription factor. The active SREBP then translocates to the nucleus and binds to the promoter of a target gene, such as the LDL receptor, and activates transcription. Subsequently, the increased number of LDL receptors takes up circulating LDL and raise intracellular cholesterol levels, while lowering plasma cholesterol levels.

Another class of nuclear receptor transcription factors, the **liver X receptors (LXRs)**, also play a major role in hepatic lipid metabolism, and exist in two forms, **LXR-α**, abundant in liver and other tissues, including intestine, and **LXR-β**, ubiquitous in many tissues. When activated by binding of a ligand, LXRs heterodimerize with the **retinoid X receptor (RXR)** and bind to specific response elements, as well as coactivator and corepressor proteins, in the promoters of their target genes to modulate cholesterol metabolism at several sites to avoid cholesterol overload. Interestingly, LXR-α has an LXR response element in its gene promoter and can autoregulate its own transcription. The LXRs promote transcription of genes that have wide-ranging effects on cholesterol homeostasis, including efflux of cholesterol from peripheral cells, such as macrophages involved

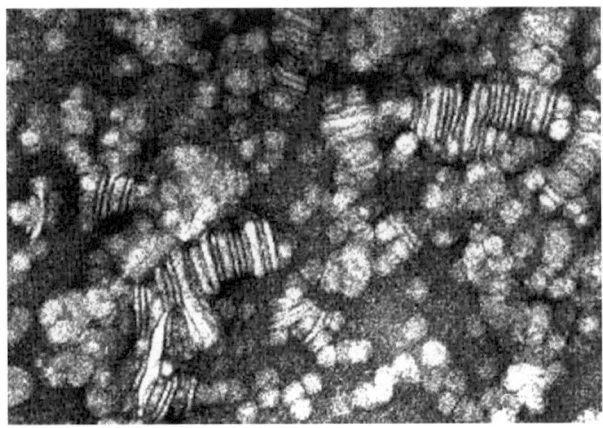

Fig. 11.7 Electron photomicrograph of negatively stained lipoprotein-X from an infant with biliary obstruction. Original magnification 100,000× (Reproduced from Williams et al. [15])

in atherosclerosis via **ATP binding cassette (ABC) proteins ABCA1** and **ABCG1**. They also regulate hepatic excretion of cholesterol into bile and intestinal absorption of cholesterol by inducing **ABCG5** and **ABCG8**. The endogenous ligands for LXRs are **oxysterols**, formed by the oxidation of cholesterol and present in proportion to the amount of intracellular cholesterol, thus enabling these receptors to function as cholesterol sensors. The gene for the rate-limiting enzyme in bile acid synthesis from cholesterol, **Cyp7α1**, is also regulated by LXRs. Recently, LXRs have been implicated in the regulation of target genes involved in glucose metabolism and inflammation. Adverse effects of LXR activation, including hepatic triglyceride overproduction via up-regulation of fatty acid synthesis, have hampered the development of small molecule LXR agonists to prevent atherosclerosis and treat other diseases. However, recent evidence suggests that exclusive targeting of extrahepatic LXRs may bypass this problem.

4.2 Hepatic Lipid and Lipoprotein Metabolism in Liver Disease

In **cholestatic liver disease**, particularly *biliary obstruction*, hyperlipidemia occurs with often striking elevation of plasma cholesterol and phospholipid levels. Furthermore, the majority of this cholesterol is unesterified, in contrast to the normal situation in which most plasma cholesterol is esterified. Much of this excess cholesterol and phospholipid is contained in an abnormal plasma lipoprotein, **lipoprotein-X**, which accumulates in individuals with cholestasis (Fig. 11.7). Lipoprotein-X is isolated in the low density lipoprotein (LDL) density range in the ultracentrifuge and consists mainly of free cholesterol and phospholipid. The major protein components of lipoprotein-X are **albumin** and the **apo C and E peptides**. Although the presence of lipoprotein-X in plasma is a sensitive marker for cholestasis, it cannot reliably discriminate intrahepatic from extrahepatic cholestasis. Interestingly, although certain cholestatic liver diseases may result in

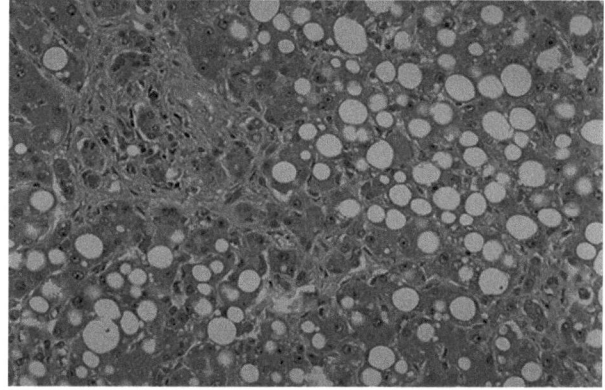

Fig. 11.8 Photomicrograph of a liver biopsy from a patient with NAFLD. Note the hepatocytes distended with large lipid droplets. Hematoxylin and eosin stain, original magnification 100×

markedly elevated lipoprotein-X levels with total serum cholesterol levels above 1,000 mg/dL, unlike LDL, this particle does not appear to be atherogenic and does not increase the risk for cardiovascular disease.

Mild to moderate **parenchymal liver disease**, such as *typical acute viral hepatitis*, usually results in an increase in plasma triglycerides, a less striking increase in cholesterol (mostly free cholesterol), and a decrease in HDL. These changes are thought to result from decreases in hepatic lipase, LCAT, and possibly peripheral lipoprotein lipase. With **severe parenchymal liver injury**, such as *fulminant hepatitis* disappearing HDL fraction portends an especially poor prognosis.

The most common plasma lipid abnormality with acute and chronic ethanol ingestion is hypertriglyceridemia contained in the triglyceride-rich very low density lipoprotein fraction (VLDL) and results from increased hepatic synthesis and secretion of VLDL. Chronic mild ethanol ingestion may result in only an elevation of plasma HDL levels. In the presence of **alcoholic hepatitis**, in addition to the abnormalities described for parenchymal liver disease, there is accumulation of an abnormal discoidal HDL particle **deficient** in *apolipoprotein A-I* and **rich** in *apolipoprotein E*. This may represent a nascent HDL particle persisting in the circulation due to **LCAT deficiency** caused by the ethanol-induced liver injury.

The current obesity epidemic in adults and children has resulted in a dramatic increase in the accumulation of lipid in the liver, termed **non-alcoholic fatty liver disease (NAFLD)** (Fig. 11.8). In many cases, the steatosis of liver cells is also associated with *inflammation*, which is then termed **non-alcoholic steatohepatitis (NASH),** and the development of cirrhosis. This disorder often accompanies the **metabolic syndrome**, a constellation of abnormalities that includes *insulin resistance, hypertension, abdominal obesity, hypertriglyceridemia, low HDL cholesterol levels*, and the presence of *abnormal, highly atherogenic small, dense LDL particles*. In the presence of insulin resistance, free fatty acids are mobilized from adipose tissue and are taken up by the liver, where they are re-esterified to triglyceride and packaged into VLDL particles. These particles are secreted and result in hypertriglyceridemia. However, overproduction often overwhelms the secretory pathway, and triglyceride accumulates in the hepatocytes. Local events, such as lipid peroxidation coupled with a systemic inflammatory state, may result in hepatic inflammation (the so-called "**second hit**") to produce **NASH** and progressive liver disease with **cirrhosis** and, in some cases, the development of **end-stage liver disease**, requiring transplantation, or **hepatocellular carcinoma**.

Recently, the contribution of the **intestinal microbiome** to progression of NAFLD and other liver diseases has been more clearly defined. The liver's location at the interface between portal blood from the gut, which carries bacteria and bacterial products, such as lipopolysaccharide (LPS) and metabolites, that have breached the intestinal barrier, and the rest of the body, places it in an ideal position to react to perturbations in the composition of the microbiome. The liver contains **immune cells**, such as the Kupffer cell, and **innate immune response receptors** that function in the initiation and modulation of inflammation, which plays a role in diverse pathologic processes, including the metabolic syndrome, progression of NAFLD to NASH, autoimmune liver diseases and others associated with immune dysregulation. **Future therapies targeting the gut microbiome** may result in prevention or amelioration of these diseases.

5 Role of the Liver in Substrate/Energy Processing and Distribution During the Fed, Postabsorptive, and Fasting States

The liver plays a pivotal role as a metabolic distribution center for the body in the face of very different metabolic conditions characterizing the **fed, postabsorptive,** and **fasting states**. In the **fed state**, nutrients absorbed from the small intestine are abundant. *Glucose* is readily available as an energy substrate for brain and red blood cells, which have no capacity for storage of glucose as glycogen. In **muscle**, the glucose is taken up for *storage as glycogen* when needed for energy production. However, unlike liver glycogen, muscle glycogen does not contribute to maintenance of blood glucose levels. In the **liver**, the incoming glucose is stored as glycogen. When glycogen stores are saturated, glucose is used for *fatty acid synthesis*. As previously discussed, these fatty acids are esterified into triglycerides and incorporated into VLDL, which are secreted and supply fatty acids to peripheral tissues. Also as previously discussed, dietary lipid is distributed to peripheral tissues and liver by intestinal chylomicron metabolism. These events are summarized in Fig. 11.9.

In the **postabsorptive fasting state**, the influx of dietary glucose is interrupted, and the liver releases glucose from glycogen to maintain blood glucose levels between meals for use by the brain. *Amino acids* released by muscle are substrates for *hepatic gluconeogenesis*. Also, adipose tissue releases fatty acids into the circulation, rather than storing them as triglycerides. Oxidation of glucose, as

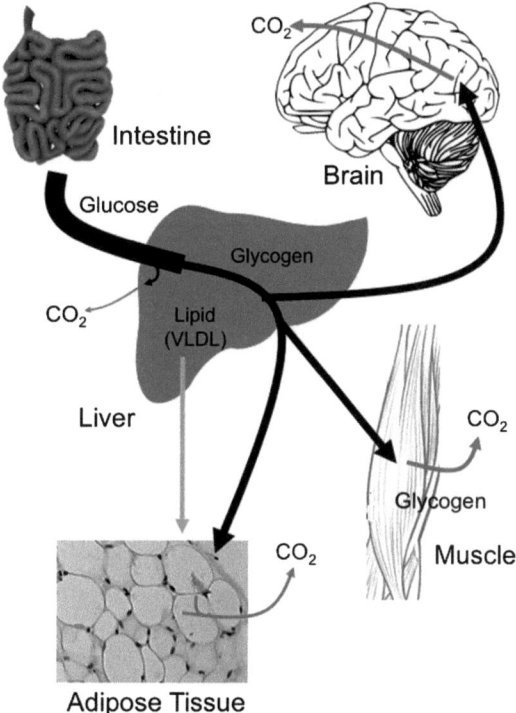

Fig. 11.9 Glucose metabolism in the fed state. Glucose is absorbed from the gut. A small fraction is used within the liver for immediate hepatic energy needs and for lipid and glycogen storage. The rest is utilized by muscle, brain, and other tissues (Adapted with permission from Van Thiel [13])

an energy source begins to decline and fatty acid oxidation by liver and muscle increases. This scheme is summarized in Fig. 11.10.

With **prolonged fasting**, hepatic glycogen stores are depleted within 48 h in adults and more rapidly in children and infants. **Gluconeogenesis** in the liver becomes more important as an energy source for brain and red blood cells. *Amino acids from muscle* are the predominant substrates for this gluconeogenesis. As fasting continues, fatty acids released from adipose tissue are oxidized to *ketones* by the liver to supply alternative energy to the central nervous system. The factors regulating these metabolic adaptations include *liver sinusoidal glucose concentrations* and *hormones* (insulin, glucagon, and catecholamines).

Glucose is readily taken up by the hepatocyte predominately by the glucose transporter, GLUT2, that is regulated by glucose levels via SREBP-1c, and the intracellular glucose concentration is generally the same as the sinusoidal concentration. The velocity of the conversion of glucose to glucose-6-phosphate by glucokinase is dependent therefore on the **sinusoidal glucose concentration**. The amount of the enzyme glucokinase is transcriptionally regulated by insulin concentrations. Therefore, during and immediately after a meal, insulin and sinusoidal glucose levels are high, and glucose is rapidly taken up, phosphorylated, and incorporated into glycogen for storage and used as a substrate for fatty acid and amino acid synthesis when glycogen stores are fully repleted. **Increasing sinusoidal concentrations of**

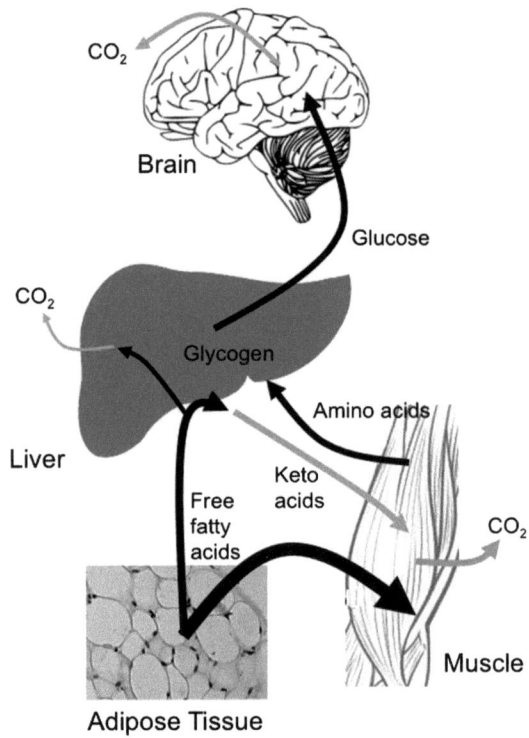

Fig. 11.10 Energy metabolism in the fasting state. Glucose is formed from amino acids released from muscle and from hepatic glycogen. The bulk of the energy requirement is satisfied as a result of lipolysis and fatty acid oxidation (Adapted with permission from Van Thiel [13])

glucose result in activation of glycogen synthetase; this incorporates glucose into glycogen, and decreased activity of the phosphorylase that breaks down glycogen to release glucose. This process is reversed when sinusoidal glucose concentrations are low. **Glycogen synthetase and phosphorylase** are activated by enzyme protein dephosphorylation and phosphorylation, respectively. A **cAMP-responsive protein kinase** is responsible for phosphorylating and thereby inactivating glycogen synthetase directly. This same protein kinase activates a phosphorylase kinase which in turn phosphorylates and activates phosphorylase. Regulatory factors involved in this complex scheme involve *glucagon* and *epinephrine*, as well as hormones and neural stimuli that affect *intracellular calcium concentrations*.

6 Role of the Liver in Vitamin Metabolism

6.1 Vitamin A

Vitamin A is essential for normal **vision**, normal **reproductive** function, and **maintenance of epithelia** throughout the body, including a role in processing cell surface glycoprotein receptors. Dietary vitamin A from *animal sources* is present in the intestine primarily as **retinyl esters**, which undergo cleavage prior to absorption of the free retinol. **Carotenoids** from *plant sources* are converted to **retinal**, then to **retinol**, prior to intestinal absorption. **Retinol** is a fat-soluble vitamin, requiring bile salt solubilization for absorption. Although a portion of carotenoids is absorbed intact, this fraction does not represent functional vitamin A. The absorbed retinol is esterified with a fatty acid in the enterocyte, and the resultant retinyl esters are incorporated into intestinal chylomicrons. Ultimately, the retinyl esters enter the hepatocyte with chylomicron remnant uptake. After uptake, hydrolysis, and reesterification by the hepatocyte, retinyl palmitate (over 90 % of total body reserve) is stored mainly in hepatic non-parenchymal fat-storing (stellate) cells. Retinol is mobilized as the **free alcohol (retinol)** bound to **retinol binding protein (RBP)**, which is synthesized by the hepatocyte. Nutritional retinol status regulates hepatic RBP secretion. Also, an **intracellular retinol binding protein (CRBP)** is present in hepatocytes and fat-storing cells. Retinol binding protein circulates complexed with prealbumin **(transthyretin)**, which also transports thyroid hormones and functions to deliver retinol to peripheral tissues. **Excess vitamin A intake** results in stellate cell overload with activation of these cells and production of excess collagen resulting in *liver fibrosis*.

6.2 Vitamin D

Vitamin D is essential for normal **calcium metabolism** and maintenance of proper **bone mineralization**. Vitamin D_3 is produced in the *skin* and absorbed from *dietary sources*, requiring bile salts for absorption. **Vitamin D3 (cholecalciferol)** is produced in the skin from the precursor 7-dehydrocholesterol in a reaction requiring ultraviolet light. **Vitamin D_2 (ergocalciferol)** is found to a limited extent in *plant products* and is the major form of *dietary vitamin D supplementation*. Vitamin D_3 is abundant in *fish oils* and *eggs*. Although vitamin D is incorporated into chylomicrons, which are secreted into intestinal lymphatics, a significant proportion is absorbed directly into the portal circulation. In the blood, vitamin D binds to **vitamin D binding protein (DBP)** for transport to liver for 25-hydroxylation. After hepatic 25-hydroxylation, 25-OH vitamin D is transported bound to DBP to kidney for final hydroxylation to **1,25-$(OH)_2$ vitamin D**. The liver does not significantly store vitamin D. Reduced levels of 25-OH vitamin D may be seen in liver disease because of reduced absorption of vitamin D with cholestasis and/or reduced hepatic synthesis of DBP. Liver disease generally does not result in significantly reduced vitamin D hydroxylation. Regulation of vitamin D metabolism occurs primarily at the level of the *kidney*. Production of 1,25-$(OH)_2$ vitamin D is enhanced by parathyroid hormone and inhibited by elevated blood levels of calcium and phosphorus. The **metabolically active 1,25-$(OH)_2$ vitamin D** regulates intestinal absorption of dietary calcium and recruits stem cells to mature into osteoclasts, which mobilize calcium from bone.

6.3 Vitamin K

Vitamin K is produced by *intestinal bacteria* and requires bile acids and pancreatic secretions for the most efficient absorption. Other significant dietary sources of vitamin K include *liver, eggs, butter, cheese,* and *green, leafy vegetables*. Vitamin K is incorporated into chylomicrons in the enterocyte and is ultimately delivered to the liver in chylomicron remnants. In the liver, vitamin K is incorporated into VLDL, which are secreted from the liver with a significant proportion of the VLDL being converted to LDL, allowing distribution of vitamin K throughout the body. The liver synthesizes the **vitamin K-dependent clotting factors,** *prothrombin* and *factors VII, IX, and X,* as well as *protein C and S*. Vitamin K is required for the **posttranslational formation of γ-carboxyglutamic acid** from specific glutamic acid residues on these proteins during hepatic synthesis. Without this carboxylation, the secreted factors are not functional in the clotting cascade. Whenever one observes a coagulopathy in an individual with liver dysfunction, particularly cholestatic liver disease, vitamin K should be administered parenterally. If the coagulopathy corrects, it was likely due to *vitamin K deficiency*. If not, the coagulopathy is probably *secondary to severe liver synthetic failure*.

Clinical Correlations

Case Study 1

A three and a half month-old male infant is brought to the emergency room having a generalized seizure. The infant was born to a 22-year-old G1P0Ab0 female after an uneventful term pregnancy. The vaginal vertex delivery went smoothly, and there were no perinatal problems. The birth weight was 8 lb and 12 oz. The infant was breast-fed from birth and has had no apparent problems until approximately 20 min before arrival at the hospital when he was noted to be poorly responsive and began having generalized jerking movements, prompting the parents to call an ambulance. In the emergency room a blood glucose is found to be 13 mg/dL. The infant is given an intravenous push of glucose, and the seizure activity stops within seconds. On examination the infant is postictal, but responsive. The baby is noted to have a "doll-like" appearance with chubby cheeks, a protuberant abdomen, and thin extremities. Abdominal examination reveals a massively enlarged liver extending down into the right pelvis and the left lobe extending across the midline to the left upper quadrant. Both kidneys are detectable and felt to be enlarged on deep palpation. There is no splenomegaly.

Laboratory studies: Hematocrit 36 %; hemoglobin 11.5 g/dL; white blood cell count 3,600 cells/mm^3 (5,000–190,500) with a neutropenia reflected in the differential; 180,000 platelets/mm^3 (150,000–350,000); prothrombin time 13.0 (control 12.4 s); glucose (before glucose infusion) 13 mg/dL (60–105 mg/dL); albumin 4.5 g/dL (3.7–5.6); total bilirubin 1.3 mg/dL; direct bilirubin 0.3 mg/dL; ALT 91 IU/L (10–54), AST 85 IU/L (25–75); GGTP 41 IU/L (5–65); alkaline phosphatase 296 IU/L (150–400); blood ammonia 59 μmol/L (29–70); sodium

135 mEq/L (135–145), potassium 5.5 mEq/L (3.5–5.0), chloride 101 mEq/L (94–106); carbon dioxide 11 mEq/L (22–29); blood urea nitrogen (BUN) 21 mg/dL (5–25); creatinine 0.3 mg/dL (0.2–0.4); lactic acid 79 mg/dL (5–20); uric acid 12 mg/dL (2.0–7.0) triglycerides 920 mg/dL (less that 99); and cholesterol 280 mg/dL (less than 203). The urine is negative for ketones.

Questions:

1. **What is the most likely diagnosis in this patient?**
   Answer: The constellation of *hypoglycemia, lactic acidosis, hyperuricemia, hyperlipidemia, nephromegaly,* and *massive hepatomegaly* in an infant with *"cherubic" facies* with *normal liver synthetic function* and *minimally elevated transaminases* suggests a diagnosis of **glycogen storage disease type I (von Gierke's disease, glucose-6-phosphatase deficiency)**, an *autosomal recessively inherited disorder.* The presence of *neutropenia* would suggest further classification as **type Ib**, which is caused by **mutations in the glucose-6-phosphate translocase gene**, resulting in defective microsomal membrane transport of glucose-6-phosphate. Inability to generate glucose from glucose-6-phosphate results in a block in the final steps of the glycogenolytic and gluconeogenic pathways, which causes the *fasting hypoglycemia* observed in these patients. There is generally no ketosis in the face of profound hypoglycemia. Accumulation of glucose-6-phosphate results in increased shunting of this substrate into the glycolytic pathway with generation of lactic acid. *Hyperuricemia* is caused by both decreased renal clearance and increased production. Lactate competes for uric acid excretion by the kidney, and hepatic accumulation of phosphate esters causes increased degradation of adenine nucleotides with resultant increased uric acid production. *Hyperlipidemia* is caused by increased hepatic production of triglycerides and VLDL due to overproduction of NADH, NADPH, acetyl-CoA and glycerol by the glycolytic pathway, which serve as the co-factors and substrates for lipogenesis.

   These patients are generally *hyperinsulinemic,* which results in decreased activity of peripheral lipoprotein lipase and decreased lipolysis of triglyceride-rich lipoproteins. Striking *hepatomegaly* is caused by accumulation of glycogen and triglyceride in the hepatocytes. *Renal enlargement* is found early in life, and renal disease is a late complication with *hypertension, proteinuria,* and *nephrolithiasis.* Most patients have growth retardation and delayed puberty. Diagnosis of this disorder requires **liver biopsy** with demonstration of **decreased glucose-6-phosphatase activity (type Ia)** or **deficiency of one of three microsomal translocase systems (type Ib)**. Genetic diagnosis is also available. There are other types of glycogen storage disease involving various steps in the synthesis and breakdown of hepatic and muscle glycogen. A photograph of an infant with glucose-6-phosphatase deficiency is shown in Fig. 11.11.

2. **How would you treat this patient?**
   The cornerstone of therapy is **maintenance of normoglycemia**. This may be accomplished by the *ingestion of raw cornstarch*, which acts as a timed-release form of glucose in the GI tract, or use of *continuous nasogastric drip*

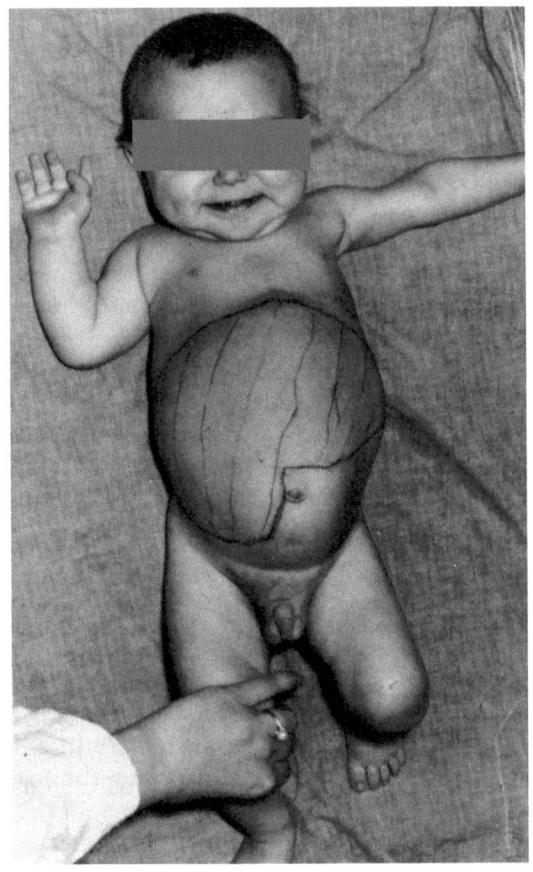

Fig. 11.11 Eight-month-old infant with glucose-6-phosphatase deficiency. Note the massive hepatomegaly and "cherubic" face (Reproduced with permission from Alagille and Odievre [1])

feedings. These measures usually result in improved growth, prevention of hypoglycemia, and lowering of lactic acid and lipid levels. The development of hepatic adenomas may necessitate *liver transplantation*.

Case Study 2

A 12-year-old female is referred for evaluation of hypercholesterolemia detected after her new family physician saw her for the first time and noticed xanthomata on her extremities. She has apparently had these lesions since early childhood. She has also previously been followed by several dermatologists for severe chronic pruritis resulting in extensive excoriations, sub-ungual hematomas, and inability to sleep at night or function in school. Further physical examination reveals short stature (<3rd percentile for height), peculiar facies with a broad forehead, deep-set eyes, and pointed chin. There is no jaundice. Cardiac examination demonstrates a grade III/VI systolic murmur compatible with pulmonic stenosis. Abdominal examination reveals a liver edge 4 cm below the right costal margin, which is firm. There is no splenomegaly, ascites, or signs of chronic liver disease or portal hypertension. Skin examination reveals diffuse excoriation from itching and extensive xanthomata on palmar creases and extensor surfaces of the upper and lower extremities. There is a family history of the deceased father having similar symptoms, although much milder.

Laboratory studies: Hematocrit 43 %; hemoglobin 14.5 g/dL; white blood cell count 9,600 cells/mm^3 (4,500–13,500); 220, 000 platelets/mm^3 (150,000–350,000); prothrombin time 15.1 s (control 12.4 s); albumin 4.7 g/dL (3.7–5.6); total bilirubin 1.6 mg/dL; direct bilirubin 1.1 mg/dL; ALT 112 IU/L (10–30), AST 102 IU/L (15–30); GGTP 660 IU/L (14–25); alkaline phosphatase 550 IU/L (50–375); blood ammonia 43 μmol/L (29–57); fasting total serum bile acids 310 μmol/L (0–6); sodium 135 mEq/L (135–145), potassium 4.2 mEq/L (3.5–5.0), chloride 99 mEq/L (94–106); carbon dioxide 25 mEq/L (22–29); blood urea nitrogen (BUN) 39 mg/dL (5–25); and creatinine 0.6 mg/dL (0.5–1.0); total serum cholesterol 1,140 mg/dL (<202); triglycerides 212 mg/dL (<125), and HDL cholesterol 25 mg/dL (32–70).

Questions:

1. **What is the most likely diagnosis in this patient?**
   The **Alagille syndrome** or **arteriohepatic dysplasia** consists of the findings of *cholestasis* with *intrahepatic ductal paucity, short stature, peculiar facies, peripheral pulmonic stenosis* or more *complex heart lesions, failure of complete vertebral body* or *arch fusion* resulting in **"butterfly" vertebral bodies** on X-ray, *posterior embryotoxon* on slit lamp eye exam, and *renal abnormalities*. This disorder is inherited as an *autosomal dominant disorder* with variable penetrance, as well as a high spontaneous mutation rate, accounting for the frequent finding of an asymptomatic or minimally symptomatic parent or lack of one or more characteristic features in probands. The disorder is due to **mutations of the *JAG1* or *NOTCH2* genes** that are involved in intercellular signaling during embryonic development. Although cholestasis may be severe, most patients do not develop cirrhosis and liver failure. However, some patients eventually require liver transplantation. **Two striking features** of this disease are the *severe cholestasis* with markedly elevated serum bile acids and *pruritis* in the face of a normal or only slightly elevated bilirubin and the resulting marked hypercholesterolemia with xanthoma formation. The majority of the serum cholesterol is unesterified and contained in lipoprotein-X. These patients also often have low HDL cholesterol levels. It is interesting that even though these patients may have serum cholesterol levels many fold higher than individuals with familial hypercholesterolemia and elevated LDL cholesterol, premature atherosclerosis is not observed. However, a patient with primary biliary cirrhosis and severe cholestasis with an extremely high lipoprotein-X level has been reported with complications from a resultant **hyperviscosity syndrome with neurologic and cardiovascular impairment**. Also, these patients have a **vasculopathy** that may result in blood vessel rupture and spontaneous bleeding, and may be fatal, especially with an intracranial site of bleeding.

2. **How would you treat this patient?**

   Chronic administration of oral ursodeoxycholic acid appears to result in improvement in the cholestasis and pruritis, as well as the serum lipid profile, in some patients. This non-toxic, water-soluble bile acid acts as a **choleretic** to increase bile flow and after chronic administration may make up 40 % or more of the bile acid pool, displacing more toxic bile acids, which may contribute to the liver disease. Some patients may achieve the same result by **chronic external biliary diversion** using a *cholecystojejunal conduit* or *ileal diversion* for bile acid pool depletion. **Fat-soluble vitamin supplementation** is a necessity with particular attention to vitamin E. Severe vitamin E deficiency may develop in these patients with resultant neurologic symptoms, including *weakness, opthalmoplegia, ataxia, areflexia,* and *decreased vibratory and proprioceptive sensation*. This patient's prothrombin time corrected to normal within 24 h after a **parenteral injection of vitamin K**.

Case Study 3

A 48-year-old white male is referred for evaluation of the recent onset of chest pain with exertion. He works as an investment banker, has a very sedentary lifestyle with no regular exercise and smokes two packs per day. He eats out frequently with high-fat and high-sugar foods. He drinks an average of 1–2 beers per day. On physical examination his weight is 88 kg with a BMI of 27.8 (overweight range) and blood pressure of 152/103. His waist to hip ratio is 1:4 (normal <1.0), indicating abdominal obesity. The cardiovascular exam is otherwise normal. The abdominal exam is normal except for hepatomegaly with the liver edge 5 cm below the right costal margin.

Laboratory studies: Hematocrit 46 %; hemoglobin 15.3 g/dL; white blood cell count 10,600 cells/mm^3 (4,500–13,500); 289, 000 platelets/mm^3 (150,000–350,000); prothrombin time 11.1 s (control 12.4 s); fasting glucose 160 mg/dL (<120); glycosylated hemoglobin 9 % (<6 %); fasting insulin 48 μU/mL (<25), C-reactive protein 3.2 mg/dL (<1); albumin 4.2 g/dL (3.7–5.6); total bilirubin 1.1 mg/dL; direct bilirubin 0.4 mg/dL; ALT 199 IU/L (10–30), AST 180 IU/L (15–30); alkaline phosphatase 89 IU/L (25–100); total serum cholesterol 260 mg/dL (<202); triglycerides 493 mg/dL (<150), and HDL cholesterol 26 mg/dL (>40); LDL cholesterol 146 mg/dL (<100); apo B 209 mg/dL (<125).

Questions:

1. **Is this patient at high risk for serious health problems?**

   Yes, he fits the criteria for **metabolic syndrome:** *abdominal obesity*; *hypertension*; *insulin resistance* that is now manifest as frank *type 2 diabetes*, as indicated by elevated glucose, glycosylated hemoglobin and insulin levels; and *hypertriglyceridemia* and *low HDL cholesterol levels*, as well as *elevated LDL cholesterol with apo B levels* (a reflection of LDL particle number) consistent with highly atherogenic small, dense LDL particles. He also has *elevated CRP levels*, a marker for inflammation, which may be a component of the

metabolic syndrome. His **chest pain** may be due to ***angina from atherosclerotic coronary artery disease***, and the patient needs a comprehensive evaluation by a cardiologist. Finally, the h**epatomegaly** and **elevated transaminase levels** are consistent with the presence of ***NAFLD***.
2. **How would you approach making the diagnosis of NAFLD?**
Other causes of liver disease, including viral hepatitis, alcoholic liver disease, A-1-antitrypsin deficiency, hemochromatosis, and Wilson disease, should be ruled out. **Imaging of the liver with CT or MRI** (Chap. 10) may be used to detect *fatty liver*. **Serum markers for fibrosis or elastography** may be helpful in detecting *advanced liver fibrosis or cirrhosis*. Finally, a **percutaneous liver biopsy** will help confirm *steatosis*, as well as *degree of fibrosis* and the *presence of inflammation*, which would indicate progression to NASH.
3. **How would you treat this patient?**
Treatment of metabolic syndrome encompasses comprehensive lifestyle changes, including **weight reduction, regular exercise**, and **dietary reduction of total calories** with particular attention to **eliminating sugar**, especially fructose-containing foods, and refined carbohydrates. Effective medications are available to treat insulin resistance, type 2 diabetes and hypertension. Treatment with **vitamin E supplementation**, as an antioxidant, reduces serum transaminase levels and improves liver histology (but not fibrosis). However, at higher doses this therapy may be associated with significant adverse effects. **Insulin sensitizers**, such as *thiazolidinediones*, may also be an effective therapy, but may also be associated with serious adverse effects with long-term use. Recent data suggest that the intestinal microbiota may play a role in the development of NAFLD, and **alteration of the microbiome by the use of probiotics** may prove to be a useful therapeutic approach. **Bariatric surgery for morbid obesity** improves insulin resistance and diabetes and may also improve NAFLD, but more study is needed to define associated risks.

Further Reading

1. Alagille D, Odievre M (1979) Liver and biliary tract disease in children. Wiley, New York, p 212
2. Asrih M, Jornayvaz FR (2013) Inflammation as a potential link between nonalcoholic fatty liver disease and insulin resistance. J Endocrinol 218(3):R25–R36
3. Black DD (2005) Chronic cholestasis and dyslipidemia: what is the cardiovascular risk? J Pediatr 146(3):306–307
4. Brown MS, Goldstein JL (1976) Receptor-mediated control of cholesterol metabolism. Science 191:150–154
5. Ducheix S, Montagner A, Theodorou V, Ferrier L, Guillou H (2013) The liver X receptor: a master regulator of the gut-liver axis and a target for non alcoholic fatty liver disease. Biochem Pharmacol 86(1):96–105
6. Ghishan FK (2012) In: Boyer TD, Manns MP, Sanyal AJ (eds) Hepatology: a textbook of liver disease, 6th edn. Elsevier Saunders, Philadelphia, p 1168

7. Glickman RM, Sabesin SM (1994) In: Arias IM, Boyer JL, Fausto N, Jakoby WB, Schachter DA, Shafritz DA (eds) The liver: biology and pathobiology, 3rd edn. Raven Press, New York, p 393
8. Henao-Mejia J, Elinav E, Thaiss CA, Licona-Limon P, Flavell RA (2013) Role of the intestinal microbiome in liver disease. J Autoimmun 46:66–73
9. Hicks J, Wartchow E, Mierau G (2011) Glycogen storage diseases: a brief review and update on clinical features, genetic abnormalities, pathologic features, and treatment. Ultrastruct Pathol 35(5):183–196
10. Norata GD, Tibolla G, Catapano AL (2013) Targeting PCSK9 for hypercholesterolemia. Annu Rev Pharmacol Toxicol 54:273–293
11. Perlmutter DH (2011) Alpha-1-antitrypsin deficiency: importance of proteasomal and autophagic degradative pathways in disposal of liver disease-associated protein aggregates. Annu Rev Med 62:333–345
12. Shao W, Espenshade PJ (2012) Expanding roles for SREBP in metabolism. Cell Metab 16(4):414–419
13. Van Thiel DA (1988) In: Arias IM, Jacoby WB, Popper H, Schachter D, Shafritz DA (eds) The liver: biology and pathobiology, 2nd edn. Raven Press, New York, p 1012
14. Van Thiel DA (1988) In: Arias IM, Jacoby WB, Popper H, Schachter D, Shafritz DA (eds) The liver: biology and pathobiology, 2nd edn. Raven Press, New York, p 1010
15. Williams GJ, Whitington PF, Weidman SW, Black DD, Sabesin SS (1985) Correctable plasma lipoprotein abnormalities in infants with choledochal cysts. Pediatr Res 19:240
16. Zakim D (1990) In: Zakim D, Zakim D, Boyer TD (eds) Hepatology: a textbook of liver disease, 2nd edn. W. B. Saunders Co., Ltd, Philadelphia

Chapter 12
Biotransformation, Elimination and Bile Acid Metabolism

Dennis D. Black

1 Introduction

The liver detoxifies and excretes both **exogenous** xenobiotics, such as drugs and environmental toxins, and **endogenous** substances, including ammonia and bilirubin. Efficient processing and elimination of these substances prevent cellular injury, including CNS toxicity in the case of ammonia and bilirubin. **Hepatic synthesis, secretion and re-uptake of bile acids** is an exquisitely regulated process that allows modulation of cholesterol metabolism and efficient absorption of the products of fat digestion, as well as compensatory trafficking of bile acids out of the hepatocyte in the face of bile acid retention as occurs in cholestasis. **Nuclear receptor transcription factors** tightly regulate these processes and offer potential therapeutic targets for new approaches to treatment of a variety of diseases. This chapter reviews these important functions of the liver, underscoring another aspect of its central role in human physiology.

2 Urea Synthesis and Ammonia Metabolism

The urea cycle (Fig. 12.1) **is responsible for clearing the majority of the body's excess** nitrogen and producing arginine for protein synthesis or conversion to urea. All five enzymes are located primarily in the **zone 1 periportal hepatocytes**. The first two enzymes are in the *mitochondrial matrix* and the other three are in the *cytosol*. As shown in Fig. 12.1, **ammonia** is the common substrate for entry into the cycle. Ammonia is derived from several sources in the body, including the degradation of *amino acids*, *amines*, and *nucleic acids*. Ammonia is also a

D.D. Black, M.D. (✉)
Department of Pediatrics, University of Tennessee, Memphis, TN, USA
e-mail: dblack@uthsc.edu

P.S. Leung (ed.), *The Gastrointestinal System: Gastrointestinal, Nutritional and Hepatobiliary Physiology*, DOI 10.1007/978-94-017-8771-0_12,

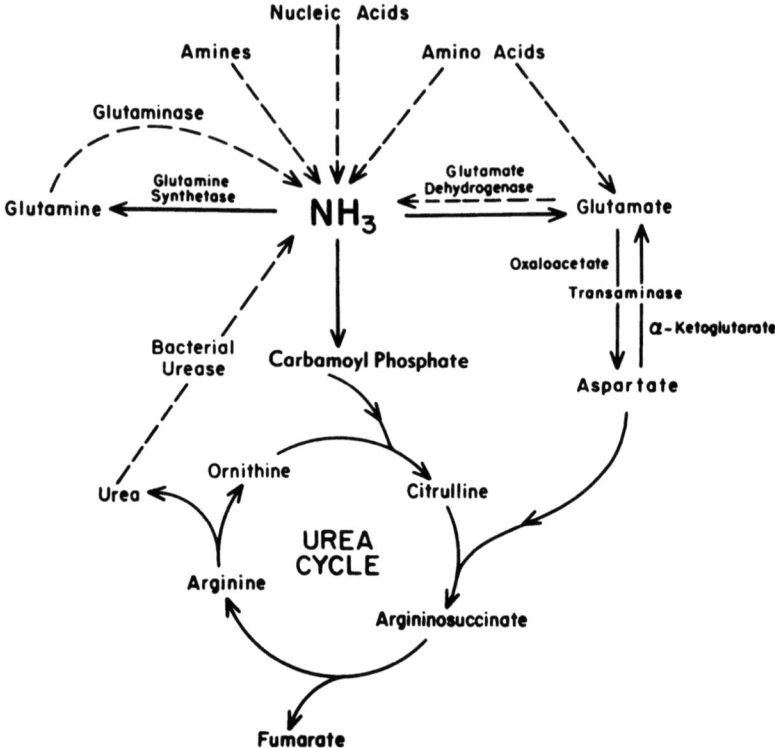

Fig. 12.1 General scheme of ammonia metabolism in the liver. *Solid arrows*, pathways of ammonia utilization; *broken arrows*, pathways of ammonia formation (Reproduced with permission from Ampola [1])

byproduct of bacterial degradation of urea in the gastrointestinal tract, which diffuses into the bowel lumen from the bloodstream. Urease-producing bacteria in the lumen of the colon split the urea, as well as luminal amino acids, into CO_2 and ammonia. The ammonia then diffuses back into the bloodstream into the portal circulation and is cleared by the liver, predominantly on the first pass. **Glutamate** is produced in large quantities from the catabolism of amino acids. Glutamate serves as a major source of urea by undergoing **transamination** in the liver to aspartate, which then enters the urea cycle. Although several organs produce urea, the **liver** is the only quantitatively important organ. **Skeletal muscle** is capable of ammonia detoxification by the reaction of glutamate with ammonia to produce glutamine. Ammonia clearance by skeletal muscle is the reason for the higher ammonia concentration in arterial than in venous blood. In skeletal muscle, protein catabolism generates large amounts of ammonia resulting in formation of glutamate by transamination, and the glutamate is converted to alanine. Alanine is transported to the liver, where glutamate is regenerated by transamination of alanine to pyruvate. Ammonia is then released from the glutamate and enters the urea cycle.

The pyruvate is a substrate for gluconeogenesis to produce glucose, which may circulate back to the muscle for energy production.

The urea cycle is subject to both short- and long-term regulation. **Short-term regulation** is accomplished by intracellular substrate levels. In **long-term regulation**, total hepatic content of the five urea cycle enzymes is proportional to protein intake and rate of urea excretion. All five urea cycle enzymes may be induced by glucagon and glucocorticoids. The rate-limiting enzyme in vivo is probably carbamylphosphate synthetase I.

Ammonia is toxic to the central nervous system, and in disorders characterized by **hyperammonemia**, there is increased movement of ammonia across the blood–brain barrier. The cells most susceptible to ammonia toxicity are the **astrocytic** or **glial cells**. Astrocytes are the site of ammonia detoxification in the central nervous system with glutamine production from **glutamate**, a major excitatory neurotransmitter in the brain. In fact, cerebrospinal fluid glutamine levels parallel the severity of neurologic dysfunction. As the level of glutamate declines, it is replenished by condensation of ammonia with α-**ketoglutarate. Depletion of α-ketoglutarate** results in *impaired mitochondrial glucose oxidation*. Ammonia may also adversely affect neurotransmitter function and directly interfere with cellular energy production. As the degree of hyperammonemia increases, neurologic function progressively deteriorates with onset of deepening coma and finally development of cerebral edema, which is often a terminal event.

Inherited defects in the urea cycle enzymes may result in *hyperammonemia* and *neurologic dysfunction*, often to a severe degree. Inherited defects in all five enzymes of the cycle, as well as N-acetylglutamate synthetase (which produces an essential cofactor for carbamyl phosphate synthetase), have been described. There may be a spectrum of severity, depending on the specific defect in the enzyme and whether there is partial residual function. In severe cases, *seizures*, *coma*, and *death* may occur in the neonatal period, if untreated.

3 Hepatic Encephalopathy

Hepatic encephalopathy may be classified as two general types. In patients with **cirrhosis**, **portal hypertension**, and **large portosystemic shunts**, ammonia from intestinal sources is not cleared effectively on first-pass through the liver. Also, the relief of portal hypertension by surgical creation of a **portosystemic shunt** or placement of **a transjugular intrahepatic portosystemic shunt** (**TIPS**) may precipitate encephalopathy. This ineffective clearance, combined with generally poor hepatic function, may result in chronic encephalopathy, punctuated by acute exacerbations often precipitated by gastrointestinal bleeding from esophageal varices (resulting in an increased intestinal ammonia load), infection, excessive protein intake, or overzealous diuresis to treat ascites. Chronic pathologic changes in astrocytes in the central nervous system may be present. This type of encephalopathy is referred to as

portosystemic encephalopathy. The other category of encephalopathy is observed in the patient with acute, severe liver disease (usually *fulminant hepatitis*) who develops progressive encephalopathy due primarily to severe liver dysfunction with inadequate ammonia detoxification. Acute liver failure with encephalopathy often rapidly progresses to cerebral edema and death. The cerebral edema is thought to be a direct result of ammonia toxicity. In fact, the edema may actually be a result of the osmotic effect of ammonia-induced elevation of brain glutamine levels.

It should be stressed that in encephalopathy associated with liver disease, other factors may contribute to the central nervous system dysfunction in addition to hyperammonemia. In fact, the degree of hyperammonemia does not always correlate with the degree of encephalopathy, suggesting that other mechanisms may be involved. In acute liver failure, other neurotoxic substances, such as *mercaptans* and *short-chain fatty acids*, may play a role. There is evidence that hepatic encephalopathy may result in part from the action of endogenous substances that bind to the **GABA-benzodiazepine receptor complex** in the brain. This is supported clinically by the observation that portosystemic encephalopathy and encephalopathy associated with acute liver failure may improve transiently with treatment with benzodiazepine receptor antagonists. However, there are limited data to support their clinical use. There may be increased entry of ammonia into the central nervous system due to increased permeability of the blood–brain barrier and conversion of positively charged ammonium ion to ammonia, which more readily crosses the blood–brain barrier, secondary to changes in blood pH. Accumulation of glutamine in the brain results in enhanced central nervous system uptake of tryptophan, which is converted into both neuroactive and neurotoxic metabolites, including *serotonin* and *tryptamine*.

Clinically, the presentation of the *first stage* of hepatic encephalopathy may be subtle. There may be **personality changes** and **inversion of sleep patterns**. The patient may exhibit a **short attention span** and **easy forgetfulness**. With portosystemic encephalopathy, it is now recognized that very mild impairment may affect social skills and emotional behavior, and evaluation of driving skills may be a practical index to diagnose what is termed **minimal hepatic encephalopathy** (**MHE**). An early physical sign of hepatic encephalopathy is **asterixis**. This sign represents an inability to maintain posture and is usually elicited by asking the patient to dorsiflex the hand with the arm extended. After several seconds a positive sign is characterized by the hand falling forward followed by a brisk recovery, thereby resulting in the typical "flapping" tremor. This sign is dependent upon the patient being able to understand and follow directions, and is highly variable over short periods of time, even in the same individual. Asterixis is not specific for hepatic encephalopathy and may be elicited in other disorders, including uremia. In *stage 2*, the individual becomes **lethargic and disoriented** with other abnormal reflexes in addition to asterixis. In *stage 3*, the patient is **somnolent, but arousable**. However, when awake the patient is unable to communicate or perform mental tasks such as calculation. *Stage 4 encephalopathy* is characterized by *total* **inability to arouse the patient** accompanied by a **positive Babinski sign** and **decerebrate** posturing, and usually immediately precedes the development of **cerebral edema and death**. Abnormal changes on electroencephalogram (EEG) are detectable as early as stage 1 and generally worsen in parallel with clinical progression to stage 4.

Treatment of hyperammonemia should be directed toward reducing the production of ammonia and enhancing its clearance. **Moderate protein restriction** had been recommended in the past. However, since restriction results in enhanced catabolism of endogenous protein, and a significant positive effect on outcomes has yet to be proven, such restriction is not currently recommended. **Branched-chain amino acid supplementation** may have a modest beneficial effect, but is prohibitively expensive. Attention should be directed to sources of gastrointestinal bleeding, such as esophageal varices, which may result in a large intestinal ammonia load. **Lactulose** is a non-absorbable carbohydrate administered enterally and is fermented by colonic bacteria to form organic acids, which lower the luminal pH. Since only ammonia in the form of NH_3 is readily absorbed, an increase in the NH_4^+/NH_3 ratio results in trapping of ammonia in the gut and elimination in the feces. Also, the osmotic property of lactulose results in a significant laxative effect and may effectively clear ammoniagenic substances from the gastrointestinal tract. The non-absorbable antibiotic, **rifaximin**, is used to suppress intestinal ammonia-forming bacteria. In urgent situations, high ammonia levels may also be cleared by hemodialysis or exchange transfusion. Use of external liver support systems has generally not significantly impacted outcomes. The **molecular absorbent recirculating system** (**MARS**) has shown some benefit in more severe cases. In cases of encephalopathy associated with liver failure, liver transplantation may result in prompt resolution of severe encephalopathy, providing that cerebral edema has not developed.

4 Hepatic Detoxification

The liver is responsible for the biotransformation and clearance of a variety of endogenous and exogenous toxic compounds. Exogenous organic substances entering the body that have no nutritive value for energy production, structural function, or enzyme cofactor function are called **xenobiotics**. The liver is capable of numerous reactions which generally serve to make the substance to be eliminated more polar and water soluble to subsequently be excreted in bile or urine (**Phase I reactions: oxidoreductases and hydrolases**) and to make the substance less toxic (**Phase II reactions: transferases**). As will also be demonstrated, these reactions do not always result in detoxification, but may actually form toxic intermediary metabolites. Apical canalicular excretion of these biotransformed molecules into bile for elimination or basolateral membrane transport into the circulation to be cleared by the kidney occurs via **specific ATP-binding cassette** (**ABC**) **transporters** and is termed **Phase III transport**.

4.1 Phase I Reactions

The enzyme systems responsible for **oxidation and reduction** reactions are components of Phase I biotransformation. **Dehydrogenases** catalyze hydrogen transfer from a substrate to a hydrogen receptor, frequently a pyridine nucleotide. Examples include *alcohol dehydrogenases*, cytosolic enzymes, which catalyze NAD-dependent oxidation of alcohols to aldehydes or ketones by the following reaction:

$$RCH_2OH + NAD^+ \rightarrow RCHO + NADH + H^+$$

Aldehyde dehydrogenases are present in *mitochondrial, microsomal,* and *cytosolic hepatocyte fractions*. The production of **aldehydes from alcohols** in the above reaction is *reversible*. However, conversion of the **aldehydes to ketones** by aldehyde dehydrogenases is essentially *irreversible*:

$$RCHO + NAD^+ + H_2O \rightarrow RCOOH + NADH + H^+$$

Another dehydrogenase is **dihydrodiol dehydrogenase**, which catalyzes the $NADP^+$-dependent oxidation of metabolites of polycyclic aromatic hydrocarbon compounds.

Hepatic reductases include *aldehyde and ketone reductases, quinone reductases, nitro and nitroso reductases, azoreductases, N-oxide reductases,* and *sulfoxide reductases*. Substrates for these enzymes are drugs and xenobiotics, including warfarin, aflatoxin, adriamycin, chloramphenicol, and dimethylsulfoxide. **Oxidases** catalyze transfer of electrons from substrate to oxygen, generating either hydrogen peroxide or superoxide anions, and include *aldehyde oxidases* and *monoamine oxidases*.

The hepatic **mono-oxygenases** are microsomal enzymes that catalyze reactions in which one atom of oxygen is incorporated into product and the other is reduced to water:

$$R + O_2 + NADPH + H^+ \rightarrow RO + H_2O + NADP^+$$

The **cytochrome P450-dependent mono-oxygenases** have been the most intensively studied. Although these enzymes are found in many tissues throughout the body, they are most abundant in the *liver*. Within the hepatocyte this system is located primarily in the *smooth endoplasmic reticulum*. It is also found in the *nuclear membrane*, where this fraction is probably more important in carcinogenesis. The system derives its name from "P" which stands for "pigment" and "450" which indicates the absorption wavelength in nm at which a maximum occurs with the binding of carbon monoxide under anaerobic conditions. This mono-oxygenase system consists of a **flavoprotein** and a family of **hemoproteins**. The scheme for the function of this complex is shown in Fig. 12.2. In the *first step*, the oxidized

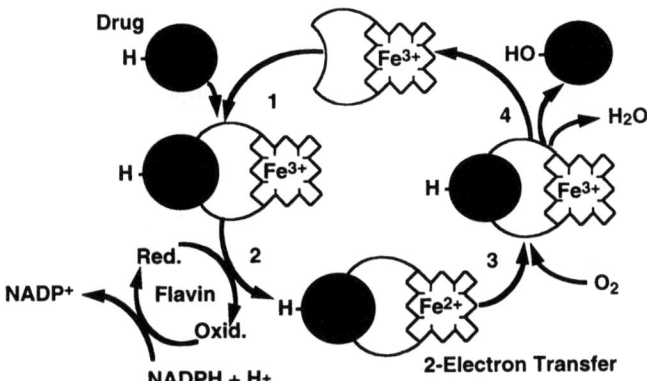

Fig. 12.2 The cytochrome P450-dependent mono-oxygenase redox cycle. In the first step, the oxidized (Fe^{3+}) form of the enzyme binds the reduced drug substrate. In the rate-limiting second step, the complexed cytochrome is reduced to the Fe^{2+} form by a flavoprotein, NADPH-cytochrome P450 reductase, with NADPH as the ultimate electron donor. In step three, the reduced cytochrome-drug complex combines with molecular oxygen to yield a ternary intermediate in which the iron is again oxidized to the Fe^{3+} state. In the fourth step, the ternary complex decomposes to yield the free oxidized cytochrome, the oxidized form of the drug, and water

(Fe^{3+}) form of the enzyme binds the reduced substrate. In the *rate-limiting second step*, the complexed cytochrome is reduced to the Fe^{2+} form by a flavoprotein, NADPH-cytochrome P450 reductase, with NADPH as the ultimate electron donor. In *step three*, the reduced cytochrome-drug complex combines with molecular oxygen to yield a ternary intermediate in which the iron is again oxidized to the Fe^{3+} state. In the *fourth step*, the ternary complex decomposes to yield the free oxidized cytochrome, the oxidized form of the substrate, and water. The entire cycle involves a **two-electron transfer** and **consumption of one mole of NADPH** and **one mole of molecular oxygen per mole of drug reduced**. The components of this system are integral membrane proteins. Reconstitution of the proteins in vitro requires the presence of phospholipid for a catalytically active complex. These mono-oxygenases are subject to genetic variation and can be induced or inhibited by xenobiotics, which may affect the metabolism of certain drugs.

Hydrolytic reactions involve substrates that contain ester linkages. **Hydrolases** cleave ester linkages with addition of a molecule of water:

$$RCOOR + H_2O \rightarrow RCOOH + ROH$$

The products of this hydrolysis, an acid and an alcohol, are more water-soluble than the parent ester. In addition to increasing water solubility, this reaction often results in detoxification of the xenobiotic. The reaction products may also be substrates for Phase II reactions. Xenobiotics subject to hydrolysis include aspirin, meperidine, malathion, and nerve gas.

4.2 Phase II Reactions

The general scheme for conjugation reactions is as follows:

$$\text{Substrate} + \text{Donor-A} \rightarrow \text{Substrate-A} + \text{Donor}$$

One of the most common conjugation reactions is **glucuronidation**. Glucuronic acid is transferred from **UDP-glucuronic acid** (**UDPGA**) to the hydroxyl or carboxyl group on the acceptor substrate molecule to form a β-glucuronide. Glucuronidation generally not only enhances water solubility, but also detoxifies the compound. Substrates for this reaction include *acetaminophen, morphine, chloramphenicol, furosemide, sulfisoxazole,* and *valproic acid*. As discussed subsequently, bilirubin, an endogenous waste product of heme catabolism, is a major substrate for conjugation with glucuronide. The glucuronidation capacity of the liver depends on the rate of synthesis of UDPGA from UDP-glucose, which requires adequate carbohydrate reserves. Another Phase II conjugation reaction is **sulfation**. Conjugation with sulfate generally results in compounds that are less toxic and more water-soluble. Substrates for this reaction include *alcohols, bile acids,* and *phenolic compounds*.

Glutathione (**GSH**) is a tripeptide (γ-glutamyl-cysteinyl-glycine) which participates in substitution reactions that are catalyzed by **glutathione S-transferases**, whereby the GSH displaces a group on a xenobiotic, which generally has an electrophilic center. Electrophilic compounds are particularly toxic to cells because of their propensity to react spontaneously with -SH groups on proteins or amine groups on nucleic acids, leading to cell necrosis or carcinogenesis. Substrates for this reaction include nitroglycerin, the diuretic ethacrynic acid, certain insecticides, and sulfobromophthalein (BSP). The GSH conjugates are generally excreted into bile. However, they may be reabsorbed in the intestine and subsequently taken up by the kidney, where the enzyme γ-**glutamyl-transpeptidase** (**GGTP**) transfers the γ-glutamyl group of the conjugate to an amino acid, leaving the cysteine-glycine conjugate. A dipeptidase then removes the glycine, producing the **cysteine conjugate**. This conjugate may be excreted in the urine or acetylated to **mercapturic acid** in the liver or kidney. The mercapturic acids may be excreted in urine, and to a lesser extent, in bile. Under certain conditions, such as acetaminophen overdosage to be discussed subsequently, the GSH stores may be depleted and the potential for toxicity greatly increases.

Methylation is another Phase II reaction, which serves mainly to detoxify and actually may render some substrates less water-soluble. In this reaction the methyl group of S-adenosylmethionine is transferred to an acceptor xenobiotic that contains a polyphenol, amine, or sulfhydryl group. Substrates for this reaction include *norepinephrine, epinephrine, dopamine,* and *tyramine* and other catachols after hydroxylation by the cytochrome P450 system.

5 Bioactivation and Hepatocellular Injury

Although the biotransformation reactions just described generally function to detoxify and enhance the elimination of xenobiotics, there are instances in which the substrates are actually rendered *more toxic or carcinogenic*. There are **five classes of such biotransformation**. The xenobiotic may be transformed into a **stable, but toxic metabolite**. Examples include *acetonitrile* and *methoxyflurane*. A *second class* involves transformation into a **reactive, electrophilic metabolite**. *Acetaminophen* and *furosemide* are in this class. ***Third***, the substrate may be converted to a **free radical**, as in the case of *carbon tetrachloride*. ***Fourth***, the biotransformation reaction may lead to the formation of **reduced oxygen metabolites**, as occurs with *paraquat* and *quinones*. ***Finally***, the biotransformation reaction itself may generate a **metabolic derangement** within the cell, as in the case of *galactosamine-induced liver injury*, where there is a depression of uracil nucleotide dependent biosynthesis of nucleic acids, glycolipids, glycoproteins, and glycogen, which results in cell organelle injury.

Review of the biochemical events occurring in acetaminophen toxicity serves as an excellent example of these metabolic pathways. These events are summarized in Fig. 12.3. When taken in appropriate doses, most of the acetaminophen is

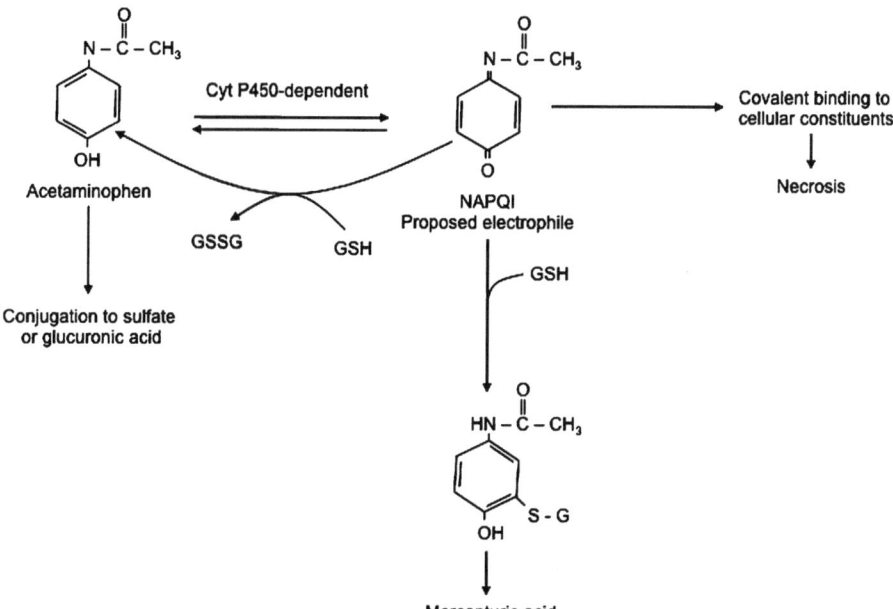

Fig. 12.3 Hepatic metabolic pathways in acetaminophen overdose. *GSH* glutathione, *NAPQI* N-acetyl-*p*-benzoquinone imine (Reproduced with permission from Vessey [20])

conjugated with sulfate and glucuronic acid and eliminated. About 5-15 % is converted to a reactive, electrophilic intermediate, **N-acetyl-p-benzoquinone imine (NAPQI)**, which reacts with sulfhydryl groups in proteins. The NAPQI is normally immediately conjugated with glutathione before it can do any damage to the cell and is converted into a mercapturic acid for mainly renal elimination. However, in the case of an overdose the conjugation pathway is saturated and available glutathione is quickly depleted, allowing accumulation of significant amounts of NAPQI. The NAPQI causes cell injury and death resulting in the hepatic necrosis seen with *acetaminophen overdose*, which can result in liver failure and death. Obviously, one therapeutic strategy would be **repletion of the liver's glutathione stores**. Although this cannot be done directly, another successful approach is to administer **N-acetylcysteine (Mucomist[R])**, which restores glutathione levels by enhancing synthesis. If administered early enough in the course of intoxication, this intervention can effectively prevent significant liver injury.

6 Hepatic Bilirubin Metabolism

Bilirubin is a pigment derived from the catabolism of **heme**. It is a toxic waste product, which undergoes hepatic biotransformation and excretion in bile. Bilirubin has an **open-chain tetrapyrrole structure** with **eight side chains** (Fig. 12.4). The IX-α isomer is the most abundant naturally occurring isomer of bilirubin. From the chemical structure of bilirubin, one would predict a water-soluble compound at

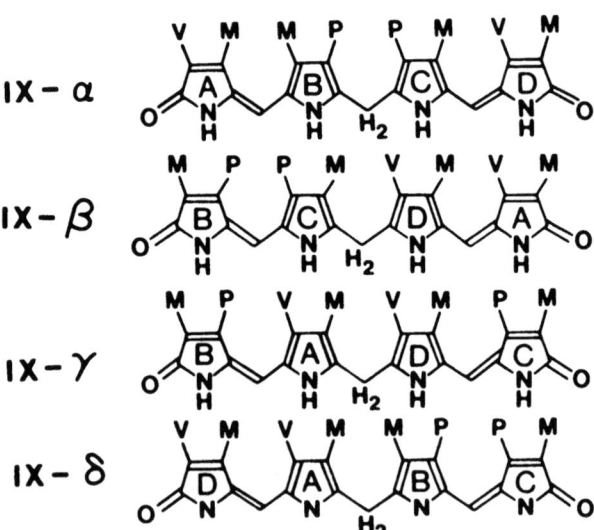

Fig. 12.4 The four isomeric forms of bilirubin. The IX-α isomer is the most abundant naturally occurring form. (M, CH_3; V, $CH_2 = CH_2$; P, CH_2-CH_2-COOH.) (Reproduced with permission from Odell [14])

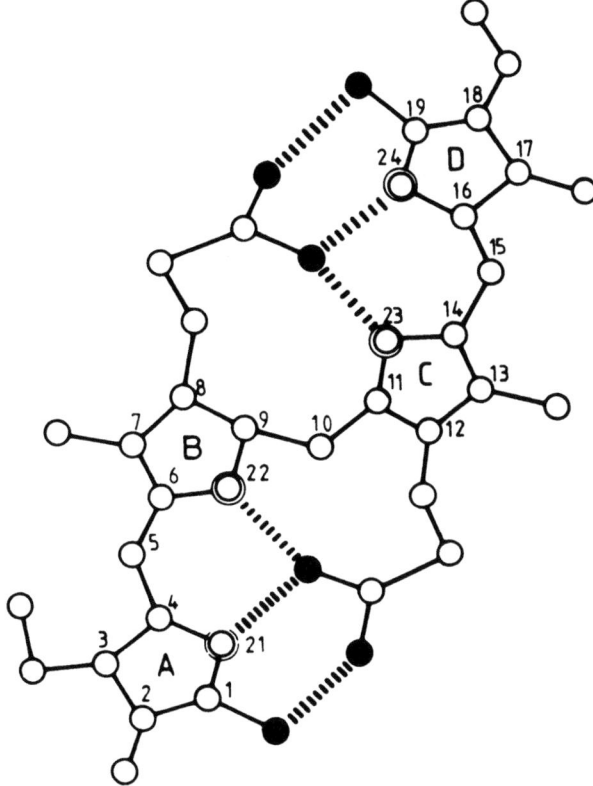

Fig. 12.5 Molecular structure of internally hydrogen-bonded bilirubin IX-α. Open, double-contoured, and filled circles represent carbon, nitrogen, and oxygen atoms, respectively. Hydrogen atoms are not shown. The six intramolecular hydrogen bonds are indicated by broken lines (Reproduced with permission from Blanckaert and Fevery [2])

physiological pH. However, in reality, unconjugated bilirubin has only **very limited water solubility**. This phenomenon may be explained by the presence of *extensive internal hydrogen bonding* within the molecule as shown in Fig. 12.5. This results in the involvement of all of the polar groups of the molecule in the hydrogen bonding, so that none are available for hydration.

Normally, serum bilirubin levels are very low, and unconjugated bilirubin predominates. In normal individuals, men have slightly higher serum bilirubin levels (0.6 mg/dL) than females (0.4 mg/dL), and any level above 1.0 mg/dL is considered abnormal. *Jaundice* with the appearance of the pigment in sclerae, skin, and mucus membranes generally is noted when the bilirubin level exceeds 3 mg/dL in adults and 6 mg/dL in infants.

The major site of bilirubin toxicity is the **central nervous system**, particularly in the *newborn infant*. Bilirubin inhibits central nervous system RNA and protein synthesis, uncouples oxidative phosphorylation, and inhibits ATPase activity in brain mitochondria. Clinically, bilirubin toxicity can produce severe central nervous system dysfunction resulting in permanent injury or death. Accumulation of bilirubin in the brain, primarily in the *basal ganglia, hippocampus,* and *cochlear and cerebellar nuclei*, with associated toxicity is termed **kernicterus**.

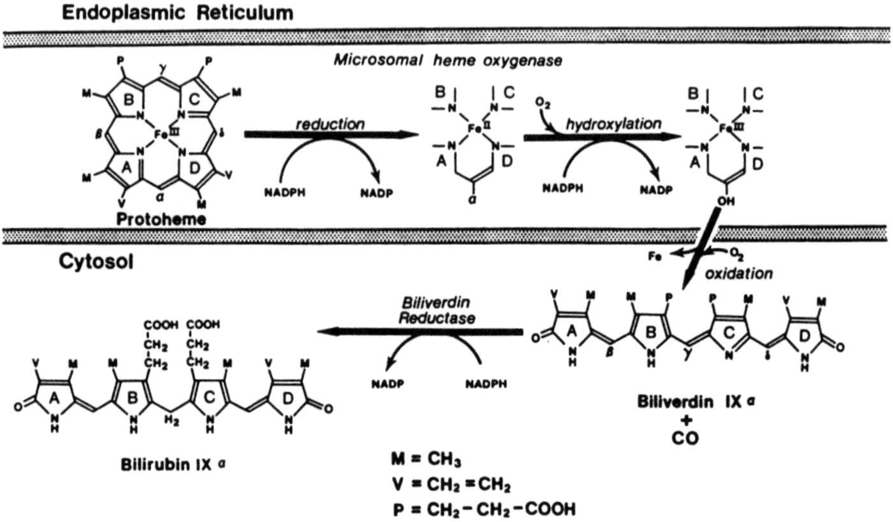

Fig. 12.6 Pathways of biliverdin production from heme by chain-opening and bilirubin production by reduction of biliverdin (Reproduced with permission from Odell [15])

Bilirubin is the major degradation product of heme in mammals. Degradation of hemoglobin from senescent red blood cells by reticuloendothelial cells in the spleen accounts for 80 % of bilirubin production. The other 20 % is derived from ineffective erythropoiesis in the bone marrow, as well as the turnover of heme-containing proteins and myoglobin. The series of reactions responsible for the conversion of heme into bilirubin is shown in Fig. 12.6. The first step is the conversion of heme to biliverdin mediated by the enzyme **heme oxygenase**, which exists as *three isoforms*. The activity of this microsomal enzyme is highest in the *spleen* and *Kupffer cells* in the *liver* and requires NADPH and oxygen. This is **the only reaction in the body that produces carbon monoxide**, which is excreted in expired air and may be quantitated as an *index of bilirubin production*. **Carbon monoxide** is also an important signaling molecule and functions as a potent vasodilator. Interestingly, **biliverdin** is a non-toxic, water-soluble compound and is the end-product of heme catabolism in birds and other non-mammalian vertebrates. However, in mammals biliverdin is converted to bilirubin, which is toxic and requires energy-dependent biotransformation for excretion. Conversion of biliverdin to bilirubin also occurs in reticuloendothelial cells by **biliverdin reductase**, a cytosolic enzyme that requires NADPH, and utilizes NADH at lower pH. Both bilirubin and biliverdin may have beneficial effects as *anti-oxidants* under certain conditions and may affect progression of atherosclerosis.

After its formation in reticuloendothelial cells, bilirubin is released into the bloodstream, where it is transported bound to **albumin** via one primary and two secondary binding sites. This binding prevents the diffusion of free bilirubin into tissues, particularly the central nervous system, thereby preventing toxicity.

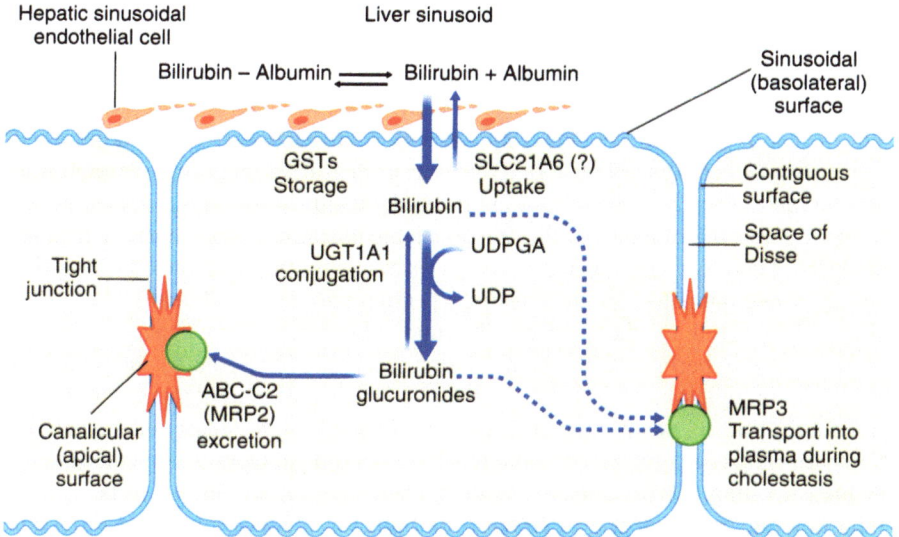

Fig. 12.7 Summary of hepatic transport and metabolism of bilirubin. During its carriage in plasma, bilirubin is strongly, but reversibly, bound to albumin. In hepatic sinusoids, this complex comes in direct contact with the basolateral domain of the hepatocyte plasma membrane through fenestrae of the specialized hepatic endothelial cells. At the hepatocyte surface, dissociation of the albumin–bilirubin complex occurs, and bilirubin enters hepatocytes by a specific uptake mechanism(s). A fraction of the bilirubin is also derived from catabolism of hepatocellular heme proteins. Storage within the hepatocyte is accomplished by binding of bilirubin to a group of cytosolic proteins, glutathione-*S*-transferases (GSTs) (also termed ligandin or Y-protein). Binding to these proteins keeps bilirubin in solution and inhibits its efflux from the cell, thereby increasing the net uptake. Conjugation of bilirubin in the endoplasmic reticulum is catalyzed by bilirubin-uridinediphosphoglucuronate glucuronosyltransferase (UGT1A1), forming bilirubin monoglucuronide and diglucuronide. Both conjugates may bind to GSTs in the cytosol. Conjugation is obligatory for efficient transport across the bile canaliculus. Bilirubin glucuronides are secreted across the bile canaliculus by an energy-consuming process mediated by MRP2 (previously termed cMOAT). This process is normally rate-limiting in bilirubin throughput, and is shared by other organic anions, but not bile salts (Reproduced with permission from Elsevier [17])

However, other ligands such as ***drugs*** (salicylates, sulfonamides, furosemide, etc.) and ***fatty acids*** can displace bilirubin from albumin binding sites and precipitate toxicity.

At the hepatocyte plasma membrane, bilirubin enters the hepatocyte via a specific uptake mechanism, as shown in Fig. 12.7. Once inside the cell, bilirubin binds to a group of cytosolic carrier proteins, **glutathione-*S*-transferases (GSTs)**, also termed **ligandin** or **Y protein**, keeping the bilirubin in solution and within the cell. The bilirubin is then conjugated with glucuronide by the microsomal enzyme, **bilirubin UDP-glucuronosyltransferase (UGT1A1)**. The resulting bilirubin monoglucuronide and diglucuronide are excreted at the canalicular membrane as a component of bile by the **ATP-dependent transporter MRP2 (ABCC2)**.

The transcription factors, **constitutive androstane receptor** (**CAR**) and **pregnane X receptor** (**PXR**), regulate the various proteins involved in bilirubin uptake, glucuronidation and secretion. Conjugation renders bilirubin *non-toxic*, *water-soluble*, and *non-absorbable* from the gastrointestinal tract. In the distal small intestinal and colonic lumen, the conjugated bilirubin is converted to **urobilinogens** by bacteria and eliminated in stool. However, a small fraction of these tetrapyrroles may be reabsorbed and re-excreted in bile, and the kidney excretes a small fraction.

In the fetus, the **placenta** is the only available mechanism for clearance of waste products. Therefore, conjugated bilirubin excreted by the liver into the intestinal lumen is deconjugated by an intestinal β-**glucuronidase** and reabsorbed to be ultimately cleared by the placenta. Normal infants may experience a transient unconjugated hyperbilirubinemia in the first week of life caused by immaturity of bilirubin conjugation, as well as enhanced enterohepatic circulation of bilirubin, potentially due to postpartum persistence of intestinal β-glucuronidase activity. Newborn infants may develop unconjugated hyperbilirubinemia as a consequence of breastfeeding. There appears to be a factor in certain breast milks, which enhances the enterohepatic circulation of bilirubin leading to the unconjugated hyperbilirubinemia of *"breast milk jaundice"*.

Disorders of bilirubin metabolism may occur at several of the metabolic steps just described. *Unconjugated hyperbilirubinemia* may occur when there is an overproduction of bilirubin from red blood cell hemolysis, as may occur in the newborn with a maternal-fetal blood group incompatibility or an extravascular accumulation of blood, such as a hematoma. There may be defective bilirubin conjugation. In the *Crigler-Najjar syndrome*, there is a genetic defect in UGT1A1. In autosomal recessively inherited **type I** of this syndrome, there is *complete absence* of enzyme activity, resulting in severe hyperbilirubinemia and often death from **kernicterus**. In **type II** there is *reduced, but not absent*, transferase activity. This residual activity is inducible by **phenobarbital treatment**, and this form of the disorder usually has a more favorable prognosis, although kernicterus can occasionally occur.

The *Gilbert syndrome* is a benign condition, which is relatively common (5–10 % of the Caucasian population) and results in mild elevation of the serum unconjugated bilirubin level. This elevation and associated visible icterus is often precipitated or exacerbated by fasting, stress, or illness. This **autosomal dominant** condition results from a genetic defect in the **promoter of the UGT1A1 gene** resulting in reduced transcription and moderate impairment of conjugation activity. Other benign disorders of bilirubin metabolism include the Dubin-Johnson and Rotor syndromes. The *Dubin-Johnson syndrome*, caused by genetic defects in the **canalicular MRP2 transporter**, is characterized by the presence of intermittent mild jaundice with a predominantly conjugated hyperbilirubinemia. There is also accumulation of a dark pigment in the liver primarily in lysosomes. *Rotor syndrome* resembles Dubin-Johnson syndrome except the liver pigment is lacking, and recent evidence suggests presence of simultaneous defects in **both basolateral membrane organic anion transporting polypeptides OATP1B1 and OATP1B3**. Both conditions exhibit an **autosomal recessive** mode of inheritance.

In conditions such as ***intra- and extrahepatic cholestasis and hepatitis***, there is failure to excrete conjugated bilirubin, which is regurgitated into the bloodstream resulting in conjugated hyperbilirubinemia. Unlike unconjugated bilirubin, conjugated bilirubin is filtered by the kidney and excreted in the urine resulting in the dark urine, which is characteristic of these disorders. Another serum bilirubin fraction found in cholestatic liver disease is **delta-bilirubin**, which consists of conjugated bilirubin covalently bound to albumin.

There are several therapeutic modalities used to treat unconjugated hyperbilirubinemia, particularly in the neonate. **Phototherapy**, which involves exposure of the individual to light (the *blue portion* of the spectrum appears to be most effective), results in production of water-soluble photooxidation products, as well as photoisomerization of unconjugated bilirubin with disruption of the internal hydrogen bonding. This results in increased water solubility and elimination of these isomers in bile and urine. This treatment is *less effective in older children and adults* (such as those with Crigler-Najjar syndrome, type 1) due to the *higher body mass to surface area ratio* in these individuals. In extreme cases, **exchange transfusion** or **plasmapheresis** may be employed to rapidly reduce the unconjugated bilirubin level. Ultimately, patients with Crigler-Najjar syndrome, type 1, require **liver transplantation** to correct the hepatic enzymatic defect. The enterohepatic circulation of bilirubin may be interrupted by **frequent feeding**. Also, **oral administration of substances that bind bilirubin**, such as agar, will interrupt the enterohepatic circulation, but has not achieved wide use. **Pharmacologic induction of UGT1A1** may reduce serum bilirubin levels. Substances such as *Sn-protoporphyrin*, which block the first enzyme involved in bilirubin production, heme oxygenase, have shown some effectiveness, but are not widely used. Experimental treatments currently under development include **hepatocyte transplantation** and **gene therapy**.

7 Hepatic Bile Acid Metabolism

Bile formation and excretion are major functions of the liver. Bile is *secreted* by the **hepatocyte canaliculus**, undergoes *modification* in the **bile ducts**, is *concentrated and stored* in the **gallbladder**, and is ultimately *excreted* into the **duodenum**. Biotransformed xenobiotics and endogenous waste products, such as bilirubin, as well as bile acids and biliary lipids, are secreted in bile. Of the total solutes in bile, bile acids constitute 67 %; phospholipids, 22 %; cholesterol, 4 %; proteins, 4.5 %; and bilirubin, 0.3 %. Bile acids are **amphiphilic** molecules, that is, one side of the molecule is *hydrophilic* and the other is *hydrophobic*. This property of bile acids allows them to act as **detergents** in the solubilization of the products of lipid digestion to facilitate intestinal absorption. In an aqueous environment, bile acids are in a **monomeric** form until they reach their *critical micellar concentration*, approximately 1-5 mM, when they aggregate as **micelles** and incorporate cholesterol and phospholipid (Fig. 12.8). Once in the gut lumen,

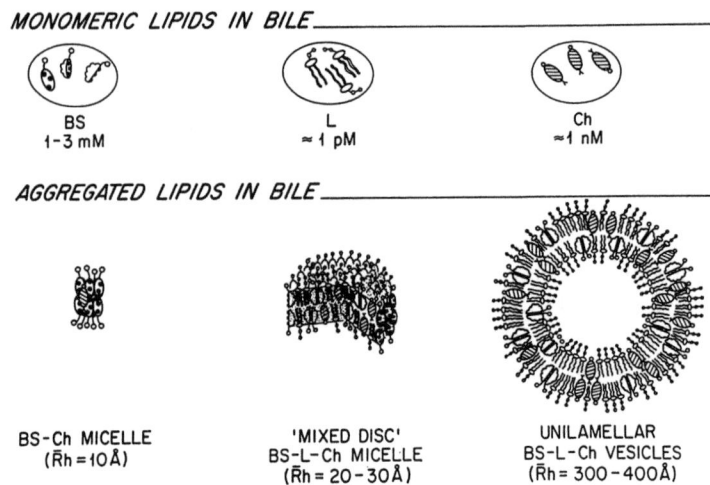

Fig. 12.8 Molecular models of the major biliary lipid molecules with depictions of the average sizes and structures of simple and mixed micelles found in human bile. BS, bile salt; L, lecithin; Ch, cholesterol; Rh, mean hydrodynamic radius; Å, Ångstrom unit. Monomeric solubilities are given in SI units. With typical gallbladder lipid compositions, simple bile salt and mixed bile salt-lecithin micelles coexist in a ratio of about 1:5. The site of attachment for cholesterol molecules on simple micelles is on the exterior (hydrophilic) surface. In mixed micelles cholesterol is solubilized within the micelles. The unilamellar vesicles/micelles ratio depends on the bile salt and cholesterol content of the bile, being greatest in bile with low bile salt and high and high cholesterol content (Reproduced with permission from Carey and Duane [4])

products of dietary lipid digestion, including monoacylglycerols, long-chain free fatty acids, lysophospholipids, and fat-soluble vitamins, are incorporated to form mixed micelles to allow these products to efficiently traverse the diffusion barrier of the unstirred water layer and gain access to the enterocyte microvillus membrane for absorption.

In addition to their role in lipid absorption, bile acids have other functions as well. They participate in **cholesterol balance**, since bile acids are synthesized from cholesterol in a tightly regulated fashion. Although the focus of this chapter will be the role of bile acids as regulatory molecules for their synthesis in liver and their enterohepatic circulation, bile acids are also important signaling molecules in the regulation of **gut motility** and **carbohydrate metabolism**. Bile acid mixed micelles have an impact on the gut microbiome through **bacteriostatic action** in the *proximal intestine* and **induction of genes** in the *distal intestine* that inhibit microbial growth. Proper regulation of the composition of gallbladder bile, including the concentration of bile acids, prevents the formation of gallstones.

The **primary bile acids** synthesized in the liver are **cholic acid** and **chenodeoxycholic acid**. They are synthesized from cholesterol as shown in Fig. 12.9. The *rate-limiting step* in bile acid synthesis is catalyzed by the microsomal enzyme, **cholesterol 7α-hydroxylase** (**CYP7A1**). A minor (10 %) pathway in humans

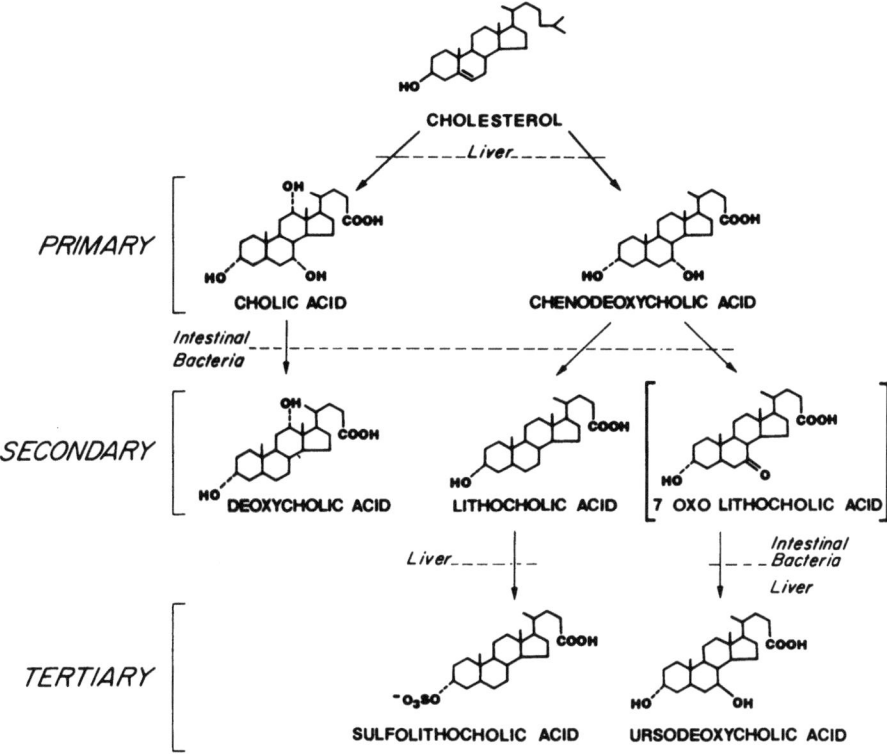

Fig. 12.9 Major primary, secondary, and tertiary (modified secondary) bile salts of humans with sites of synthesis and metabolism (Reproduced with permission from Carey and Duane [5])

involves sterol **12α-hydroxylase** (**CYP8B1**, not shown) that regulates the amount of cholic acid synthesized and thereby *determines the ratio of cholic acid to chenodeoxycholic acid*, which modulates the composition and hydrophobicity of the bile acid pool. Prior to canalicular secretion, bile acids are conjugated with either **glycine** or **taurine**, which lowers their pKa resulting in more effective lipid solubilization in the duodenum. **Taurine conjugates** predominate in infants, and **glycine conjugates** are the most abundant later in life. In addition to the primary bile acids, cholic acid and chenodeoxycholic acid, which are synthesized by the hepatocyte, there are secondary and tertiary bile acids in the bile acid pool (Fig. 12.9). **Secondary bile acids** are produced by biotransformation of primary bile acids by intestinal bacteria. *Cholic acid* is converted to **deoxycholic acid**, *chenodeoxycholic acid* is converted to both **lithocholic acid** and **7-oxo-lithocholic acid**. **Tertiary bile acids** are formed by both the liver and intestinal bacteria from secondary bile acids. *Lithocholic acid* is sulfated to form **sulfolithocholic acid**, and *7-oxo-lithocholic acid* is converted to **ursodeoxycholic acid**. Sulfolithocholic acid is poorly absorbed and rapidly lost from the enterohepatic circulation.

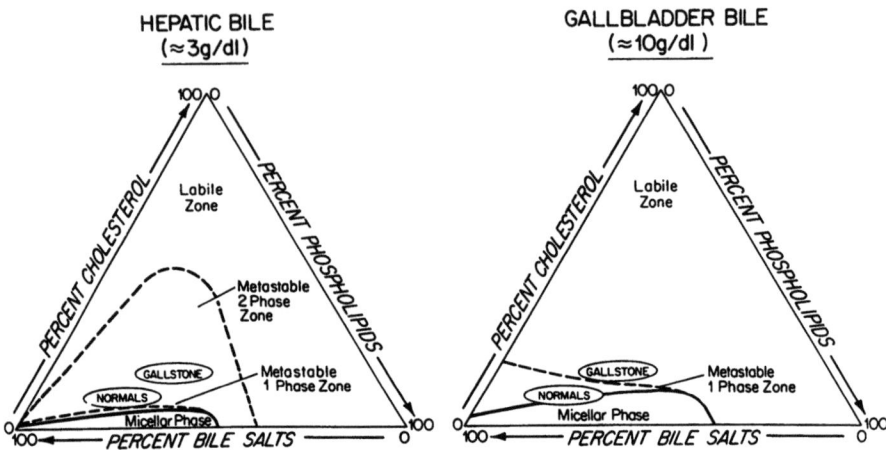

Fig. 12.10 Relative lipid compositions as functions of micellar, one-phase metastable (supersaturated micelles), and two-phase metastable (supersaturated micelles plus vesicles) zones of human hepatic and gallbladder biles (Reproduced with permission from Carey and Duane [6])

After canalicular secretion, bile undergoes modification as it passes through the ductular system. Biliary tract epithelium is capable of absorption and secretion of electrolytes and water by active and passive processes. The hormone **secretin** is capable of stimulating secretion by bile ducts and ductules. The gallbladder concentrates bile five to tenfold by active absorption of cations and chloride/bicarbonate ions coupled with passive water movement. During fasting the gallbladder stores and concentrates bile. Post-prandially, the gallbladder contracts in response to the release of **cholecystokinin** and empties its contents into the duodenum. The relative percentages of bile acids, cholesterol, and phospholipid are normally maintained in a relatively narrow range in bile (Fig. 12.10). If this balance is disturbed such that the percentage of cholesterol exceeds this range, cholesterol gallstones may form in the gallbladder, where bile is normally concentrated and there is relative stasis. Also, in conditions characterized by increased bilirubin production, such as hemolytic anemias, there is a propensity to form bilirubin gallstones.

Bile acids undergo an **enterohepatic circulation** as summarized in Fig. 12.11. After participation in lipid absorption, bile acids are actively absorbed in the distal small intestine via an **ileal high-affinity transporter**. A small fraction of bile acids may be passively absorbed in the proximal small intestine, particularly under conditions of abnormally low intraluminal pH. The absorbed bile acids are secreted into portal blood where they are bound to albumin and lipoproteins. There is greater than 90 % extraction of the bile acids from the portal blood on the first pass through the liver via a **high-affinity hepatocyte membrane transporter** (Fig. 12.11). Therefore, under normal conditions, the concentration of bile acids in peripheral blood is extremely low. These bile acids may be re-secreted into bile and make repeated passes through the enterohepatic circulation. In adults, it is estimated that

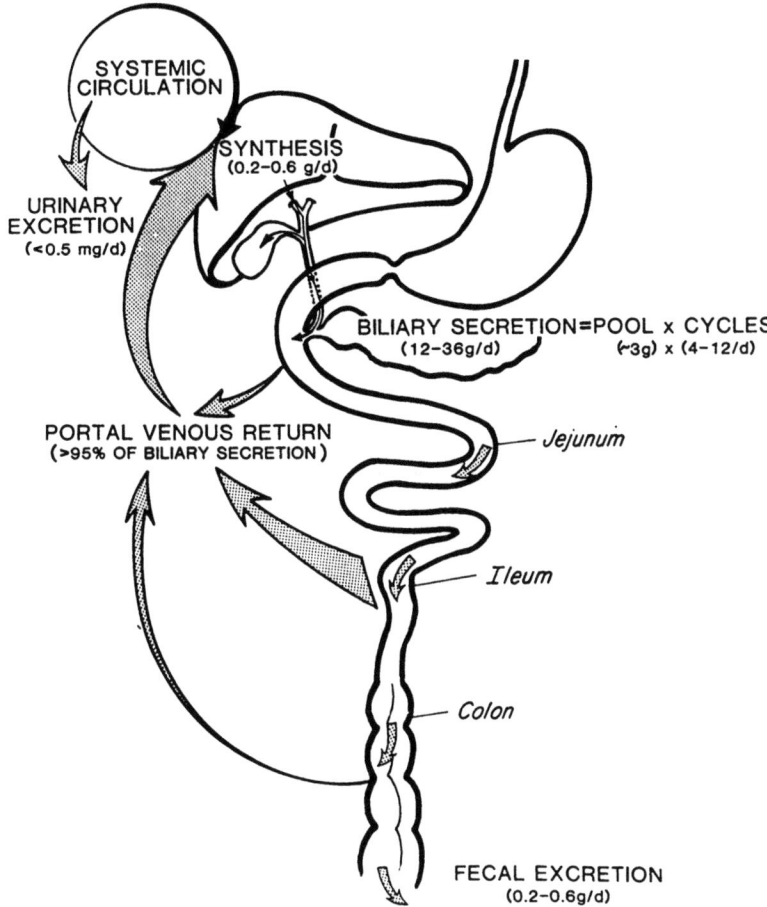

Fig. 12.11 Enterohepatic circulation of bile salts showing typical kinetic values for healthy human adults. (Reproduced with permission from Carey and Duane [5])

the bile acid pool circulates 8–10 times per day and approximately two times with each meal. Conservation of bile acids during this cycling is highly efficient, and only small amounts of the bile acid pool (400–600 mg/day) are lost each day into stool. This small amount is important, however, since this represents the *only* route by which cholesterol (the precursor of bile acid synthesis) is eliminated from the body. This loss is compensated for by an increase in hepatic synthesis.

The various transporters involved in hepatocyte basolateral bile acid uptake and canalicular secretion, modification of bile by the bile ducts, and ileal uptake of bile acids have been identified (Fig. 12.12). Their nomenclature, function and associated diseases caused by genetic mutations are shown in Table 12.1. At the *canalicular membrane*, conjugated bile acids are secreted by **BSEP**. The other major components of bile, phosphatidylcholine and cholesterol, are secreted via the **phospholipid**

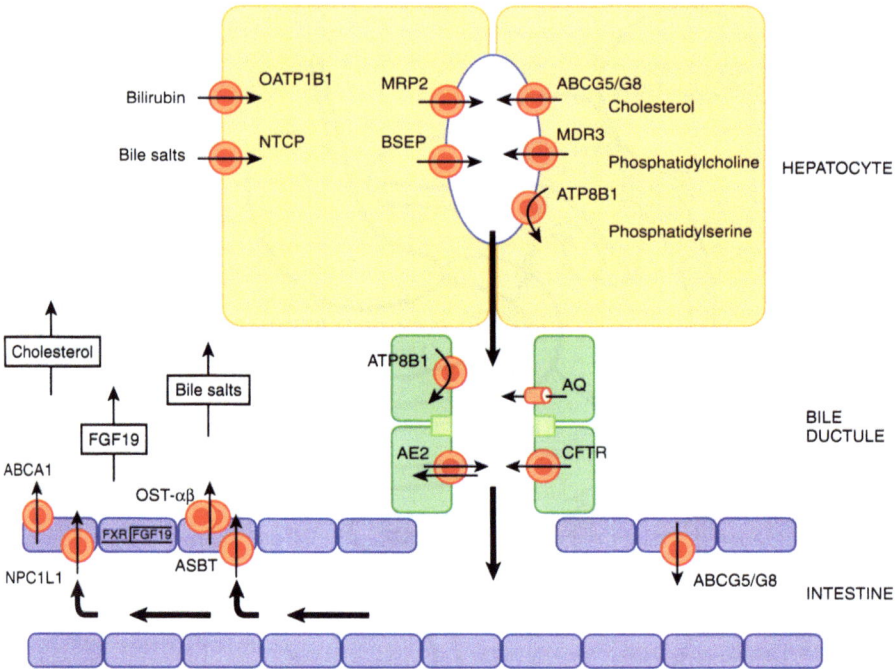

Fig. 12.12 Human liver transporter proteins involved in bile formation. Transporter proteins located in the basolateral membrane are responsible for the uptake of bile acids (NTCP), bulky organic anions, uncharged compounds (OATPs), and cations (OATPs, OCT1). Transporter proteins located in the canalicular membrane are responsible for the biliary secretion of compounds such as bile acids, phosphatidylcholine, cholesterol, bilirubin conjugates, and oxidized and reduced glutathione. These transporter proteins comprise the bile salt transporter, BSEP, the phosphatidylcholine translocator, MDR3, the anionic conjugate transporter, MRP2, and the multidrug transporter, MDR1 (not shown). The organic anion transporters, MRP3, MRP4, and OST-α/β, are present at very low levels in normal hepatocytes but are up-regulated during cholestasis. ABCG5/G8 are two half-transporters (half the molecular mass of regular ABC transporters) and as a heterodimer function as cholesterol and plant sterol transporters. Fibroblast growth factor 19 (FGF19) is synthesized in the terminal ileum upon binding of bile acids to FXR. Niemann-Pick C1-Like 1 (NPC1L1) protein is the intestinal cholesterol transporter. Both bile acids and cholesterol participate in an enterohepatic cycle (Reproduced with permission from Elsevier [12])

flippase, **MDR3**, and the **heterodimer**, **ABCG5/ABCG8**, respectively. Anionic conjugates, including conjugated bilirubin, are secreted by **MRP2**. The gene ATP8B1 encodes a protein termed **FIC1**, an aminophospholipid transporter, that is defective in genetic cholestasis syndromes. The electrolyte and water content in bile is modified in the bile ducts after secretion through multiple transport processes including *water movement* via **aquaporin (AQ)**, *chloride secretion* by **CFTR** and *chloride/bicarbonate exchange* by **AE2**. After participation in solubilization of the products of lipid digestion, bile acids are actively taken up by *the luminal membrane of ileal enterocytes* by **ASBT** and secreted from the *basolateral* aspect

Table 12.1 Human liver transport proteins

Name	Gene	Localization	Transport function	Phenotype when defective
NTCP	SLC10A1	H-BL	BS	
ASBT	SLC10A2	CH-A, E-A	BS	Bile salt diarrhea
OCT1	SLC22A1	H-BL, E-BL	OC	
OAT2	SLC22A7	H-BL	OA	
OATP	SLC21A3	H-BL	BS, OA, OC	
OATP2	SLC21A6	H-BL	B, BS, OA, OC	
OATP8	SLC21A8	H-BL	BS, OA, OC	
OATP-B	SLC21A9	H-BL	OA	
OATP-1B1	SLC1B1	H-BL	OA	Rotor syndrome
OATP-1B3	SLC1B3	H-BL	OA	Rotor syndrome
ABCA1	ABCA1	H-BL, E-BL	Phospholipid	High-density lipoprotein deficiency, Tangier type
MDR1	ABCB1	H-A, CH-A	Drugs, chemotherapeutics	
MDR3	ABCB4	H-A	PC	PFIC3, ICP, intrahepatic gallstones
BSEP	ABCB11	H-A	BS	PFIC2, BRIC2
MRP1	ABCC1	H-BL, CH-BL	OA, OA conjugates	
MRP2	ABCC2	H-A, E-A	OS, OA conjugates	Dubin-Johnson syndrome
MRP3	ABCC3	H-BL, CH-BL	OA conjugates	
MRP4	ABCC4	H-BL	BS sulfates	
MRP6	ABCC6	H-BL	Peptides, endothelin receptor antagonist BQ123	Pseudoxanthomaelasticum
CFTR	ABCC7	CH-A	Chloride	Cystic fibrosis
ABCG2	ABCG2	H-A	Chemotherapeutics, chlorophyll metabolites, protoporphyrin	Photosensitivity (mice)
ABCG5	ABCG5	H-A E-A	Plant sterols	Sitosterolemia
ABCG8	ABCG8	H-A, E-A	Plant sterols	Sitosterolemia
FIC1	ATP8B1	H-A, CH-A, E-A	Aminophospholipid translocation	PFIC1, BRIC1, ICP
WND	ATP7B	H-INT	Copper	Wilson disease
NPC1L1	NPC1L1	E-A, H-A	Cholesterol	Inhibited by ezetimibe

Adapted from Jansen et al. [12]
A apical, *B* bilirubin, *BL* basolateral, *BS* bile salts, *CH* cholangiocytes, *E* enterocytes, *H* hepatocytes, *ICP* intrahepatic cholestasis of pregnancy, *INT* intracellular, *MDR* multidrug resistance, *OA* organic anions, *OC* organic cations, *PC* phosphatidylcholine, *PFIC* progressive familial intrahepatic cholestasis

of the cell by **OST-αβ**. The bile acids then enter the portal venous blood and are transported back to the liver, where the majority are extracted on first pass by the **hepatocyte basolateral membrane transporter**, **NCTP**, and may again participate in the enterohepatic circulation. As indicated in Fig. 12.12, biliary cholesterol also undergoes **enterohepatic recycling**.

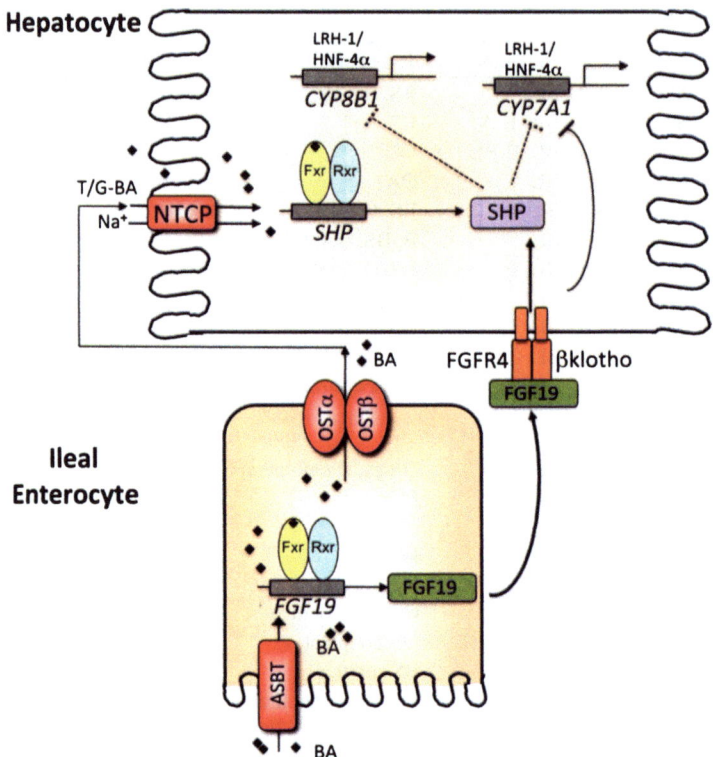

Fig. 12.13 Regulation of bile acid synthesis via the enterohepatic circulation. In the liver, bile acids activate FXR, which induces SHP expression. SHP then inhibits LRH-1 and HNF-4α activation of CYP7A1 and CYP8B1. In the major physiological pathway, intestinal bile acids are taken up by the ASBT and activate FXR to induce FGF19 expression. FGF19 is then carried to the liver where it signals via a receptor tyrosine kinase, the FGFR4:βKlotho complex, and the MAPK/ERK1/2 pathway to inhibit CYP7A1 transcription. At the liver, FGF19 may also signal via FGFR4:βKlotho to stabilize SHP protein and further reduce CYP7A1 expression (Reproduced with permission from Dawson [9])

Integration and regulation of all aspects of the enterohepatic circulation are accomplished via nuclear receptors that modulate target gene transcription (Fig. 12.13). Ileal bile acid uptake by ASBT activates **farnesoid X receptor (FXR)** that induces expression of **fibroblast growth factor 19 (FGF19)**. Then FGF19 is transported to the liver via the portal blood and binds to a **basolateral hepatocyte membrane receptor (FGFR4/βKlotho complex)** and down-regulates bile acid synthesis through MAPK/ERK1/2 signaling and stabilization of **small heterodimer protein (SHP)** to suppress activation of transcription of the CYP7A1 gene **by liver receptor homolog-1 (LRH-1)** and **hepatocyte nuclear factor-α (HNF-4 α)**. Also, intracellular bile acids, including those extracted from portal blood, may also inhibit synthesis via FXR activation, which induces SHP expression to down-regulate CYP7A1. CYP8B1 is also regulated via SHP.

The newborn infant, especially those born prematurely, have developmental immaturity at several steps in the enterohepatic cycling of bile acids. There is **reduced hepatocyte bile acid synthesis, conjugation, uptake, and secretion**. Also, **ileal uptake of bile acids** is reduced and not demonstrable until 3 weeks of age. These factors lead to a reduced bile acid pool size in the newborn. Defective hepatic bile acid uptake and secretion lead to **slightly elevated serum bile acids** and *"physiologic cholestasis of infancy."* These deficiencies also lead to decreased intraluminal bile acid concentrations and transient fat malabsorption in the newborn, which is even more striking in the premature infant.

7.1 Disease States Characterized by Disturbances in Bile Acid Metabolism and Enterohepatic Circulation

Cholestasis

Cholestasis is defined as bile excretory failure and may be the result of a variety of disorders. Cholestasis may be broadly classified as **intrahepatic (or hepatocellular)** or **extrahepatic (or obstructive)**. **Examples of intrahepatic cholestasis** would include certain types of *drug-induced liver injury*, *inherited cholestatic syndromes* such as *Alagille syndrome*, and *idiopathic neonatal hepatitis*. **Gram negative endotoxin** inhibits bile acid transporters and may cause cholestasis during sepsis or even in infants with gram negative bacterial urinary tract infections. **Examples of extrahepatic cholestasis** include *bile duct obstruction* by gallstones, *choledochal cyst*, *primary sclerosing cholangitis*, and *biliary atresia* in infants. In chronic cholestasis there is regurgitation of bile acids and biliary lipids into the bloodstream, which leads to several abnormalities. Accumulation of bile acids in the skin causes severe itching or **pruritis**, which can be quite debilitating. As previously described, lipoprotein metabolism is perturbed, and hyperlipidemia may be quite striking, particularly cholesterol levels. An abnormal lipoprotein particle enriched in free cholesterol and phospholipid called **lipoprotein-X** accumulates in the circulation. *Cutaneous xanthomata* often result from the hyperlipidemia. Lack of adequate intraluminal bile acid concentrations results in dietary fat malabsorption, including fat-soluble vitamins. *Jaundice* and *dark urine* may be present due to conjugated hyperbilirubinemia. It is interesting that in several chronic cholestatic conditions, the serum bilirubin level may be normal or slightly elevated in the face of a markedly elevated serum bile acid level.

There is evidence that bile acids, which accumulate during cholestasis, particularly hydrophobic bile acids such as lithocholic acid, may perpetuate or exacerbate the underlying liver disease. During cholestasis associated with retention of bile acids and conjugated bilirubin within the hepatocyte, **basolateral OST-$\alpha\beta$** is up-regulated via **FXR** to transport potentially toxic bile acids out of the cell. Also, **basolateral MRP3** and **MRP4** are up-regulated by **pregnane X receptor** (**PXR**) to transport bilirubin conjugates and bile acids, respectively, out of the hepatocyte.

As would be expected, genetic defects in bile acid transporters lead to *clinical cholestatic syndromes* (Table 12.1). **Progressive familial intrahepatic cholestasis (PFIC)** comprises a group of these disorders that has led to identification of some of these transporters and better understanding of their function. Patients with PFIC are characterized by the presence of cholestasis early in life that is progressive usually leading to need for transplantation. Pruritis, often severe, and growth retardation are features.

Type 1 PFIC

Type 1 PFIC results from **mutations in the gene encoding FIC1 (ATP8B1)**, a protein that translocates aminophospholipid from the outer to inner leaflets of the canalicular membrane. Type 1 PFIC is also associated with *diarrhea* thought to be due to the expression of **defective FIC1** in the small intestine. This type of PFIC was formerly known as **Byler disease** after the Amish kindred in whom it was first described. Mutations in ATP8B1 also may result in a much milder phenotype, *benign recurrent intrahepatic cholestasis* (*BRIC*) characterized by recurrent episodes of mild to severe cholestasis with intervening asymptomatic periods.

Type 2 PFIC

Type 2 PFIC is due **to mutations in the bile salt export pump, BSEP (ABCB11)**, and is similar to type 1, but with more rapid progression to need for transplantation.

Type 3 PFIC

Type 3 PFIC is caused by **mutations in the gene for MDR3 (ABCB4)**, a canalicular phospholipid flippase responsible for secretion of phospholipid into bile. The resultant abnormal bile devoid of phospholipid permits bile acids to cause injury to hepatocytes and cholangiocytes resulting in *cholestasis*. Progression may be variable depending on the specific mutation, although chronic cholestasis, marked fibrosis and liver failure usually result. Mutations in ABCB4 may also be related to *intrahepatic cholestasis of pregnancy* and *gallstone formation*. A subset of patients with CFTR mutations and cystic fibrosis also develop a form of liver disease called *focal biliary cirrhosis*, and some mutations in this gene may be responsible for the development of *primary sclerosing cholangitis* in the absence of the classic phenotype of cystic fibrosis.

7.2 Treatment of Cholestasis due to Bile Transporter Mutations

Patients with cholestasis due to bile transporter mutations represent challenges to management of their disease. **Oral administration of ursodeoxycholic acid**, a non-toxic, water-soluble bile acid, may result in improvement through its ability to increase bile flow and displace toxic bile acids in the bile acid pool. This bile acid also has *anti-inflammatory* and *anti-apoptotic* properties. It is a minor tertiary bile acid in humans, but is the major bile acid in bears; thus the name. Another approach is to **orally administer a bile acid binding resin**, such as *cholestyramine*, or perform a **surgical external biliary diversion**. Both of these measures chronically eliminate bile acids from the body and reduce the size of the bile acid pool. However, response to both of these treatments is variable, and compliance with bile acid sequestrants is often poor and biliary diversion may be associated with complications.

Ileal Dysfunction or Resection

Individuals with inflammatory disease of the ileum, such as *Crohn's disease*, or an ileal resection, as in infants requiring resection secondary to necrotizing enterocolitis, have defective ileal bile acid uptake and bile acid malabsorption. Bile acids may enter the colon and cause a *secretory diarrhea* mediated by a **cAMP-dependent mechanism**. With increased fecal loss of bile acids, there is a compensatory increase in bile acid synthesis, and the bile acid pool size is maintained. These patients may be treated with **oral administration of bile acid binding resins**, which bind the malabsorbed bile acids to allow elimination in stool and prevent the associated diarrhea. If the inflammatory process is very severe or the amount of resected ileum is extensive, the bile acid loss may exceed the capacity of the liver to increase synthesis to compensate for the loss. In this situation there is a reduction is the bile acid pool size and resultant reduction in the intraluminal bile acid concentration, often to below the critical micellar concentration. This results in *malabsorption* of *dietary fat* and *fat-soluble vitamins*.

Intestinal Blind Loop Syndrome

In conditions that result in *stasis of proximal intestinal contents* such as blind loop syndrome, **bacterial overgrowth** may occur resulting in deconjugation of bile acids by primarily anaerobic bacteria. Unconjugated bile acids are readily passively absorbed leading to a deficient intraluminal bile acid concentration and fat malabsorption. These unconjugated bile acids may also exert a toxic effect on the intestinal mucosa, leading to exacerbation of malabsorption and diarrhea. **Oral administration of broad-spectrum non-absorbable antibiotics** may result in clinical improvement.

Clinical Correlations

Case Study 1

A 22-year-old male comes to your office with the complaint of noticing "yellow eyes" intermittently throughout his life, generally associated with an acute febrile illness with decreased appetite and food intake. Apparently, blood tests had been performed in childhood, and his parents had been told that he had "mild hepatitis", since the only abnormality reported was an elevated bilirubin. He otherwise has been healthy with normal growth and development and no other complaints. There is no history of blood exposure or exposure to anyone with hepatitis. Family history is significant for similar symptoms in his father who was killed in a traffic accident at age 32 years. The present problem prompting a visit to you is the patient's inability to obtain health insurance coverage for a pre-existing "recurrent hepatitis". The physical examination is normal. No jaundice is noted. There is no hepatosplenomegaly or ascites. There are no findings to suggest chronic liver disease or cirrhosis.

Laboratory studies: Hematocrit 46 %; hemoglobin 15.1 g/dL; reticulocyte count 1.1 % (0.8–2.5); white blood cell count 7,900 cells/mm^3 (4,500–11,000); 220,000 platelets/mm^3 (150,000–350,000); prothrombin time 12.2 s (control 12.4 s); albumin 4.9 g/dL (3.7–5.6); total bilirubin 1.8 mg/dL (0.1–1.0); direct bilirubin 0.2 mg/dL (0–0.2); ALT 22 IU/L (10–30), AST 24 IU/L (15–30); GGTP 21 IU/L (0–25); alkaline phosphatase 61 IU/L (30–100); urinalysis is negative for bilirubin and shows normal urobilinogen.

Questions:

1. **What is this patient's most likely diagnosis?**
   Answer: The presence of a ***non-hemolytic*** (normal Hgb, reticulocyte count, and urine urobilinogen) ***unconjugated hyperbilirubinemia*** in an adult, which is exaggerated with the occurrence of an acute illness, would strongly suggest the diagnosis of the **Gilbert syndrome**. This is especially true with the positive family history, since there appears to be autosomal dominant inheritance.
2. **Why is the patient not icteric now?**
   Answer: Even though individuals with the Gilbert syndrome have an exaggeration of their indirect hyperbilirubinemia with acute illness or fasting, the baseline level is generally above normal, as in this individual. However, adults do not generally demonstrate visible jaundice until the serum bilirubin level reaches 2.0 mg/dL. Fasting or illness is thought to elevate the bilirubin level further by reducing the availability of glucose for formation of UDP-glucuronic acid by UDP-glucose dehydrogenase, thereby reducing bilirubin conjugation. However, the precise mechanism is not known.
3. **How would you confirm the diagnosis?**
   Answer: The Gilbert syndrome is generally considered a diagnosis of exclusion with mild unconjugated hyperbilirubinemia in an otherwise healthy individual without evidence of hemolysis or structural liver disease. A past approach has

involved measuring bilirubin levels in the parents after a 48-h fast to uncover the affected parent. However, this approach has not been shown to be reliable. In this case, the presumably affected parent is deceased. In most situations, the present evaluation would be adequate. However, the measurement of the ratio of monoglucuronide to diglucuronide bilirubin conjugates in duodenal bile may be confirmatory if an increased ratio of monoconjugates to diconjugates is found. This same finding is also present in the Crigler-Najjar syndrome, type II. Genetic testing is also available for the Gilbert syndrome mutation.

4. How would you counsel this patient?
   Answer: He should be reassured that the Gilbert syndrome is a completely **benign disorder** and should not affect his health or activity in any way.

Case Study 2

A 28-year-old female has a history of Crohn's disease involving the terminal ileum for the past 8 years. Six months ago she underwent resection of an unknown length of her terminal ileum secondary to intractable inflammation and enterocolic fistula formation. Her recovery from surgery was uneventful. She initially had an ileostomy, which was later reconnected. She has continued to have watery diarrhea with 6–7 large watery stools per day. No mucus or blood has been noted in the stool. Otherwise, she has been afebrile and has had no other symptoms. A contrast radiographic study of the area of the anastomosis shows no evidence of recurrent disease. A colonoscopy with ileoscopy also shows normal mucosa without evidence of recurrent Crohn's disease which is confirmed by biopsies. Physical examination is unremarkable except for well-healed abdominal surgical scars.

Laboratory studies: Hematocrit 30 %; hemoglobin 10.2 g/dL (RBC indices are macrocytic); reticulocyte count 0.4 % (0.8–2.5); white blood cell count 5,400 cells/mm^3 (4,500–11,000); 233, 000 platelets/mm^3 (150,000–350,000); erythrocyte sedimentation rate 4 mm/h (0–20); albumin 4.2 g/dL (3.7–5.6); serum iron 150 µg/dL (50–170); serum folate 2.2.ng/L (1.8–9.0); serum vitamin B_{12} 19 pg/mL (30–785); stool negative for white blood cells, occult blood, Giardia-specific antigen, Clostridium difficile toxin; stool culture negative for enteric pathogens.

Questions:

1. **What is the most likely cause of the diarrhea and how would you confirm it?**
   Answer: A recurrence of Crohn's disease proximal to the surgical anastomosis is ruled out by **radiographic studies** and **endoscopy with biopsy**. Additionally, the ESR and serum albumin are normal. **Stool studies and cultures** are negative and make an infectious etiology unlikely. The fact that the patient is vitamin B_{12}-deficient with associated megaloblastic anemia suggests that a significant amount of ileum, the site of absorption of this vitamin, was resected. Therefore, deficient ileal absorption of bile acids with a resulting colonic secretory diarrhea is a strong possibility. A therapeutic trial of ***oral cholestyramine*** (a bile acid binding resin) is instituted and results in resolution of the diarrhea.

2. **What is another complication of deficient ileal absorption of bile acids?**
   Answer: If the liver is unable to compensate for the chronic loss of bile acids from the bile acid pool by increased synthesis, **bile acid deficiency** will result. This deficiency can result in a luminal bile acid concentration below the critical micellar concentration with resultant **fat and fat-soluble vitamin malabsorption**.

Case Study 3

A 16-year-old white female is brought to the emergency room after admitting to her parents that she attempted suicide by ingesting an unknown number of acetaminophen tablets approximately 6 h ago. The patient is otherwise healthy, but does have a past history of depression and has been seeing a psychologist. Her physical examination is unremarkable. She is afebrile with normal vital signs. She is anicteric and in no distress. She is alert and oriented. Her abdomen is soft and non-distended. Her liver span is 10 cm by percussion without splenomegaly. Her neurological examination is normal without asterixis.

Laboratory studies: Hematocrit 38 %; hemoglobin 18.1 g/dL; white blood cell count 9,900 cells/mm^3 (4,500–11,000); 305, 000 platelets/mm^3 (150,000–350,000); prothrombin time 11.0 s (control 12.4 s); albumin 4.6 g/dL (3.7–5.6); total bilirubin 0.8 mg/dL (0.1–1.0); direct bilirubin 0.2 mg/dL (0–0.2); ALT 96 IU/L (10–30), AST 110 IU/L (15–30); GGTP 21 IU/L (6–29); alkaline phosphatase 230 IU/L (61–264); urinalysis is negative for bilirubin and shows normal urobilinogen; toxicology screen positive for fluoxetine, which the patient takes for depression, and acetaminophen.

Questions:

1. **What is the next step?**
   Answer: An acetaminophen level is measured and is 200 µg/mL 8 h post estimated ingestion. The **Rumack-Matthew nomogram** is consulted to determine the risk for significant hepatotoxicity and need for treatment based on serum level and time post-ingestion. The patient's level is in the range for toxicity, and treatment is indicated. The time frame for effective gastric lavage and administration of activated charcoal is past. **N-acetylcysteine** is administered by nasogastric tube per protocol that includes a loading dose and every 4 h dosing based on body mass for a total of 72 h. The patient is admitted with an **intravenous line for hydration**, and is closely monitored with vital sign and neurological checks, as well as serial laboratory testing.
2. **What is the expected clinical course?**
   Answer: The clinical course of acetaminophen intoxication may be divided into **four phases**:

   Phase 1, 30 min to 24 h after ingestion, during which there may be **minimal or no symptoms** and **subclinical rise in transaminases**.
   Phase 2, 18-72 h post-ingestion, during which patients may have **nausea** and **vomiting, right upper quadrant pain**, and **decreased urine output**.

Phase 3, *hepatic phase*, 72-96 h post-ingestion, is associated with **continued nausea** and **vomiting, abdominal pain** and **tender liver**. **Hepatic necrosis** may become obvious with *markedly elevated transaminases, jaundice, coagulopathy, hypoglycemia* and *hepatic encephalopathy*. **Acute renal failure** may occur, followed by **death from multiorgan failure. Transplantation** should be considered.

Phase 4, *recovery phase*, 4 days to 3 weeks post-ingestion, and patients who survive usually have **complete resolution of symptoms** and **hepatic recovery**.

In the case of this patient, her liver tests peaked at between 72 and 96 h post ingestion with AST 12,000 IU/L and ALT 10,700 IU/L, bilirubin total 3.7 mg/dL, direct 2.9 mg/dL, prothrombin time 15.5 s, and normal blood ammonia level. The patient experienced nausea, vomiting and abdominal pain, but remained alert with no signs of hepatic encephalopathy. She subsequently made an uneventful recovery.

3. **To what do you attribute the patient's recovery?**
   Answer: The features of this case that contributed to full recovery included a **young, otherwise healthy patient**, a **single acute ingestion**, and **rapid institution of N-acetylcysteine therapy** to restore hepatic glutathione levels to deplete toxic NAPQI levels. The **acetaminophen level** and **Rumack-Matthew nomogram** are useful in guiding therapy, but are not helpful if there is multiple chronic dosing or the time of ingestion is unknown. Studies of the use other **markers for potential hepatotoxicity**, such as serum levels of the degradation products of acetaminophen protein adducts, are ongoing and may prove useful for clinical management.

Further Reading

1. Ampola MG (1994) In: Arias IM, Boyer JL, Fausto N, Jakoby WB, Schachter DA, Shafritz DA (eds) The liver: biology and pathobiology, 3rd edn. Raven Press, New York, p 366
2. Blanckaert N, Fevery J (1990) In: Zakim D, Boyer TD (eds) Hepatology: a textbook of liver disease, 2nd edn. W. B. Saunders Co., Ltd, Philadelphia, p 263
3. Brites D (2012) The evolving landscape of neurotoxicity by unconjugated bilirubin: role of glial cells and inflammation. Front Pharmacol 3:88
4. Carey MC, Duane WC (1994) In: Arias IM, Boyer JL, Fausto N, Jakoby WB, Schachter DA, Shafritz DA (eds) The liver: biology and pathobiology, 3rd edn. Raven Press, New York, p 728
5. Carey MC, Duane WC (1994) In: Arias IM, Boyer JL, Fausto N, Jakoby WB, Schachter DA, Shafritz DA (eds) The liver: biology and pathobiology, 3rd edn. Raven Press, New York, p 722
6. Carey MC, Duane WC (1994) In: Arias IM, Boyer JL, Fausto N, Jakoby WB, Schachter DA, Shafritz DA (eds) The liver: biology and pathobiology, 3rd edn. Raven Press, New York, p 749
7. Carey EJ, Lindor KD (2012) Current pharmacotherapy for cholestatic liver disease. Expert Opin Pharmacother 13(17):2473–2484
8. Dawson PA (2011) Role of the intestinal bile acid transporters in bile acid and drug disposition. Handb Exp Pharmacol 201:169–203

9. Dawson PA (2012) In: Johnson LR (ed) Physiology of the gastrointestinal tract, 5th edn. Academic Press, New York, p 1467
10. Imam MH, Gossard AA, Sinakos E, Lindor KD (2012) Pathogenesis and management of pruritus in cholestatic liver disease. J Gastroenterol Hepatol 27(7):1150–1158
11. Jacquemin E (2012) Progressive familial intrahepatic cholestasis. Clin Res Hepatol Gastroenterol 36(Suppl 1):S26–S35
12. Jansen LMJ, Beuers U, Oude Elferink RPJ (2012) In: Boyer TD, Manns MP, Sanyal AJ (eds) Hepatology: a textbook of liver disease, 6th edn. Elsevier Saunders, Philadelphia, p 49
13. Montagnese S, De Pitta C, De Rui M et al (2013) Sleep-wake abnormalities in patients with cirrhosis. Hepatology 59:705–712
14. Odell GB (1980) Neonatal hyperbilirubinemia. Grune and Stratton, New York, p 5
15. Odell GB (1980) Neonatal hyperbilirubinemia. Grune and Stratton, New York, p 4
16. Poh Z, Chang PE (2012) A current review of the diagnostic and treatment strategies of hepatic encephalopathy. Int J Hepatol 2012:480309
17. Roy-Chowdhury J, Roy-Chowdhury N (2012) In: Boyer TD, Manns MP, Sanyal AJ (eds) Hepatology: a textbook of liver disease, 6th edn. Elsevier Saunders, Philadelphia, p 1087
18. Sticova E, Jirsa M (2013) New insights in bilirubin metabolism and their clinical implications. World J Gastroenterol: WJG 19(38):6398–6407
19. Trauner M, Baghdasaryan A, Claudel T et al (2011) Targeting nuclear bile acid receptors for liver disease. Dig Dis 29(1):98–102
20. Vessey DA (1990) In: Zakim D, Boyer TD (eds) Hepatology: a textbook of liver disease, 2nd edn. W. B. Saunders Co., Ltd, Philadelphia, p 228
21. Wagner M, Zollner G, Trauner M (2011) Nuclear receptors in liver disease. Hepatology 53(3):1023–1034

Part IV
Review Examination

Chapter 13
Multiple Choice Questions

Dennis D. Black, Eugene B. Chang, Po Sing Leung, and Michael D. Sitrin

Chapter 1

1. Guanylin is a recently identified gut hormone produced by the ileum and colon. Which of the following statement regarding this peptide regulator is correct?

    (a) It stimulates a specific G-kinase receptor, resulting in increases in epithelial cGMP and net secretion.
    (b) It is secreted by antral G-cell and activates gastric secretion.
    (c) It is made by intestinal smooth muscle cells.
    (d) It increases cGMP, inhibiting net secretion.
    (e) It activates soluble guanylate cyclase.

2. Cholera toxin's effects are mediated by:

    (a) Increases in cytosolic calcium.
    (b) ADP-ribosylation of the G_s alpha subunit, rendering it persistently active.
    (c) A membrane guanylate cyclase receptor.

(d) A paracrine effect.
(e) Mucosal invasion and destruction.

3. Gastrin stimulates acid secretion by:

   (a) A paracrine effect
   (b) A juxtacrine action
   (c) A neurocrine action
   (d) An endocrine action
   (e) An autocrine action

4. Which of the following gut hormones is not involved in the regulation of the initial phase of digestion?

   (a) Guanylin
   (b) Secretin
   (c) Gastrin
   (d) Cholecystokinin
   (e) Gastric inhibitory polypeptide (GIP)

5. Which of the following statements is correct?

   (a) Secretin is made and secreted by ileal endocrine mucosal cells.
   (b) CCK is secreted by the stomach.
   (c) Gastrin is made by antral G-cells.
   (d) GIP is produced by the liver.
   (e) GIP stimulates gastric acid secretion.

6. All of the following belong to gut peptide hormones, except:

   (a) Somatostatin
   (b) Gastrin
   (c) Substance P
   (d) Enkephalins
   (e) Nitric oxide

7. Which of the following statements is incorrect?

   (a) Intestinal functions are only possible through inputs from vagal or spinal neurons.
   (b) Enteric neurons are organized into networks that can produce programmed responses.
   (c) Enteric neurons integrate gut function.
   (d) Peptide neurotransmitters are made by enteric neurons.
   (e) The intestine is one of the most innervated organ systems in the body.

8. Mark one of the following parings of gut endocrine cells with its correct secretions:

   (a) S cell – Somatostatin.
   (b) G cell – Secretin.

(c) I cell – Gastrin.
(d) D cell – Cholecystokinin.
(e) L cell – Glucagon like-peptide-1.

9. Which of the following statements is most likely to best describe the feature of enteric nervous system?

    (a) It works independently of the autonomous nervous system.
    (b) It does not have motor neurons.
    (c) It contains only cholinergic nerves.
    (d) Its myenteric nerves control gut secretion and local blood flow.
    (e) Its submucosal nerves control gut peristalsis and segmentation.

10. Select one of the following gut peptide hormones that best pair with its respective action:

    (a) Gastric inhibitory peptide – Inhibits pancreatic ductal secretion.
    (b) Cholecystokinin – Inhibits pancreatic enzyme secretion.
    (c) Gastrin – Inhibit intestinal electrolyte secretion.
    (d) Secretin – Inhibits gastric acid secretion.
    (e) Motilin – Inhibits bile secretion.

11. Regarding the regulation of gut secretion and motility, there are diverse peptides that act as endocrine, paracrine and neurocrine regulators. Which of the following is a paracrine regulator?

    (a) Gastrin.
    (b) Secretin.
    (c) Motilin.
    (d) Somatostatin.
    (e) Gastrin releasing peptide.

12. Migrating motor complex is an interdigestive gut contraction which propels food residue and controls bacterial overgrowth in our gastrointestinal tract. Which of the following peptide hormone is responsible for this gut motility?

    (a) Motilin.
    (b) Cholecystokinin.
    (c) Gastrin releasing peptide.
    (d) Gastrin inhibitory peptide.
    (e) Vasoactive intestinal peptide.

Answers: (1) a; (2) b; (3) d; (4) a; (5) c; (6) e; (7) a; (8) e; (9) a; (10) d; (11) d; (12) a

Chapter 2

1. The following are true of neural innervation of the intestine except:

    (a) Preganglionic sympathetic nerves from the spinal cord synapse at paravertebral ganglia, from which post-ganglionic fibers project to the intestine.

(b) Post-ganglionic sympathetic fibers project to the intestine following the celiac, superior mesenteric, and inferior mesenteric arteries in their respective distributions.
(c) The neurons of the enteric nervous system project from the spinal cord.
(d) The enteric nervous system has "hard-wired" circuits which produce patterned responses.
(e) The enteric nervous system has inhibitory, stimulatory, and connecting neurons.

2. All of the following statements are true regarding the basic electrical rhythm except:

(a) The depolarization phase is due to opening of voltage-sensitive K channels.
(b) BER frequency is not affected by neurohumoral stimulation.
(c) BER are not always associated with motor contraction.
(d) BER originate from gut pacemakers found throughout the GI tract.
(e) Spike potentials during depolarizing phases of the BER determine motor contraction.

3. Which of the following agents inhibits intestinal smooth muscle contraction?

(a) Acetylcholine
(b) Serotonin
(c) CCK
(d) Gastrin
(e) Somatostatin

4. Which of the following statements concerning gut smooth muscle movement is incorrect?

(a) Segmentation mixes intestinal contents.
(b) Tonic contractions characterize intestinal sphincters.
(c) Peristalsis is primarily found in the esophagus and stomach.
(d) Increases in smooth muscle cAMP stimulate contraction.
(e) Secondary peristalsis in the esophagus is initiated by luminal distention.

5. Migrating motor complexes are characterized by the following statements except:

(a) Cyclic periodicity of approximately 90 min.
(b) Stimulated by meals.
(c) Wavelike contraction starting from stomach to terminal ileum.
(d) Interdigestive motor contractions that sweep the bowel of luminal contents.
(e) Motilin may play a role in initiating MMCs.

6. Which of the following statements concerning the colonic motor functions is incorrect?

(a) Longitudinal muscles are continuous sheaths enveloping the colonic wall.
(b) Segmentations and haustrations are the major motor activities of the small and large bowels.

(c) There is an increasing gradient of BER frequency from proximal to distal.
(d) Rectal and anal functions are distinct from colonic motor functions.
(e) The relaxation of the internal anal sphincter is involuntary.

7. Gastric emptying is regulated at a rate optimal for digestion and absorption of a meal. Which one of the following statements concerning factors that will increase the rate of gastric emptying of a meal?

   (a) Increasing the fat content of the meal.
   (b) Increasing the size of food particles of the meal.
   (c) Increasing the hypotonicity of the meal.
   (d) Increasing the volume of the meal.
   (e) Increasing the acidity of the meal.

8. Slow waves are the intrinsic and spontaneous electrical activity of the gut smooth muscle. Which of the following statements concerning slow wave activity in small intestinal smooth muscle is true?

   (a) It directly triggers muscle contractions.
   (b) It always contains spike or action potentials.
   (c) It has a lower frequency in proximal compared to distal regions of the gastrointestinal tract.
   (d) It occurs at approximately 90-min intervals.
   (e) It sets the maximum frequency of small intestinal contractions.

9. Which of the following statements concerning propulsion of colonic contents into the rectum is true?

   (a) It is caused by migrating motor complex.
   (b) It causes the internal anal sphincter to relax.
   (c) It causes involuntary contraction of the external anal sphincter.
   (d) It is under voluntary control.
   (e) It initiates defecation.

10. There are two major gastric functions, namely gastric motility and secretion. Which of the following statements concerning gastric function is true?

    (a) Gastric emptying is stimulated by cholecystokinin.
    (b) Gastric emptying is increased by the arrival of acid in the duodenum.
    (c) Acid secretion is stimulated by secretin.
    (d) Acid secretion is inhibited by gastrin.
    (e) Acid secretion is inhibited by hydrogen ions in the gastric lumen.

11. The function of lower esophageal sphincter (LES) is to allow food bolus entering into the stomach and prevent the reflux of gastric content back to the lower part of esophagus. Which of the following statements concerning LES is true?

    (a) LES is able to be anatomically identified as a structure.
    (b) Abnormally high LES pressure leads to gastro-esophageal reflux disease.

(c) There is a positive pressure gradient between the abdomen and thorax that tends to promote reflux of gastric contents into the esophagus.
(d) LES tone is decreased by acetylcholine.
(e) LES tone is increased by ethanol and smoking.

12. Which of the following statements concerning receptive relaxation of the stomach is true?

    (a) It is not affected by vagotomy.
    (b) It is triggered by relaxation of the lower esophageal sphincter.
    (c) It results from decreased contractile activity of antral smooth muscle.
    (d) It results in a low intragastric pressure during a meal.
    (e) It depends entirely on the enteric nervous system.

Answers: (1) c; (2) a; (3) e; (4) d; (5) b; (6) a; (7) d; (8) e; (9) b; (10) e; (11) c; (12) d

Chapter 3

1. The following statements are concerning about the gastric function. Which of the following is true?

   (a) The endocrine portion of the stomach is found in the body and fundus.
   (b) Mucous neck cells are predominantly responsible to produce bicarbonate.
   (c) Chief cells make and secrete acid.
   (d) Parietal cells are located in the upper part of gastric pits.
   (e) Mast cells regulate acid secretion through an endocrine pathway.

2. Mucus cells have all but one of the following properties:

   (a) Secretion of bicarbonate.
   (b) Mucin production.
   (c) Intrinsic factor production.
   (d) Protect the gastric mucosa against autodigestion.
   (e) Increased secretion following prostaglandin E stimulation.

3. Which one of the following gut hormones works through an endocrine pathway?

   (a) Somatostatin
   (b) Histamine
   (c) Bombesin
   (d) Gastrin
   (e) Guanylin

4. The proton pump, or H^+/K^+-ATPase, is the major mediator of gastric acid secretion. Which of the following statements is incorrect?

   (a) For the proton pump to function, its activity must be coupled to K^+ and Cl^- efflux.
   (b) The proton pump is an electroneutral transporter.

(c) Direct regulation of proton pump activity determines the rate of gastric acid secretion.
(d) The activation of the proton pump is associated with insertion of tubulovesicles into the canalicular membrane.
(e) Proton pump inhibitors covalently modify the proton pump.

5. The three most important physiological activators of gastric acid secretion are:

   (a) Secretin, GIP, and VIP.
   (b) Gastrin, GIP, histamine.
   (c) Acetylcholine, GIP, histamine.
   (d) Gastrin, serotonin, histamine.
   (e) Gastrin, acetylcholine, histamine.

6. Which one of the following statements is true?

   (a) The effects of acetylcholine and gastrin are synergistic.
   (b) Gastrin stimulates increases in cellular cAMP.
   (c) Histamine activates gastric acid secretion through type 1 histamine receptors.
   (d) Histamine potentiates the effects of gastrin and acetylcholine.
   (e) H_2-receptor blockade is not an effective therapy for peptic ulcer disease because the stimulatory actions of gastrin and acetylcholine remain unblocked.

7. The following statements concerning the cephalic phase are correct except:

   (a) It is abolished by vagotomy.
   (b) It can be activated by conditioned responses.
   (c) It is in part mediated by increased absorption of amino acids.
   (d) It is associated with increased gastric acid secretion and motility.
   (e) It accounts for 30–40 % of the meal-stimulated gastric acid secretion response.

8. Which of the following statements is correct?

   (a) Pepsin accelerates the conversion of pepsinogen to pepsin.
   (b) Pepsin primarily digests vegetable proteins.
   (c) Pepsinogen is converted to pepsin prior to secretion into the lumen of the gastric pit.
   (d) Proton pump inhibitors have no effect on the conversion of pepsinogen to pepsin.
   (e) Pepsinogen secretion does not parallel acid secretion.

9. Which of the following is the least important contributing factor for causing peptic ulcer disease?

   (a) *Helicobacter pylori*
   (b) NSAIDs

(c) Diabetes
(d) Smoking
(e) Gastrinoma

10. The following are true about *Helicobacter pylori* except:

    (a) The organism expresses urease, allowing it to survive the harsh milieu of the stomach.
    (b) *H. pylori* causes peptic disease in all patients it infects.
    (c) *H. pylori*-related ulcer disease is treated with antibiotics and acid suppression.
    (d) This organism causes the most common world-wide infection.
    (e) *H. pylori* is not an invasive organism.

11. Which of the following statements concerning acidification of the gastric content to pH 2 is true?

    (a) It will lead to decreased release of secretin.
    (b) It will lead to inhibition of acid secretion via a vago-vagal reflex.
    (c) It will lead to inhibition of acid secretion via release of somatostatin.
    (d) It will lead to decreased conversion of pepsinogen to pepsin.
    (e) It will lead to increased secretion of gastrin.

12. Which of the following statements concerning acidification of the duodenal lumen to pH 3 is true?

    (a) It will lead to inhibition of pancreatic bicarbonate secretion.
    (b) It will lead to inhibition of pepsinogen secretion.
    (c) It will lead to inhibition of pancreatic enzyme secretion.
    (d) It will lead to inhibition of bile production.
    (e) It will lead to inhibition of gastric emptying.

Answers: (1) d; (2) c; (3) d; (4) c; (5) e; (6) d; (7) c; (8) a; (9) c; (10) b; (11) c; (12) e

Chapter 4

1. One therapeutic target for treating chronic pancreatitis is to reduce stimuli of pancreatic functions, i.e. keep the pancreas in its resting state, as the release of digestive enzymes would injure the pancreas further. All but one of the following measures would be consistent with this therapeutic goal:

    (a) Fasting the patient.
    (b) Administration of octreotide, a long-lived and potent analog of somatostatin.
    (c) Oral administration of pancreatic proteases.
    (d) Placing the patient on a high protein diet.
    (e) Suppressing gastric acid secretion.

13 Multiple Choice Questions 335

2. Vagotomy or surgical interruption of the vagal nerve affects all of the following except one. Which one of the answers would not be affected by vagotomy?

   (a) Meal-stimulated acinar cell secretion.
   (b) CCK release from duodenal mucosal endocrine cells.
   (c) Bicarbonate response to acid load in the duodenum.
   (d) Enteropeptidase activation of trypsinogen.
   (e) Cephalic phase of meal-stimulated pancreatic functions.

3. At high flow rates of pancreatic secretion, all but one of the following statements is false?

   (a) The potassium concentration of pancreatic juice significantly increases.
   (b) Pancreatic juice is bicarbonate rich.
   (c) Increased admixture of ductular fluid changes the composition of pancreatic juice.
   (d) Na concentration of pancreatic juice remains unchanged.
   (e) Ductular cell CFTR is activated.

4. Which of the following transporters is not required for ductular bicarbonate secretion.

   (a) Luminal membrane Cl^-/HCO_3^- exchanger.
   (b) CFTR.
   (c) Basolateral Na^+/K^+-ATPase
   (d) Basolateral Na^+/H^+ exchanger
   (e) Apical or luminal membrane Na^+ channel

5. Which one of the following statements is false?

   (a) Increases in cytosolic calcium in acinar cells stimulate zymogen granule release.
   (b) Nitric oxide is an important mediator of acinar cell secretion.
   (c) CCK's major physiological effect in stimulating acinar cell secretion involves direct activation of basolateral CCK receptors.
   (d) Vagal nerves mediate CCK's stimulatory effects on acinar cells.
   (e) Secretin is the primary regulator of digestive enzyme secretion.

6. Which of the following is not a proteolytic enzyme?

   (a) Trypsin
   (b) Colipase
   (c) Elastase
   (d) Chymotrypsin
   (e) Carboxypeptidase

7. Which one of the following statements is not true?

   (a) Lipase is involved in fat digestion.
   (b) Endopeptidases cleave proteins at internal sites.

    (c) Amylase is important for digesting carbohydrates.
    (d) Nucleases digest RNA.
    (e) Proteolytic enzymes make up 20 % of all digestive enzymes secreted by the pancreas.

8. Regarding the combination of stimuli that will produce the highest rate of pancreatic bicarbonate secretion, which of the following statement is true?

    (a) Secretin plus histamine.
    (b) Cholecystokinin plus acetylcholine.
    (c) Secretin plus gastrin.
    (d) Cholecystokinin plus gastrin.
    (e) Secretin plus acetylcholine.

9. Regarding the pancreatic enzyme secretion, which of the following statements is true?

    (a) They are all secreted as inactive proenzymes or zymogens.
    (b) They are all secreted by pancreatic acinar cells.
    (c) They are synthesized in response to cholecystokinin.
    (d) They pass into the colon after digestion.
    (e) They are important for only protein digestion.

10. Regarding the salivary secretion, which of the following statements is true?

    (a) It is produced at high volumes relative to the mass of the glands.
    (b) It is usually hypertonic.
    (c) It is primarily regulated by hormones.
    (d) It is unaffected by the treatment with atropine.
    (e) It has a lower concentration of sodium as the rate of flow increases.

11. Regarding the functions of pancreas, which of the following statements is true?

    (a) Stimulation of the vagus nerve causes a predominantly watery secretion from the pancreas.
    (b) Secretion of cholecystokinin from the duodenum causes a watery secretion from the pancreas.
    (c) Pancreatic exocrine secretion contains enzymes which are essential in the digestion of fats.
    (d) Damage to the pancreas leading to fat malabsorption may result in deficiency of vitamin B_{12}.
    (e) The secretions of the pancreas have a pH of about 7.

12. Regarding the pancreatic exocrine secretion, which of the following statements is true?

    (a) Pancreatic enzyme secretion is largely controlled by the parasympathetic nervous system.
    (b) Pancreatic bicarbonate secretion is stimulated by gastrin.
    (c) Most secretion occurs during the gastric phase of digestion.

(d) Pancreatic amylase is not essential since amylase is also present in saliva.

(e) The intestinal mucosa produces an enzyme that activates pancreatic proteolytic enzymes.

Answers: (1) d; (2) d; (3) a; (4) e; (5) c; (6) b; (7) e; (8) e; (9) b; (10) a; (11) c; (12) e

Chapter 5

1. Which one of the following statements concerning intestinal function is true?

    (a) Mucosal permeability is greater in the colon than in the small intestine.
    (b) Paneth cells originate from the proliferative zone and migrate to villus tips.
    (c) The colon is the site of greatest fluid absorption.
    (d) Intestinal secretions make up the majority of the daily fluid load presented to the intestine.
    (e) Solidification of stool begins in the cecum.

2. All of the following statements are true except one. Which one is incorrect?

    (a) Villus cells are mature absorptive epithelial cells.
    (b) Mucosal permeability in the villus region is higher than in the crypt regions.
    (c) Brush-border hydrolases are predominantly expressed by villus cells.
    (d) Active anion secretion is primarily found in crypt cells.
    (e) The turnover rate of intestinal epithelial cells is 3–5 days in humans.

3. Which of the following transporters is not a carrier protein?

    (a) Facilitative Glut 2 transporter
    (b) Na^+/H^+ exchanger
    (c) $Na^+/K^+/2Cl^-$ co-transporter
    (d) SGLT1 (Na^+-glucose co-transporter)
    (e) Na^+/K^+-ATPase

4. Which of the following statements is incorrect?

    (a) Na^+-glucose cotransport is found throughout the GI tract.
    (b) Luminal membrane Na^+-H^+ exchangers are the major mediators of non-nutrient dependent intestinal Na absorption.
    (c) Na^+ channels are expressed in distal colon and rectum.
    (d) Cl^- secretion is found throughout the GI tract.
    (e) Na^+-coupled Cl^- absorption is found in ileum and colon.

5. Which of the following statements is incorrect?

    (a) Sodium glucose co-transporter (SGLT1) is inhibited by cholera toxin.
    (b) Starch based electrolyte solution are effective in the treatment of cholera.
    (c) Somatostatin promotes intestinal absorption.
    (d) Imbalances in homeostatic regulation of intestinal water and electrolyte transport can cause diarrhea.
    (e) Increases in intestinal epithelial cytosolic calcium stimulate net secretion.

6. All of the following transporters are required for active epithelial Cl⁻ secretion except:

    (a) CFTR
    (b) H^+/K^+-ATPase
    (c) Basolateral membrane K channels
    (d) $Na^+/K^+/2Cl^-$ cotransporter
    (e) Na^+/K^+-ATPase

7. Which of the following statements concerning short chain fatty acids (SCFAs) is correct?

    (a) SCFAs are primarily found in ingested foods and are absorbed in the small intestine.
    (b) SCFAs promote colonic proliferation and differentiation.
    (c) SCFAs stimulate colonic water absorption.
    (d) Antibiotic treatment would have no effect on the availability of SCFAs.
    (e) Glucose, not SCFAs, is the preferred metabolic substrate of colonocytes.

8. Which of the following statements concerning diarrhea is correct?

    (a) Cystic fibrosis patients are very susceptible to the effects of cholera toxin.
    (b) Diabetic diarrhea is caused by the development of cholinergic neuropathy.
    (c) Heat stable enterotoxin (ST_a) of *E. Coli* binds to GM_1 ganglioside receptors and increases cellular cGMP.
    (d) Octreotide, as a somatostatin analog, is effective in treating diabetic diarrhea.
    (e) Abnormal motility is the primary cause of most diarrheal diseases.

9. The major function of the small intestinal villus cell is digestion, absorption and secretion. Which of the following statements concerning the villus cells that line in the upper small intestine is correct?

    (a) They produce an enzyme that is able to break down starch.
    (b) They produce a hormone that activates trypsinogen.
    (c) They transport fructose by a sodium-independent mechanism.
    (d) They secrete hydrogen ions in exchange for potassium ions across the brush-border membrane.
    (e) They act as stem cells to replace the worn out epithelial cells of the top of the villi after 3–6 days.

10. Enterokinase is a membrane-bound enzyme which is anchored to the brush-border of the duodenal epithelium. Which of the following statements concerning the actions of enterokinase is true?

    (a) It is directly responsible for the activation of colipase.
    (b) It is directly responsible for the activation of trypsinogen.
    (c) It is directly responsible for the activation of amylase.

(d) It is directly responsible for the activation of cholecystokinin.
(e) It is directly responsible for the activation of pepsinogen.

11. Which of the following statements concerning the partial removal of the distal ileum is true?

    (a) It will increase the bile salt levels in the hepatic venous blood.
    (b) It will increase the bile salt levels in the portal vein.
    (c) It will increase the rate of bile acid secretion by hepatocytes.
    (d) It will increase the bile acid absorption across the distal ileum.
    (e) It will increase bile acid synthesis by hepatocytes.

12. Concerning intestinal solute and water transport, which of the following statements is true?

    (a) Peptide absorption is directly linked to sodium uptake.
    (b) Fructose absorption is directly linked to sodium uptake.
    (c) Potassium is secreted in the large intestine.
    (d) Sodium absorption is via passive transport throughout the small and large intestine.
    (e) Water absorption is via active transport throughout the small and large intestine.

Answers: (1) d; (2) b; (3) e; (4) a; (5) a; (6) b; (7) b; (8) d; (9) c; (10) b; (11) e; (12) c

Chapter 6

1. Which of the following statements regarding starch digestion is true?

    (a) Salivary and pancreatic amylases are the products of the same gene.
    (b) Amylase digests amylose, but not amylopectin.
    (c) Amylase cleaves α-1,6-linkages in amylopectin.
    (d) The major products of amylose digestion are maltose and maltotriose.
    (e) α-limit dextrins are products of amylose digestion.

2. The brush-border membrane enzyme that can hydrolyze the α-1,6 bond in an α-limit dextrin is:

    (a) Glucoamylase
    (b) Sucrase
    (c) Amylase
    (d) Lactase
    (e) Isomaltase

3. Which of the following is not a property of the glucose transporter SGLT1?

    (a) Two sodium ions enter the enterocyte for each molecule of glucose transported.
    (b) This absorptive process is electrogenic (net charge transfer across the membrane).

(c) Is the major brush-border membrane fructose transporter.
(d) Sodium increases the transporter affinity for glucose.
(e) The energy for glucose accumulation is derived from the transmembrane electrochemical Na^+ gradient.

4. The carrier for transporting glucose and other monosaccharides across the enterocyte basolateral membrane is:

    (a) SGLT1
    (b) Glut-2
    (c) Glut-5
    (d) Sucrase
    (e) Glucoamylase

5. Which of the following is a brush-border membrane fructose transporter?

    (a) SGLT1
    (b) Glut-5
    (c) PEPT1
    (d) Lactase
    (e) Galactose

6. Which of the following statements regarding human lactase is true?

    (a) Lactase levels are low in the neonatal small bowel.
    (b) Lactase is not the rate-limiting step in intestinal lactose absorption.
    (c) Children with congenital lactose deficiency have hypoglycemia as the major clinical symptom.
    (d) Persistence of lactase in the adult small intestine is a genetically determined trait.
    (e) Lactase is not synthesized in the intestine as a pro-protein.

7. Which of the following statements is true?

    (a) Fructose is better absorbed when given as sucrose rather than as an equivalent amount of the monosaccharide.
    (b) Patients with fructose malabsorption will have little rise in breath hydrogen after given fructose orally.
    (c) Transport of glucose by SGLT1 inhibits incorporation of Glut-2 into the enterocyte brush-border membrane.
    (d) Starches that resist efficient digestion will be converted to long-chain fatty acids by intestinal bacteria.
    (e) The composition and amount of fiber in the diet has little effect on intestinal flora.

8. Pepsins in the stomach are:

    (a) Essential for efficient protein digestion and absorption.
    (b) Active mainly in the jejunum.
    (c) Synthesized in the gastric parietal cells.

(d) Synthesized as inactive pre-proenzymes.
(e) Products of a single gene.

9. Which of the following statements is false?

   (a) Pancreatic proteases are secreted as proenzymes.
   (b) Activation of pancreatic proteases is initiated by cleavage of trysinogen by the brush-border enzyme enteropeptidase.
   (c) Pancreatic secretions contain a variety of endo- and exopeptidases.
   (d) Proline-containing peptides are resistant to pancreatic enzymes.
   (e) Free amino acids are the only products of pancreatic peptidases.

10. Which of the following statements regarding brush-border membrane peptidases is true?

    (a) Only 3–4 brush-border membrane peptidases have been identified.
    (b) Brush-border membrane peptidases hydrolyze oligopeptides generated by intraluminal digestion to free amino acids and di- and tripeptides.
    (c) Aminopeptidase N is directly inserted into the brush-border membrane.
    (d) All brush-border membrane peptidases are carboxypeptidases.
    (e) Brush-border membrane peptidases are not regulated during development or cellular differentiation.

11. Which of the following is a property of the di-tripeptide carrier PEPT1?

    (a) Is a potassium coupled transporter.
    (b) PEPT1 can transport a myriad of possible di- and tripeptides generated from protein digestion.
    (c) Is one of many di-tripeptide carriers that have been identified in the small intestine.
    (d) Is normally present in high levels in human colon.
    (e) Can transport long polypeptide chains.

12. Which of the following statements is true?

    (a) Multiple transport systems for the uptake of amino acids have been identified in the small intestinal brush-border membrane.
    (b) All small intestinal brush-border membrane amino acid transporters are sodium-dependent.
    (c) Lactase persistence in adults appears to have an X-linked mode of inheritance.
    (d) Activities of oligosaccharidases are greater in the terminal ileum than in the jejunum.
    (e) The small intestinal brush-border does not contain peptidases than can hydrolyze peptide bonds containing proline.

Answers: (1) d; (2) e; (3) c; (4) b; (5) b; (6) d; (7) a; (8) d; (9) e; (10) b; (11) b; (12) a

Chapter 7

1. Which of the following is a property of gastric lipase?

   (a) Is rapidly degraded by pepsin.
   (b) Preferentially hydrolyzes the ester bond in position 3 of triglyceride, generating fatty acids and diglycerides.
   (c) In adult humans accounts for 75 % of triglyceride hydrolysis.
   (d) Contributes little to triglyceride hydrolysis in neonates.
   (e) In humans is produced mainly by surface epithelial cells in the stomach.

2. Which of the following is a characteristic of pancreatic lipase?

   (a) Is activated by a pH of less than 4.0.
   (b) Binding of pancreatic lipase to a lipid droplet is inhibited by colipase.
   (c) Is secreted by the pancreas in a pro-enzyme (zymogen) form.
   (d) Is an interfacial enzyme that is most active at an oil-water interface.
   (e) Completely hydrolyzes triglyceride to fatty acids and glycerol.

3. Which of the following statements is false?

   (a) A patient with congenital deficiency of pancreatic lipase will absorb no triglyceride.
   (b) The normal human secretes a great excess of pancreatic lipase.
   (c) A normal human bile salt-activated lipase catalyzes the complete hydrolysis of triglyceride to fatty acids and glycerol.
   (d) The pancreatic lipase related proteins PLRP-1 and -2 may play a role in neonatal triglyceride digestion.
   (e) CCK stimulates pancreatic secretion of pancreatic lipase.

4. Which of the following statements is true?

   (a) In mixed micelles bile salts are oriented with the polar aspect pointing towards the interior.
   (b) In a patient with complete biliary obstruction and no intraluminal bile salts no triglyceride will be absorbed.
   (c) Micelles diffuse across in unstirred water layer faster than monomeric fatty acids in solution.
   (d) Human milk contains a bile-salt activated lipase.
   (e) Colipase is produced by enterocytes.

5. Which of the following proteins facilitates fatty acid uptake across the intestinal brush-border membrane?

   (a) CD36
   (b) Colipase
   (c) B-monoglyceride
   (d) Secretin
   (e) Trypsin

6. Which protein is responsible for most of the triglyceride synthesized in the intestine?

    (a) Intestinal fatty acid binding protein (I-FABP)
    (b) Liver fatty acid binding protein (L-FABP)
    (c) Diacylglycerol acyltransferase 1 (DGAT1)
    (d) Apolipoprotein B
    (e) Bile-salt dependent lipase

7. Which of the following proteins is involved in the assembly of chylomicrons in the enterocyte?

    (a) ApoB100
    (b) LDL receptor
    (c) Lecithin-cholesterol acyltransferase (LCAT)
    (d) Liver fatty acid binding protein (L-FABP)
    (e) Apo B48

8. Which of the following statements is false?

    (a) In the Golgi apolipoprotein AI and other lipids are added to the chylomicrons.
    (b) In the Golgi apolipoproteins are glycosylated.
    (c) Golgi organelles containing chylomicrons fuse into secretory vesicles that fuse with the basolateral membrane and release their contents into the extracellular space.
    (d) Chylomicrons enter intestinal lacteals and are transported through intestinal lymphatics into the circulation.
    (e) Anderson/chylomicron-retention disease is caused by a failure of intestinal triglyceride synthesis.

9. Medium chain triglycerides are:

    (a) Hydrolyzed by pancreatic lipase more slowly than long-chain triglycerides.
    (b) A major component of chylomicrons.
    (c) Absorbed well even in patients with cholestasis.
    (d) A major component of dietary fat.
    (e) An excellent source of essential fatty acids.

10. Which of the following statements is true?

    (a) Dietary fat is the major stimulus of secretin release.
    (b) CCK stimulates gall bladder contraction and relaxation of the sphincter of Oddi delivering concentrated bile into the intestine.
    (c) Secretin is the major stimulus of pancreatic lipase secretion.
    (d) Pancreatic lipase is most active at a pH of 3.
    (e) Colipase is secreted into pancreatic fluid as the fully active protein.

11. Which of the following statements regarding medium chain triglyceride absorption is false?

    (a) Medium chain fatty acids are directly released from the enterocytes into the portal circulation.
    (b) Absorbed medium chains fatty acids are rapidly taken up the liver and other tissues and used as an energy source.
    (c) Can be a useful source of calories in patients with malabsorption.
    (d) Medium-chain triglycerides contain fatty acids of 6–12 carbon chain lengths.
    (e) Medium chain triglycerides are not hydrolyzed by gastric lipase.

12. Which of the following statements is true?

    (a) The major pathway for intestinal triglyceride synthesis involves the generation of diglycerides from fatty acyl-CoA and β-monoglyceride by monoglyceride acyltransferase (MGAT) enzymes.
    (b) The major pathway for intestinal triglyceride synthesis uses phosphatidic acid which is dephosphorylated to produce diacylglycerol.
    (c) ApoB-mRNA editing refers to the production of ApoB from ApoAI.
    (d) ApoAI is found in high density lipoproteins (HDL), but not chylomicrons.
    (e) The most common cause of abetalipoproteinemia is a mutation in the gene for apolipoprotein AIV

Answers: (1) b; (2) d; (3) a; (4) d; (5) a; (6) c; (7) e; (8) e; (9) c; (10) b; (11) e; (12) a

Chapter 8

1. Which of the following statements regarding cholesterol absorption is true?

    (a) More cholesterol is typically present in the diet than is secreted in bile.
    (b) Dietary and biliary cholesterol form a single pool in the intestinal lumen.
    (c) Almost 100 % of luminal cholesterol is absorbed.
    (d) Pancreatic lipase is the major enzyme involved in cholesteryl ester hydrolysis.
    (e) Cholesterol in mixed micelles is in rapid equilibrium with cholesterol in monomolecular solution.

2. Which of the following is the target of the cholesterol absorption inhibitor ezetimibe?

    (a) Niemann-Pick C1-Like 1 (NPC1L1)
    (b) ABCG5
    (c) ABCG8
    (d) Pancreatic lipase
    (e) LXRα

3. Which of the following is involved in cholesterol efflux from the enterocyte into the intestinal lumen?

    (a) Niemann-Pick C1-Like 1 (NPC1L1)
    (b) ABCG5/ABCG8
    (c) Microsomal triglyceride transfer protein (MTTP)
    (d) LDL receptor
    (e) LXRα

4. Which of the following is the principal enzyme responsible for cholesterol esterification in the intestine?

    (a) Niemann-Pick C1-Like 1 (NPC1L1)
    (b) ABCG5/ABCG8
    (c) Microsomal triglyceride transfer protein (MTTP)
    (d) Scavenger receptor class B type 1 (SR-B1)
    (e) Acyl CoA:cholesterol acyl transferase 2 (ACAT2)

5. Which of the following statements regarding bile acid absorption is true?

    (a) Bile acids are secreted by the liver almost exclusively as unconjugated forms.
    (b) Most conjugated bile acids are absorbed in the proximal jejunum.
    (c) In the ileum a high capacity active transport system absorbs more than 95 % of the secreted bile acids.
    (d) Bile acid uptake in the ileum occurs via a H^+-bile acid co-transporter.
    (e) The ileal bile acid transporter transports unconjugated bile acids more efficiently than conjugated bile acids.

6. Which of the following is responsible for most of the bile acid transport across the basolateral membrane into the circulation?

    (a) Ileal binding acid protein (IBABP)
    (b) ASBT
    (c) OSTα/β
    (d) Phospholipase A_2
    (e) Microsomal triglyceride transfer protein (MTTP)

7. Which of the following statements is false?

    (a) Dietary retinyl esters are hydrolyzed prior to intestinal absorption.
    (b) β-carotene uptake across the brush-border membrane is mediated by the scavenger receptor class B, type 1 (SR-B1).
    (c) In the enterocyte β-carotene is cleaved forming retinal.
    (d) Lecithin-retinol acyltransferase (LRAT) esterifies CRBP(II)-bound retinol.
    (e) All dietary carotenes are sources of vitamin A.

8. Which cells in the liver contain most of the vitamin A?

    (a) Hepatocytes
    (b) Kupffer cells
    (c) Endothelial cells
    (d) Stellate cells
    (e) Lymphocytes

9. Which of the following statements is true?

    (a) Vitamin D is present in the diet mainly in an esterified form.
    (b) The liver is the principal site for formation of 1,25(OH)$_2$ vitamin D
    (c) Formation of 24,25(OH)$_2$ vitamin D is the initial step in conversion of vitamin D to its active form.
    (d) Proteins involved in cholesterol absorption, such as SR-B1, CD-36, and NPC1L1 also facilitate vitamin D$_3$ uptake across the brush-border membrane.
    (e) Intestinal absorption of 25(OH) vitamin D$_3$ and 1,25(OH)$_2$ vitamin D$_3$ is less efficient than absorption of vitamin D$_3$.

10. Which of the following statements regarding vitamin E is false?

    (a) Dietary vitamin E is comprised of multiple tocopherols and tocotrienols.
    (b) The richest sources of vitamin E are vegetable oils.
    (c) Bile salts facilitate vitamin E absorption.
    (d) Mutations in α-tocopherol-transfer protein (α-TTP) cause vitamin E deficiency by impairing intestinal absorption of vitamin E.
    (e) In plasma, vitamin E is found in several classes of lipoproteins.

11. Vitamin K:

    (a) Regulates transcription of genes coding for clotting factors.
    (b) Is a co-factor involved in the post-transcriptional modification of proteins.
    (c) Is involved in hydroxylation of certain lysine residues in clotting factors.
    (d) In the liver is mostly in the phylloquinone form.
    (e) Is not incorporated into chylomicrons, but is released from enterocytes directly into the portal blood.

12. The anti-coagulant warfarin:

    (a) Inhibits dithiol-dependent vitamin K-epoxide reductase and vitamin K-reductase.
    (b) Blocks intestinal vitamin K absorption.
    (c) Inhibits vitamin K-dependent transcription of clotting factor genes.
    (d) Accelerates vitamin K catabolism and biliary excretion.
    (e) Blocks the conversion of menaquinone to phylloquinone.

Answers: (1) e; (2) a; (3) b; (4) e; (5) c; (6) c; (7) e; (8) d; (9) d; (10) d; (11) b; (12) a

Chapter 9

1. Which of the following describes the function of most water-soluble vitamins?

    (a) Co-enzymes in biochemical reactions
    (b) Regulators of gene transcription
    (c) Transport proteins
    (d) Substrates for oxidation-reduction reactions
    (e) Anti-oxidants

2. Which of the following is not a characteristic of water-soluble vitamin absorption?

    (a) The vitamins are often present in the diet as complex forms that require digestion in the intestinal lumen or at the brush-border membrane prior to transport across the intestinal epithelium.
    (b) At low concentrations, transport across the brush-border membrane typically occurs by membrane carriers, active transport systems or membrane binding proteins and receptors.
    (c) Extensive metabolism of water-soluble vitamins occurs with the enterocyte.
    (d) Bile acids are required for solubilization of water-soluble vitamins prior to intestinal transport.
    (e) At high doses water-soluble vitamins may be absorbed by passive diffusion.

3. Which of the following factors does not commonly affect intestinal absorption of minerals and trace elements?

    (a) Intraluminal pH
    (b) Redox state of the metal
    (c) Formation of chelates in the intestinal lumen
    (d) Cholesterol content of the meal
    (e) Digestion of proteins that is associated with the metal.

4. Which of the following significantly decreases intestinal copper absorption?

    (a) Glycine
    (b) Glucose
    (c) Zinc
    (d) Sodium
    (e) Potassium

5. Which of the following statements regarding folate absorption is true?

    (a) Folate is present in the diet mainly as the monoglutamate form.
    (b) At low doses, folate is absorbed principally by passive diffusion.
    (c) Folates are transported across the basolateral intestinal protein by the reduced folate carrier (RFC).
    (d) Dietary polyglutamyl folates are digested to the monoglutamate form prior to intestinal absorption.
    (e) Vitamin B_{12} is a competitive inhibitor of folate absorption.

6. Which of the follow statements is false?

    (a) Sulfasalazine is an inhibitor of folylpolyglutamate deconjugation.
    (b) Sulfasalazine is an inhibitor of intestinal monoglutamyl folate transport.
    (c) At physiologic concentrations, folic acid is largely reduced and methylated or formylated within the enterocyte.
    (d) Folic acid, reduced folates, and methotrexate are transported by different brush-border membrane carriers.
    (e) Intestinal bacteria synthesize folates.

7. A 70-year-old man has a total gastrectomy for gastric adenocarcinoma. He subsequently develops anemia and neurologic symptoms, and is found to be severely vitamin B_{12} deficient. Which of the following is the major reason for his vitamin B_{12} deficiency?

    (a) Poor dietary vitamin B_{12} intake
    (b) Lack of Intrinsic Factor
    (c) Deficiency of a duodenal transport protein
    (d) Lack of gastric acid
    (e) Folic acid deficiency

8. The receptor for the Intrinsic Factor-vitamin B_{12} complex in the ileum is:

    (a) Transcobalamin II
    (b) Haptocorrin
    (c) Multidrug resistance protein 1
    (d) Cubilin
    (e) Vitamin D receptor

9. Which of the following would increase intestinal iron transport?

    (a) Iron-deficiency anemia
    (b) A high serum hepcidin level
    (c) A high duodenal ferritin content
    (d) Inflammatory states
    (e) Riboflavin

10. Which of the following statements regarding hereditary hemochromatosis is false?

    (a) The most common mutation causing this disorder is in the iron transporter DMT-1.
    (b) The primary treatment for this disease is periodic blood donation until the body iron burden is reduced.
    (c) Hemochromatosis is characterized by excessive iron absorption in spite of iron overload.
    (d) Hemochromatosis is usually caused by mutations in the HFE gene.
    (e) Hepcidin mutations can also cause iron overload.

13 Multiple Choice Questions

11. Vitamin D regulates which of the following:
    (a) Calcium uptake across the brush-border membrane
    (b) Calcium transport from the brush-border to the basolateral enterocyte membrane
    (c) The calcium transporting ATPase in the basolateral membrane
    (d) All of the above
    (e) None of the above

12. Which of the following statements is true?
    (a) At high dietary intakes, vitamin D is needed for calcium absorption.
    (b) Uptake of calcium from the intestinal lumen into the enterocyte does not require metabolic energy.
    (c) Lactose inhibits intestinal calcium absorption.
    (d) Knockout of the gene for intestinal calbindin-D completely blocks intestinal calcium absorption.
    (e) Transport of calcium across the basolateral membrane occurs by incorporation of calcium into lipoproteins.

Answers: (1) a; (2) d; (3) d; (4) c; (5) d; (6) d; (7) b; (8) d; (9) a; (10) a; (11) d; (12) b

Chapter 10

1. All of the following are true of the hepatic acinus EXCEPT:
    (a) The portal triad is the center of the acinus.
    (b) The terminal hepatic venules are at the periphery of the acinus.
    (c) Zone 1 hepatocytes are the closest to the center of the acinus.
    (d) Zone 1 cells are the most prone to develop hypoxic injury.
    (e) Zone 1 cells are most active in oxidative energy metabolism.

2. Which liver test would be most sensitive in evaluating hepatocyte necrosis?
    (a) Total serum bilirubin level
    (b) Serum transaminases (ALT and AST)
    (c) Serum 5' nucleotidase
    (d) Serum alkaline phosphatase
    (e) Liver ultrasound

3. Which liver test may provide the most reliable information on the degree of liver fibrosis?
    (a) Elastography
    (b) Serum GGTP
    (c) Prothrombin time
    (d) MRCP
    (e) Liver biopsy

4. What is the most practical and accurate method to measure portal venous pressure?

   (a) Splenic pulp pressure
   (b) Hepatic venous wedged pressure
   (c) Doppler ultrasound
   (d) Hepatic artery wedged pressure
   (e) Observation of esophageal varices

5. All of the following are complications of portal hypertension EXCEPT:

   (a) Esophageal varices
   (b) Portal hypertensive gastropathy
   (c) Hypersplenism
   (d) Renal sodium wasting
   (e) Ascites

6. All of the following are important in the development of ascites EXCEPT:

   (a) Increased aldosterone and renal sodium retention
   (b) Increased *effective* plasma volume
   (c) Increased circulating NO and other vasodilators
   (d) Decrease in systemic vascular resistance
   (e) Increased hydrostatic pressure in the splanchnic vascular bed

7. Which of the following is NOT true of the hepatorenal syndrome?

   (a) May be precipitated by overzealous diuresis or paracentesis for ascites.
   (b) Development of oliguria with decreased GFR and preservation of tubular function.
   (c) The kidney exhibits cortical vasoconstriction.
   (d) Affected kidneys do not function when transplanted to another individual.
   (e) Likely due to local and circulating vasoactive and other factors.

8. All of the following may be a part of the management of esophageal varices EXCEPT:

   (a) Blood transfusion
   (b) Endoscopic banding of the varices
   (c) Administration of an oral diuretic
   (d) Intravenous infusion of octreotide
   (e) Portosystemic shunting

9. All of the following are signs of hyperestrogenism and hypogonadism in the cirrhotic patient EXCEPT:

   (a) Palmar erythema
   (b) "Spider" angiomata
   (c) Gynecomastia
   (d) Female escutcheon in male patients
   (e) Hairy pinnae

10. All of the following are examples of prehepatic portal hypertension EXCEPT:
    (a) Portal vein thrombosis
    (b) Extrinsic compression of the portal vein by a tumor
    (c) Alcoholic cirrhosis
    (d) Splenic vein thrombosis
    (e) Hepatic artery-portal vein fistula

11. Which of the following cells of the liver receives blood supply solely from the hepatic artery:
    (a) Hepatocytes
    (b) Sinusoidal endothelial cells
    (c) Stellate cells
    (d) Kupffer cells
    (e) Bile ductular cells

12. Which of the following statements is true?
    (a) Zone 3 cells are closest to the portal triad and are the first to receive oxygenated, nutrient-rich blood.
    (b) Zone 2 cells are the last to develop hypoxic injury and the first to regenerate
    (c) Zone 3 cells, located around the terminal hepatic venule, are last to receive blood, and as such, are most sensitive to any form of toxic or vascular injury. Zone 2 cells are intermediate.
    (d) Zone 3 hepatocytes are most active in glucose synthesis and release, oxidative energy metabolism, amino acid utilization, bile acid and bilirubin excretion, and ammonia detoxification.
    (e) Zone 1 hepatocytes are most active in glucose uptake and utilization and biotransformation.

Answers: (1) d; (2) b; (3) e; (4) b; (5) d; (6) b; (7) d; (8) c; (9) e; (10) c; (11) e; (12) c

Chapter 11

1. Which ONE of the following steps is NOT involved in the SREBP regulation of liver LDL receptors during statin-induced reduction of cholesterol synthesis?

   (a) Response of SREBP cleavage activation protein (SCAP) to decreased free cholesterol levels allowing SREBP precursor to move to the Golgi.
   (b) Protease cleavage of SREBP in the Golgi and productions of the active peptide.
   (c) Binding of activated SREBP to the sterol response element (SRE) in the LDL receptor gene promoter.
   (d) Sensing of cellular cholesterol content.
   (e) Down-regulation of LDL receptor gene transcription.

2. Which is NOT true of alpha-1-antitrypsin deficiency?

   (a) Retention of misfolded, polymerized and aggregated mutant protein in the hepatocyte ER contributes to liver disease.
   (b) Most individuals with the mutant PiZZ phenotype have serious liver disease.
   (c) Alpha-1-antitrypsin is a protease inhibitor produced and secreted by the liver and protects tissues from damage by endogenous proteases.
   (d) Alpha-1-antitrypsin deficiency may cause chronic liver and lung disease.
   (e) Alpha-1-antitrypsin is a glycoprotein.

3. Which ONE of the following is NOT true of the LXRs?

   (a) They are nuclear receptor transcription factors.
   (b) They regulate several key steps in cholesterol and lipoprotein metabolism.
   (c) They heterodimerize with RXR.
   (d) They are regulated by oxysterols.
   (e) They bind to mRNA.

4. Which of the following is NOT a part of the metabolic syndrome?

   (a) Abdominal obesity
   (b) Hypertension
   (c) Insulin resistance
   (d) High HDL levels
   (e) Hypertriglyceridemia

5. Which ONE of the following statements about lipoprotein lipase (LPL) is CORRECT?

   (a) LPL is bound to chylomicrons or VLDL particles when they interact with HDL.
   (b) LPL hydrolyzes triglyceride in adipose tissue.
   (c) LPL is activated by apo A-I.
   (d) LPL action produces HDL from IDL.
   (e) LPL is activated by apo C-II.

6. Which ONE of the following regulates net intestinal cholesterol absorption?

   (a) MTP
   (b) HMG CoA reductase
   (c) ABCG5/ABCG8
   (d) LCAT
   (e) CETP

7. In humans, LDL particles are derived from which ONE of the following?

   (a) Chylomicrons
   (b) VLDL
   (c) HDL

(d) Chylomicron remnants
(e) Intestinal lipid micelles

8. Regulation of glycolysis in the liver is tightly integrated with that of all of the following EXCEPT:

    (a) Gluconeogensis
    (b) Lipogenesis
    (c) Glycogen synthesis
    (d) Glycogenolysis
    (e) Protein synthesis

9. Which of the following is the rate-limiting step in glycogen synthesis?

    (a) Glycogen synthetase
    (b) Phosphorylase
    (c) Phosphofructokinase
    (d) Glucose-6-phosphatase
    (e) Galactokinase

10. Which of the following is NOT true of PCSK9?

    (a) Originally identified as involved in apoptosis of nerve cells.
    (b) Gain of function mutations are associated with elevated serum LDL levels.
    (c) Functions to target LDL receptors away from recycling to a degradative pathway.
    (d) Loss of function mutations are associated with reduced serum LDL levels.
    (e) PCSK9 is not an attractive therapeutic target for treatment of hypercholesterolemia.

11. Which of the following does NOT occur during prolonged fasting?

    (a) Hepatic glycogen stores are depleted.
    (b) Gluconeogenesis in liver becomes more important as an energy source for the brain.
    (c) Amino acids from muscle are the predominant substrates for hepatic gluconeogenesis.
    (d) Fatty acids are released from adipose tissue and are oxidized to ketones by the liver to supply alternative energy to the brain.
    (e) Incoming glucose is stored as glycogen.

12. In vitamin metabolism, the liver plays a major role in:

    (a) The storage of vitamin A
    (b) The conversion of 7-dehydrocholesterol to cholecalciferol
    (c) The storage of vitamin D
    (d) The conversion of 25-hydroxyvitamin D_3 to 1,25-hydroxyvitamin D_3
    (e) The production of factor VIII with the aid of vitamin K

Answers: (1) e; (2) b; (3) e; (4) d; (5) e; (6) c; (7) b; (8) e; (9) a; (10) e; (11) e; (12) a

Chapter 12

1. The following are all sources of ammonia in the body EXCEPT:

    (a) Amino acids
    (b) Nucleic acids
    (c) Urea
    (d) GI bleeding
    (e) Fructose

2. The following are all true of hepatic encephalopathy EXCEPT:

    (a) Blood ammonia levels closely correlate with degree of encephalopathy.
    (b) Hepatic encephalopathy may be acute or chronic.
    (c) Variceal bleeding may exacerbate encephalopathy.
    (d) Oral administration of lactulose may be used to treat encephalopathy.
    (e) Early clinical presentation may include inversion of sleep patterns.

3. All of the following are true of a xenobiotic except:

    (a) Exogenous organic substance
    (b) No nutritive value for energy production
    (c) The liver plays a minor role in their biotransformation and clearance
    (d) No necessary structural function
    (e) No function as an enzyme cofactor

4. All of the following are true of acetaminophen EXCEPT:

    (a) Normally, the toxic NAPQI produced by acetaminophen metabolism is immediately conjugated with glutathione and cleared.
    (b) With poisoning, glutathione is depleted, allowing accumulation of NAPQI with resultant cell injury and death.
    (c) The Rumack-Matthew nomogram is useful in predicting acetaminophen toxicity based on serum levels in chronic poisoning.
    (d) Administration of *N*-acetylcysteine may be effective in preventing significant hepatotoxicity.
    (e) Liver transplantation may be required in severe cases of acetaminophen poisoning.

5. Which of the following is TRUE of the Gilbert syndrome?

    (a) Mutations in the coding region of the bilirubin UDP-glucuronosyltransferase gene cause Gilbert syndrome.
    (b) Gilbert syndrome is a frequent cause of kernicterus.
    (c) Results in mild elevations of the serum unconjugated bilirubin level, often in response to fasting or illness.
    (d) Is often accompanied by the Dubin-Johnson and Rotor syndromes.
    (e) Phototherapy is frequently used to treat adults with this condition.

6. The portion of the hepatocyte membrane specialized for bile secretion into the bile ducts is the:

    (a) Basolateral membrane
    (b) Lipid raft
    (c) Canaliculus
    (d) Tight junction
    (e) Peroxisome

7. All of the following are functions of bile acids EXCEPT:

    (a) Solubilize the products of lipid digestion in mixed micelles to enhance efficiency in crossing the unstirred water layer for absorption.
    (b) Along with phospholipid, help prevent cholesterol from forming gallstones in the gallbladder.
    (c) Serve as signaling molecules in the regulation of gut motility and carbohydrate metabolism.
    (d) Have bacteriostatic effects in the proximal intestine and induce genes to inhibit microbial growth in the distal intestine.
    (e) Function to inhibit fluid and electrolyte secretion in the colon.

8. Which of the following is NOT true of human bile acid metabolism?

    (a) Cholic acid and chenodeoxycholic acid are primary bile acids.
    (b) Bile acids are conjugated with glycine and taurine.
    (c) The rate-limiting step in bile acid synthesis is catalyzed by cholesterol 7-alpha hydroxylase (CYP7A1).
    (d) Bile composition is not affected by passage through the bile ducts.
    (e) Bacteria participate in the formation of secondary and tertiary bile acids.

9. All of the following are true of the bile acid enterohepatic circulation EXCEPT:

    (a) Cholecystokinin stimulates the gallbladder to contract in response to a fatty meal.
    (b) Bile acids are actively transported from the gut lumen by ileal enterocytes via the transporter, ASBT.
    (c) Bile acids returning to liver from ileum via the portal vein are taken up by the basolateral hepatocyte transporter, NCTP.
    (d) Loss of bile acids in stool is not a significant route of loss of cholesterol from the body.
    (e) Hepatic bile acid synthesis is regulated by ileal bile acid uptake via FXR and FGF19.

10. All of the following are components of management of patients with cholestasis EXCEPT:

    (a) Administration of ursodeoxycholic acid may benefit some patients.
    (b) Supplementation with fat-soluble vitamins.
    (c) Use phototherapy to reduce bilirubin levels.

(d) Avoid medications cleared by the liver and excreted into bile.
(e) Supplementation with dietary medium-chain triglycerides that do not require bile acids for absorption.

11. Which of the following is associated with unconjugated hyperbilirubinemia?

    (a) Gilbert's disease
    (b) Alagille syndrome
    (c) Biliary atresia
    (d) Dubin-Johnson syndrome
    (e) Common bile duct stone

12. Which of the following conditions is associated with hemolytic jaundice?

    (a) Increased bilirubin in the urine
    (b) Increased plasma unconjugated bilirubin
    (c) Decreased urobilinogen in the urine
    (d) Decreased absorption of vitamin B_{12}
    (e) Clay-colored stools

Answers: (1) e; (2) a; (3) c; (4) c; (5) c; (6) c; (7) e; (8) d; (9) d; (10) c; (11) a; (12) b

Index

A
ABCA1, 181, 183, 191, 200, 201, 282
Abetalipoproteinemia, 170
Abnormal sweat test, 133
ACAT2, 181, 183
Acetylcholine, 14, 16, 24, 26, 41, 43, 44, 48, 54, 67, 68, 69, 73, 75, 90, 91, 92, 114, 149
Achalasia, 48, 49, 61
Achlorhydria, 72, 85
Acinar cells, 87, 88, 89, 90, 92, 93, 94, 100, 160, 161, 162
Acinus, 88, 89, 242, 243
Active transport, 137, 141, 185, 211, 213, 241
Acyl CoA synthase, 167
Adenosylcobalamin, 219
Adult-type hypolactasia, 144, 146, 155
Afferent neurons, 28
Afferent (sensory) neurons, 25
Aganglionosis, 59
Alagille syndrome, 291, 317
Albumin, 250, 272
Alcoholic hepatitis, 283
Alkaline phosphatase, 248
Alpha-1-antitrypsin, 273
Aminopeptidases, 150
Aminopyrine breath test, 251
Amnionless, 222
Amoxycilllin, 82
Amphiphilic, 165, 309
Amylase, 94, 101, 103, 104, 138, 139
α-Amylase, 139
Amylopectin, 138
Amylose, 138
Anemia, 63, 199, 201, 213, 215, 223, 225, 227, 231, 232, 259, 266, 267, 321
Angiography, 251

Angiotensin-converting enzyme 2, 152
Anovulation, 261
Antacids, 80
Anticholinergics, 81
Antral systole, 50, 52
Apo AI, 171, 173
Apo AIV, 171, 173, 174
ApoB, 170, 177
Apo B48, 169, 170, 171, 174, 177
Apo B100, 170, 177
Apo CII, 172
Apo D, 172
Apo E, 171, 172
Apolipoproteins, 169, 171, 172, 237, 240, 273, 278, 279
Apoptosis, 109, 279
Aquaporin, 314
Arachidonic acid cascade, 16, 116
Arteriohepatic dysplasia, 291
Arteriovenous fistula, 254
ASBT, 185, 186, 187, 314, 316
Ascites, 250, 259, 260, 261, 262, 263, 264, 265, 267, 290, 297, 320
Aspartic protease, 75
Aspirin, 83
Atherosclerotic cardiovascular disease, 159
ATP-binding cassette (ABC) transporters, 299
Atrial natriuretic factor, 16, 19, 260
Atropine, 74, 100
Autocrine pathway, 23
Autonomic neuropathy, 52, 60, 61, 127
Autonomous (extrinsic) nervous system, 8

B
Barium, 60, 133
Basic electric rhythm, 39

Benign recurrent intrahepatic cholestasis, 318
Bezoar, 60
Bidirectional retinol transport, 193
Biliary obstruction, 174, 176, 248, 249, 252, 256, 267, 282
Biliary tract, 159, 207, 248, 252
Biliverdin, 306
Biotin, 211
Bone morphogenic protein 6, 227
Botulinum toxin, 62
Bradykinin, 128
Breath-hydrogen testing, 155
Bromosulfophthalein, 250
Budd-Chiari syndrome, 251, 254, 265, 267
Butyrate, 147
Byer disease, 318

C

Calcitonin gene-related peptide, 48, 114
Calcium-calmodulin-dependent protein kinase, 19
Calmodulin, 19
Campylobacter-like organisms (CLO) test, 83
Capsaicin, 100
Carbohydrate responsive element binding protein, 281
Carbonic anhydrase, 78, 122
Carboxypeptidases, 94, 150
Carnitine acyltransferase I, 276
β-Carotene, 188, 189, 190, 193
β-Carotene-9',10'-monooxygenase, 190
β-Carotene-15,15'-monooxygenase, 190
Carrier transport proteins, 118
CCK-releasing peptide, 101
Celiac disease, 7, 32, 177
Celiac sprue, 32
Cellular retinol binding protein 1, 192
Cellular retinol-binding protein, type 2, 190
Central nervous system, 8, 25, 26, 37, 72, 100, 259, 285, 297, 298, 305, 306
Cephalic-phase, 98
Channel proteins, 117
Chenodeoxycholic acid, 185, 310
Chief cells, 65
Cholecystokinin, 9, 10, 11, 24, 44, 68, 74, 91, 99, 113, 148, 149, 160, 161, 175, 312
Cholera, 17, 129, 131, 132
Cholera toxin, 17, 129, 131
Cholestatic liver disease, 174, 197, 206, 233, 248, 282, 288, 309
Cholesterol esterase, 180
Cholesteryl ester transfer protein, 280

Cholic acid, 310
Cholinergic muscarinic receptors, 90
Chylomicron, 167, 168, 169, 170, 171, 172, 173, 174, 176, 177, 181, 183, 185, 191, 192, 193, 197, 200, 201, 203, 204, 278, 279, 280, 284, 287, 288
Chylomicron remnants, 171, 172, 191, 192, 200, 278, 279, 288
Chymotrypsin, 101, 150
Circulating natriuretic factor, 260
Cirrhosis, 207, 208, 232, 233, 240, 250, 251, 252, 254, 256, 257, 258, 259, 260, 261, 263, 264, 265, 267, 283, 291, 293, 297, 318, 320
Clarithromycin, 80, 82, 83
Cl^-/HCO_3^- exchanger, 118, 122
Colipase, 161, 162, 163, 164, 165
Collectrin, 152
Colloidal bismuth subcitrate, 82
Computed tomography, 251
Congenital lactase deficiency, 146
Congenital megacolon, 59
Constipation, 59, 60, 61
Creatine, 104, 248
Crigler-Najjar syndrome, 308, 309, 321
Crohn's disease, 7, 231
Crypt cells, 16, 23, 32, 109, 141, 230
C-terminal tetra-peptide sequence, 10
Cubilin, 222
Cutaneous xanthomata, 317
Cyanocobalamin, 218
Cyclic guanosine monophosphate, 16, 43, 91, 124
Cystic fibrosis, 71, 96, 102, 122, 133, 175, 176, 206, 207, 318
Cystic fibrosis transmembrane regulator, 71, 102, 122, 133
Cystic fibrosis transmembrane regulator (CFTR) channel, 71

D

D cells, 66, 67, 68, 74, 80
7-Dehydrocholesterol, 194, 287
7α-Dehydroxylation, 187
Delta-bilirubin, 309
Depolarization, 35, 39
Dermatitis herpetiformis, 177, 178
Detoxification, 237, 244, 296, 297, 298, 299, 301
Diabetes mellitus, 60, 142, 276
Diabetic gastroparesis, 60
1,2-Diacylglycerol, 18, 91
Diacylglycerol acyltransferases, 169

Index

Diarrhea, 4, 6, 16, 17, 30, 31, 32, 33, 61, 108, 118, 127, 129, 130, 131, 132, 133, 142, 144, 146, 147, 154, 155, 156, 157, 175, 176, 177, 178, 186, 187, 188, 205, 206, 207, 233, 318, 319, 321
Dihydrothiothreitol, 77
Dihydroxylated bile salts, 163
1,25-Dihydroxyvitamin D_3, 195, 228
24,25-Dihydroxyvitamin D_3, 198
Dipeptidases, 150
Dipeptidyl aminopeptidase IV, 151
Dipeptidyl peptidase IV, 10
Disaccharides, 138, 144, 154
Dithiol-dependent reductase, 205
Divalent metal transporter, 225, 226
Dopamine, 13, 302
Doppler ultrasonography, 254
Dorsal motor nucleus, 72, 73
Dubin-Johnson syndrome, 308
Duct cells, 87, 95, 96, 97, 101
Duct of Santorini, 88
Duct of Wirsung, 88

E

Effective plasma volume, 260
Efferent (motor) neurons, 25, 115
Elastase, 101, 150
Endopeptidases, 94, 150
Endothelin, 241
Enteric nervous system, 6, 8, 25, 37, 55, 56, 58, 113, 114, 129
Enteric (intrinsic) nervous system (ENS), 8
Enteric reflexes, 30, 112
Enterogastrones, 74
Enterokinase, 93, 94, 103, 149
Epidermal growth factor, 16, 141
Epinephrine, 13, 41, 43, 149, 274, 275, 286, 302
Esophageal varices, 257, 258, 259, 263, 264, 265, 297, 299
Esophagus, 3, 8, 29, 35, 38, 46, 47, 48, 49, 60, 61, 257, 258
Ethanol toxicity, 261
Exenatide, 10
Exendin-4, 10
Ezetimibe, 181, 182

F

Facilitated diffusion, 137, 146, 187, 189, 213, 229, 285
Farnesoid X receptor, 186, 316
FATP4, 167

Fatty stools, 103
Fecal incontinence, 61
Ferritin, 225, 226, 227, 233, 234
Ferroportin, 225, 226, 227, 234
Ferroportin 1, 225, 226
Ferrous (Fe^{2+}) iron salts, 224
Fibrinogen, 240, 250
Fibroblast(s), 115, 128
Fibroblast growth factor (FGF), 19, 23, 187, 196, 316
Folate, 211, 214, 215, 216, 217, 220, 231, 232, 321
Folate conjugase, 151
Folate deficiency, 215, 217
Folate-proton symport, 216
Folic acid, 214, 217
Free cholesterol, 169, 170, 171, 180, 181, 183, 278, 280, 282, 283, 317
Fructose, 146, 147, 157, 273
Fulminant hepatitis, 283, 298

G

Galactose, 273
Galactose elimination test, 250
Galanin, 114
Gap junctions, 36, 92
Gastrectomy, 148, 232
Gastric chyme, 99
Gastric colic reflex, 57
Gastric inhibitory peptide, 9, 74
Gastric inhibitory polypeptide, 12, 27, 49, 100
Gastric motility, 22, 27, 51, 77, 81, 161
Gastric-phase, 73, 74, 98, 99
Gastrin, 9, 10, 11, 12, 21, 27, 50, 55, 64, 66, 67, 68, 69, 72, 73, 74, 75, 76, 79, 81, 82, 84, 90, 99, 113, 148, 149, 160, 220
Gastrin-cholecystokinin family, 10
Gastrinoma, 79, 84
Gastrin releasing polypeptide, 73, 74
Gastritis, 78, 79, 223, 263
Gastro-esophageal reflux disease, 4
Gastrografin, 134
Gastro-intestinal (GI) system, 3
Gastroparesis, 4, 60
Gastroparesis/dumping syndrome, 4
G cells, 27, 66, 67, 68, 74, 79
Ghrelin, 9
Gilbert syndrome, 308, 320, 321
Glucagon, 10, 44, 54, 87, 98, 274, 275, 276, 277, 278, 285, 286, 297
Glucagon-like peptide-1(GLP-1), 10
Glucagon-like peptide-2 (GLP-2), 10, 144
Glucagon-related polypeptide, 10

Glucoamylase, 139
Glucocorticoids, 13, 144, 146, 275, 297
Glucocorticoid treatment, 32
Glucokinase, 273, 285
Gluconeogenesis, 274, 285
Glucose-6-phosphatase deficiency, 289, 290
Glucose-6-phosphate, 273, 281, 285, 289
Glucuronidation, 244, 302, 308
GLUT-2, 157
GLUT-5, 157
Glutamic acid, 202, 204, 214, 216, 217, 288
Glutamine, 154, 170, 296, 297, 298
Gluten, 32
Gluten-free diet, 178
Glycentin, 10
Glycine, 150, 151, 185, 187, 207, 302, 311
Glycogenolysis, 275
Glycogen synthetase, 275, 286
Glycolysis, 274
Glycoproteins, 76, 90, 123, 139, 273, 303
GM1-ganglioside receptors, 133
Goblet cells, 64, 109, 110
G proteins, 15, 16, 17, 18, 82
Guanylin, 16, 33, 114
Gut-associated lymphoid tissue, 5, 6, 7
Gut endocrine system, 5, 8

H
Hartnup disease, 152, 156
Heat-stable enterotoxin, 16, 19, 33, 114, 129, 131, 132
Helicobacter pylori, 79, 80, 83
Hemochromatosis, 223, 227, 233, 234, 251, 253, 263, 293
Hepatic encephalopathy, 263, 298, 323
Hepato-biliary-pancreatic GI, 3
Hepatocytes, 170, 171, 172, 187, 192, 222, 227, 239, 240, 242, 256, 267, 273, 283, 284, 287, 289, 295, 307, 314, 318
Hepatorenal syndrome, 260, 261, 264
Hepcidin, 227, 233, 234
Hephaestin, 225, 226
Hereditary hemochromatosis, 233
Hirschsprung's disease, 59, 61
Histamine, 13, 66, 67, 68, 69, 74, 79, 81, 82, 84, 85, 115, 124, 128, 220
Histidine, 13, 38, 44, 148, 224
H^+/K^+-ATPase, 70, 71, 72, 81, 82, 96, 117, 119, 221
HMG-CoA reductase, 277
H_2-receptor antagonists, 81
12α-Hydroxylase, 311

24-Hydroxylase enzyme, 198
Hypercholesterolemia, 181, 279, 290, 291
Hypergastrinemia, 79, 84, 85
Hypersplenism, 259, 266
Hypochloremia, 31, 83
Hypogonadism, 261
Hypokalemia, 31, 83, 263
Hypolactasia, 155

I
Ileum, 3, 19, 52, 54, 55, 63, 114, 118, 120, 122, 130, 133, 146, 151, 152, 167, 180, 185, 189, 191, 205, 206, 216, 217, 221, 222, 229, 231, 314, 319, 321
Immunoglobulin A, 123
Incretin, 151
Indocyanine green, 250
Inflammatory bowel disease, 7, 216
Inositol 1,4,5-triphophate, 18
Insulin-like growth factor, 12
Interfacial enzyme, 162
Intermediate-density lipoproteins, 170
Interneurons, 25, 28, 38, 58, 114
Intestinal biopsy, 155
Intestinal blind loop syndrome, 6, 319
Intestinal-phase, 99
Intestinal stasis, 61
Intestino-intestinal reflex, 51
Intravenous cholangiography, 252
Iron deficiency, 223, 224, 225, 226, 232
Islet cells, 67, 87, 88
Isoleucine, 38, 44, 148, 150, 220
Isomaltase, 139, 140, 154

J
Jaundice, 305, 317
Juxtacrine pathway, 23

K
Kallikrein, 128, 260
Kennedy pathway, 184
Kernicterus, 201, 305, 308
Ketogenesis, 244, 276
K^+/H^+ exchanger, 119
Kupffer cell., 237

L
Lamina propria, 5, 12, 23, 29, 32, 67, 115, 116, 129
Lansoprazole, 71

Laparoscopic biopsy, 253
Laparoscopic cholecystectomy, 268
Laplace's law, 257
L-arginine, 13, 91
LCAT deficiency, 283
L cells, 10
Lecithin-cholesterol acyltransferase, 171
Leloir pathway, 273
Leptin, 9, 153
Leucine, 44, 148, 150
Leydig cell dysfunction, 261
L-FABP, 168, 172
Lidocaine clearance test, 251
α-Limit dextrins, 139, 140
Linoleate, 191
Lipase, 94, 101, 160, 161, 162, 163, 164, 165, 172, 173, 175, 176, 180, 189, 193, 200, 208, 278, 279, 283, 289
Lipoprotein lipase, 278
Lipoprotein-X, 282, 291, 317
5-Lipoxygenase, 116
Lithocholic acid, 311
Liver, 3, 6, 28, 29, 31, 46, 159, 168, 170, 171, 174, 177, 182, 185, 188, 192, 193, 195, 197, 200, 201, 202, 203, 204, 206, 207, 208, 215, 226, 227, 233, 237, 238, 239, 240, 241, 242, 243, 244, 247, 248, 249, 250, 251, 252, 253, 254, 256, 257, 259, 260, 261, 262, 263, 264, 265, 266, 267, 271, 273, 274, 275, 276, 277, 278, 279, 280, 281, 282, 283, 284, 285, 287, 288, 289, 290, 291, 292, 293, 295, 296, 297, 298, 299, 300, 302, 303, 304, 306, 308, 309, 310, 312, 314, 315, 316, 317, 318, 319, 320, 322, 323
Liver X receptors, 281
Low-density lipoproteins, 170, 201
Lower esophageal sphincter, 8, 22, 27, 46, 48, 60, 62
Luteinizing hormone, 261
Lysine, 148, 150

M
Magnetic resonance cholangiopancreatography, 252
Magnetic resonance imaging, 251
Malabsorption, 32, 61, 84, 102, 103, 142, 143, 144, 153, 154, 155, 157, 159, 167, 170, 174, 176, 178, 186, 197, 199, 203, 206, 207, 208, 216, 218, 223, 231, 232, 233, 317, 319, 322
Maldigestion, 28, 32, 84, 95, 102, 103
Maltotriose, 139, 140

Mast cells, 68, 115
Medium-chain triglycerides, 173
Menaquinone-4, 204
Menaquinone-9, 204
Menaquinone(s), 203
Mesenchymal cells, 115, 116, 239
Metabolic alkalosis, 83, 118, 263
Metalloporphyrin, 225
Methionine, 44, 148, 151, 218, 220
Mevalonate, 277
Microcolon, 133, 134
Microcytic anemia, 232
Microminerals, 212
Microribonucleic acid, 253
Microsomal reductase, 190
Microsomal triglyceride transfer protein, 169, 174, 183
Microspheres, 176
Migrating motor complex, 54
Mineralocorticoids, 13
Mixed micelles, 165, 166, 176, 180, 181, 183, 184, 189, 200, 203, 207, 208, 310
Monoethylglycinexylidide, 251
β-Monoglyceride(s), 163, 165, 166, 167, 168, 176, 203
Monoglyceride acyltransferase, 168
Monosaccharides, 90, 137, 138, 141, 144, 147, 154, 155, 157, 273
Motilin, 9, 31, 55
Motor neurons, 24, 25, 28, 37, 58, 59
Mucosa, 5, 12, 13, 22, 23, 26, 29, 31, 32, 49, 63, 64, 65, 66, 67, 74, 78, 79, 81, 83, 84, 93, 98, 103, 104, 107, 109, 110, 112, 114, 115, 117, 118, 120, 123, 126, 129, 130, 132, 134, 153, 160, 161, 177, 178, 179, 225, 319, 321
Mucosa-associated lymphoid tissues, 6
Mucosal hyperplasia, 130
Mucus, 76, 77
Multidrug resistance protein 2, 187
Multidrug resistance protein 3, 187
Muscularis externa, 5
Muscularis mucosae, 5
Myenteric nerve plexus, 6

N
Na^+-bile salt cotransporter, 207
Na^+ channel, 119, 121
Na^+-dependent glucose transport, 142
NADPH-dependent pathway, 205
$NaHCO_3$ cotransporter, 122
Na^+/H^+ exchanger, 72, 118, 126, 133
Neurocrine pathway, 24

Neuropeptide Y, 44, 72, 114
Neurotensin, 31, 52, 74, 114
Neutrophils, 115, 128, 129, 215
Niacin, 156, 211
Niemann-Pick C1-Like 1, 180, 314
Nitric oxide, 13, 14, 19, 20, 24, 38, 44, 48, 54, 59, 61, 91, 128, 129, 241, 257
Nitrc oxide synthase (NOS), 91
Non-steroidal anti-inflammatory drugs, 79
Norepinephrine, 13, 24, 43, 114, 302
NPC1L1, 180, 181, 182, 197, 200, 201, 314
Nucleus tractus solitarii, 72, 73

O

OATP1B1, 308
OATP1B3, 308
Obesity, 147, 159, 252, 283, 292, 293
Octreotide, 31, 85
Oddi's sphincter, 161
Oleate, 191
omeprazole, 71
Open surgical wedge biopsy, 253
Oral rehydration, 33, 132
Oral rehydration solution, 33
Organic brain syndrome, 263
Osmotic diarrhea, 144
Osteomalacia, 194
7-oxo-lithocholic acid, 311

P

Palmitate, 191
Pancreas, 3, 29, 31, 46, 84, 87, 88, 89, 93, 94, 97, 102, 103, 104, 108, 137, 148, 149, 159, 161, 175, 176, 178, 247, 249
Pancreatic α-cells, 10
Pancreatic amylase, 139
Pancreatic lipase-related proteins, 164
Pancreatic phospholipase A_2, 183, 184
Pancreatitis, 95, 102, 104
Pancreozymin, 161
Paneth cells, 109, 114
Pantothenic acid, 211
Para-aminobenzoic acid, 214
Paracellular pathway, 117, 137, 211, 216, 228, 231, 233
Paracentesis, 260, 261
Paracrine pathway, 22
Parotid, 3
Partial triglyceride hydrolysis, 162
Passive diffusion, 109, 117, 137, 167, 180, 184, 189, 197, 211, 213, 228, 231

Pepsin, 22, 60, 63, 66, 75, 76, 77, 78, 79, 80, 82, 98, 148, 149, 160, 220
Pepsinogen, 63, 65, 66, 75, 76, 149
Peptic cells, 75
Peptic ulcer(s), 78, 79, 80, 84
Peptic ulcer disease, 4, 10, 67, 79, 80, 82, 83, 103, 263
Peptones, 75
Percutaneous needle biopsy, 253
Peristalsis, 5, 6, 8, 36, 45, 47, 48, 49, 50, 51, 56, 60, 61, 64
Peroxisomal β-oxidation, 277
Pertussis toxin, 18
Peyer's patches, 6, 7
PGE_2, 79, 115, 128
Pharynx, 3, 46, 47, 160
Phase contrast microscopy, 164
Phenylalanine, 13, 74, 102, 148, 150
Phosphatidylcholine, 183
Phosphatidylethanolamine, 183
Phosphatidylinositol, 16, 18, 40, 67, 75, 91, 124, 150, 183
Phosphatidylinositol (PI) pathway, 18
Phosphatidylserine, 183
Phospholipid flippase, 313, 318
Pit cells, 240
Plasminogen multigene family, 171
Platelet activating factor (PAF), 13, 23
Platelet-derived growth factor, 16
Polymerase chain reaction, 253
Portal alkalosis, 254
Portal blood acidosis, 254
Portal triad, 242
Portosystemic encephalopathy, 266, 298
Postrema, 72
Preganglionic efferent fibers, 38
Preproglucagon, 10
Previtamin D_3, 194
Primary sclerosing cholangitis, 252, 253, 317, 318
Programmed cell death, 109
Progressive familial intrahepatic cholestasis, 318
Prolactase, 145
Proprotein convertase subtilisin kexin type 9, 279
Prostaglandin, 30, 48, 67, 74, 75, 77, 78, 79, 83, 115, 124, 125, 128, 149, 159, 176, 187, 260
Protein kinase C, 19, 67, 91, 126, 142, 143, 153
Prothrombin, 202, 205, 206, 250, 262, 264, 267, 288, 291, 292, 320, 322, 323
Prothrombin time, 206, 250, 262, 264, 267, 288, 291, 292, 320, 322, 323

Proton pump inhibitors, 81
Proximal jejunum, 32, 146, 151, 163, 207
Pruritis, 290, 291, 292, 317
Pteroylglutamic acid, 214
Pumps, 117

R
Radionuclide imaging, 252
Receptor-mediated internalization, 171
Renin-angiotensin system, 9, 152, 260
Resident tissue macrophages, 115
Resin, 205, 206, 319, 321
Resting membrane potential, 38, 39
Retinoic acid receptor, 193
Retinoids, 188, 190, 193
Retinoid X receptor, 193, 281
Retinol binding protein, 192, 208, 287
Reverse cholesterol transport, 171, 280
Rickets, 194, 248
Rifaximin, 264, 299
Rotor syndrome, 308
Rumack-Matthew nomogram, 322, 323

S
Salivary amylase, 139
Salivary glands, 3, 46, 108
Salmonella, 129
Scavenger receptor-BI, 280
Scavenger receptor class B type 1, 182
Schilling test, 223, 232
Secretagogues, 124
Secretin, 9, 10, 11, 12, 27, 32, 49, 50, 74, 75, 91, 92, 97, 99, 101, 103, 113, 149, 161, 162, 175, 178, 221, 312
Secretory canaliculi, 69, 72, 220
Serosa, 5, 6, 117
Serous glands, 160
Serum colloid osmotic pressure, 272
SGLT-1, 10, 118, 133, 142, 143, 144, 154, 157
Shigella, 129
Short-chain fatty acids, 120, 147, 298
Signal peptidase, 90
Signal peptide, 89, 145
Signal-peptide recognition particle, 90
Site-1 protease, 281
Site-2 protease, 281
β-Sitosterolemia, 182
SLC6A19, 156
Slow waves, 38, 59
Small intestine, 3, 25, 26, 28, 29, 35, 39, 50, 51, 52, 54, 56, 57, 60, 61, 74, 76, 84, 101, 114, 118, 119, 120, 139, 144, 147, 148, 149, 153, 154, 159, 161, 162, 167, 168, 169, 172, 173, 177, 178, 179, 182, 185, 190, 207, 213, 215, 216, 218, 222, 228, 230, 231, 249, 284, 312, 318
Sodium-dependent glucose co-transporter-1, 10, 133
Somatostatin, 17, 24, 31, 44, 48, 55, 66, 67, 68, 69, 74, 80, 83, 85, 87, 98, 114, 259
Spike potentials, 38, 40, 41, 50, 54
Splenic congestion, 259
Splenomegaly, 259, 263, 265, 266, 288, 290, 322
SREBP cleavage activating protein, 281
Stearate, 191
Steatorrhea, 6, 31, 32, 159, 163, 164, 175, 176, 177, 178, 197, 207, 233
Stellate cell, 192, 240, 241, 257, 287
Sterolin-1, 182
Sterolin-2, 182
Sterol regulatory element binding proteins, 281
Stomach, 3, 8, 9, 25, 26, 27, 29, 38, 39, 40, 46, 47, 48, 49, 50, 51, 52, 54, 55, 60, 61, 63, 64, 66, 68, 72, 73, 74, 76, 77, 78, 81, 82, 87, 96, 98, 114, 148, 149, 159, 160, 161, 162, 224, 257
Stop codon, 170, 177
STRA6, 193
Sublingual, 3
Submandibular glands, 3
Submucosa, 5, 115
Submucosal nerve plexus, 5
Submucosal plexus, 36
Substituted benzimidazoles, 81
Sucralfate, 82
Sucrase, 139, 140, 147, 154
Sulfolithocholic acid, 311
Systemic mastocytosis, 115

T
Tangier disease, 183
Taurine, 151, 185, 187, 311
Taurocholate, 163, 180, 224
Teniae coli, 36, 57
Testicular atrophy, 261
TGF-α, 12, 23
TGF-β, 12, 23
Threonine, 19, 76, 77, 148, 220
Toxin A, 130
Trace elements, 212, 214
Transferrin receptor 2, 227, 234
Transforming growth factors, 12
Transjugular intrahepatic portosystemic shunting, 259

Transjugular route, 253
Transthyretin, 193, 287
Trihydroxylated bile salts, 163
Triple therapy, 80, 83
Trypsin, 93, 94, 101, 103, 140, 150, 162, 167
Tryptophan, 74, 148, 150, 155, 156, 250, 272, 298
Type 2 diabetes, 147, 151, 292, 293
Tyrosine kinase receptors, 20, 21

U
UDP-glucose, 273, 274, 302, 320
UDP-glucuronic acid, 302, 320
Ulcerative colitis, 7
Urobilinogen, 249
Ursodeoxycholic acid, 292, 311, 319

V
Vagotomy, 49, 50, 51, 103
Valine, 148, 150, 220
Valsalva maneuver, 58
Vascular endothelial growth factor, 257
Vasoactive intestinal peptide, 9, 26, 38, 44, 48, 68, 74, 91, 92, 97, 114, 125
Vasoactive intestinal polypeptide, 31, 149
Very low-density lipoproteins, 170, 200, 278

Vibrio cholerae, 132
Villus cells, 23, 109, 146, 168, 171, 186, 217
VIP-induced diarrhea, 31
VIPoma, 31
Viscous isotropic phase, 165
Vitamin D, 155, 193, 194, 195, 196, 197, 206, 228, 229, 230, 231, 233, 248, 287
Vitamin D deficiency, 194
Vitamin D_3 receptor, 196
Vitamin E deficiency, 199, 200, 201, 292
Vitamin K, 201, 202, 203, 204, 205, 206, 207, 250, 264, 288, 292
Vitamin K deficiency, 202
Vitamin K 2,3-epoxide, 205
Vitamins B_6, 211
Von Gierke's disease, 289

W
Warfarin, 202, 205, 249, 250, 300
Wedged hepatic venous pressure, 254
Wernicke-Korsakoff psychosis, 263

Z
Zollinger-Ellison syndrome, 84
Zymogen, 88
Zymogen, 75, 90, 91, 92, 93, 104, 183

MIX
Papier aus verantwortungsvollen Quellen
Paper from responsible sources
FSC® C105338

If you have any concerns about our products,
you can contact us on
ProductSafety@springernature.com

In case Publisher is established outside the EU,
the EU authorized representative is:
**Springer Nature Customer Service Center GmbH
Europaplatz 3, 69115 Heidelberg, Germany**

Printed by Libri Plureos GmbH
in Hamburg, Germany